THE HUMAN
EXPERIENCE
OF TIME
The Development of
Its Philosophic Meaning

The Publication of this work
has been aided by a grant from the
Andrew W. Mellon Foundation

But now the sight of day and night, and the months and the revolutions of the years, have created number, and have given us a conception of time, and the power of enquiring about the nature of the universe; and from this source we have derived philosophy, than which no greater good ever was or will be given by the gods to mortal man. This is the greatest boon of sight . . .

— Plato

The feeling of eternity is a hypocritical one, for eternity feeds on time. The fountain retains its identity only because of the continuous pressure of water. Eternity is the time that belongs to dreaming, and the dream refers back to waking life, from which it borrows all its structures. Of what nature, then, is that waking time in which eternity takes root?

— M. Merleau–Ponty

THE HUMAN EXPERIENCE OF TIME

The Development of Its Philosophic Meaning

CHARLES M. SHEROVER

Hunter College
City University of New York

New York: New York University Press   1975

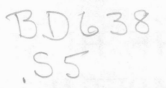

Copyright © 1975 by New York University

Library of Congress Catalog Card Number: 74-21659

ISBN: 0-8147-7759-7

Manufactured in the United States of America

To William Barrett

Preface

The genesis of this book is to be found in work done under a 1971 special summer grant by the Faculty Research Foundation of the City University of New York, which is gratefully acknowledged. It soon became apparent that much current work being done in the philosophy of time was being pursued in some ignorance of the historic sources of the questions being asked, the issues being discerned, and the problems being raised. It seemed important that an elucidation of the temporalist tradition in Western philosophy might help avoid useless labors and point us to the core of the issues facing us. This interpretive overview is, then, a first stage in my own work concerning some prime issues I see in the development of a systematic temporalism; it is hoped that it will prove to be of value to other laborers in that same vineyard.

I feel a special kind of gratitude to two persons not directly connected with this book who nevertheless have proven very important to its completion. F. Joachim Weyl, a mathematician with a philosophic perspective, gave me warm encouragement, at a crucial early period, to proceed with this exploration which goes in a direction quite different from his own. The dedication which J. T. Fraser has given to the import of the study of time, and his voiced encouragements, have been important in maintaining my own perseverance along the way.

Specific stages in the actual preparation of this book owe much to several individuals whose help has, in each case, been important. I am very indebted to John J. Blom (who has given me valuable help with special regard to the passages from Descartes), Mark N. Cohen, Charles W. Eddis, José Huertas-Jourda (who has been of vital help regarding the Husserl material), Drew Hyland and Frederick G. Weiss. Joseph McGeough worked with me as my research assistant in the early stages of the manuscript and Dennis Shea and Livia Farkas have assisted me in its later stages, particularly by preparing the Index; to each of them I extend deeply felt thanks for their aid and their many important suggestions along the way. In several ways it seems that Renate F. Murray, as usual, helped make this book possible by relieving me of otherwise time-consuming demands, and

vii

Meredith Crossman by her clerical assistance. My good friend, Bruce Wilshire joins William Barrett in always being ready to counsel with me and to provide advice that is not always congenial but is never to be ignored. I owe a vote of thanks to Hugh van Dusen for his special kindnesses, and to the folks at New York University Press (and to their resolutely anonymous adviser), for their patience, understanding, and indulgences.

This book, following a format suggested by that of Russell's book on Leibniz, is a somewhat unorthodox compendium of anthological selections together with my own connecting essays. It seems to me that many of our perplexities and our understandings of time emerge from a continuity of development that is seldom observed; its stages have so rarely been looked at together that the ample documentation provided by 'selections' became necessary. To the publishers who have made copyrighted material available for this purpose, I am, of course, grateful; specific credits have been indicated as footnotes on the first page of each selection.

Perhaps needless to say but nevertheless necessary, without any diminution of gratitude, the responsibility for what has been and has not been said within these pages is mine alone.

C.M.S.

Table of Contents

Contents

22. Royce: Time: Concept and Will   395
23. Santayana: "Sentimental Time"   407
24. Dewey: "Time and Individuality"   419

VII. *The Structure of Experiential Time*   437
25. Piaget: Developing the Concept of Time   466
26. Husserl: The Constitution of the Present   484
27. Minkowski: The Presence of the Past   504
28. Heidegger: The Priority of the Future   519

VIII. *The Open Agenda*   549
29. Collingwood: "Some Perplexities About Time"   558
30. McKeon: "Time and Temporality"   572

Notes   581
Index   593

I.

Foreword: In the Beginning

"In the beginning," Genesis tells us, "God created the heavens and the earth." The creation is presented as having required six sequential stages, the first of which was devoted to establishing the cycle of light and darkness, of day and night. Time was created by God at the outset of the creation; it was intrinsically connected with his creation as its concomitant or its form. Time was not only an initially created aspect of a created world; time was used by God for the entire process of first creating this world and then resting from this time-consuming divine labor.

If time was seen as useful for God, it was certainly necessary for men. Time was, in the development of biblical thought, not a neutral container of events, but somehow intrinsic to their essential nature. Time was to be an important guide to men concerning the propriety or appropriateness of possible alternate actions which, whatever else be true of them, had an integral temporal component. Thus, in Ecclesiastes, we are told, "For everything there is a season, and a time for every matter under heaven: a time to be born, and a time to die; a time to plant, and a time to pluck up what is planted; a time to kill, and a time to heal; a time to break down, and a time to build up. . . ."

The New Testament, although composed in a later Hellenistic world, carried this theme forward; time was portrayed as linear, as history, as the vehicle for fulfillment, as the carrier of meaning. It opens with an account of ancestral chronology, focuses on a set of historic events, and ends with prophecy. From Genesis through Revelation, there is a continuity of movement from an unrepeatable past to a yet pending future; articulated at every point is an "affirmation and realization of the possibilities of life through time rather than by cultic destruction of time in favor of eternity. . . . Each day of life was one day closer to its fulfillment, and one day further from its creation." [1]

In its beginning, the Western philosophic tradition, in contrast, did not countenance the notion of a beginning. As an aspect of the world, time was itself seen as eternal. The world was conceived as an ongoing process without beginning or end and time was the mark of the rhythm or pattern of the continuity of change within it.

Philosophy arose out of the attempt of some Ionian Greeks to understand and explain the patterns of change which they discerned in the world about them. Starting from the testimony of everyday sense perception, they realized that every physical entity was engaged in some process of change, that no perceptual object was really static. Turning from the reports of sense to the considerations of reason, they tried to understand the nature of the divergent patterns of change before them which constituted the world they experienced. Probably patterning their understanding of the regularities they beheld on the cycle of the seasons, they concluded that each of these patterns was recurrent without any genesis or finality. Explanation of particular changes was to be found in some rationally conceived principle or law which was discerned, not in the particular perception itself, but only by generalization in the thinker's reasoning. The visible object, the seen, was to be explained in terms of the thought which was literally invisible and unseen. A somehow continuing principle, reason, or law was to be discerned in the regularly recurrent changes in nature. Behind the diversity of sense experiences, there was to be a found a unity in thought. Without any denigration of the many-ness of things, without any separation of reason from nature, explanation was to be found in common unifying principles, patterns, or laws which were understood to underlay the multiplicity of particular changing things and bound the many into one.

The convention is that Thales, an inhabitant of Miletus, the port city of Ephesus in the Greek settlement of Asia Minor, initiated the philosophic quest for a unifying first rational principle of natural events. Noting that water is the one natural element which can readily be transformed, by heat or cold, into a vapor like air or a solid like earth, he apparently reasoned that it must thereby be the prime element. As such it was to be in terms of the transformations of the state of water that the patterns of changes in physical nature were to be explained or understood.

Another Miletian, Anaximander, flourished about twenty years later—about 560 B.C. Facing the question of the 'wherefrom?', the source of the plurality of things in nature, he noted that they each came into being, lasted, and then disintegrated. He sought the common 'matter' or substance out of which all particular things were generated and into which they disappeared after disintegration or death. "The Non-Limited is the original material of existing things," he concluded. And the processes of nature, which govern the patterns of generation and decay of the particular entities comprising the world, were regularly ordered and to be understood "according to necessity . . . [and] according to the arrangement of Time." [2]

Perhaps more than anyone, it was Heraclitus of Ephesus, who gave the fullest expression to this pre-Socratic Greek conception of time as the mark of ordered change in a dynamic natural world. In the fragments of his writing which have come down to us, we find these ideas forcefully enunciated and brought together in a way that suggests cardinal themes of the development which followed. Contrary to the usually facile textbook description of an anarchic apostle of unbridled change, Heraclitus recognized the pervasiveness of change in the visible world while insisting that it be understood in its complexity in terms of an apparently unchanging Law (or *Logos*), which guided all changes, and brought them into the unity of the world (see Frags. 1, 2, 45, 50). Seeing the *Logos*, which binds all diversity within the world into the unity that is the world—as encompassing the multiplicity of all the changes constituting experience—he directed attention to the essential dynamicity and continuity of change in the world of nature to which we belong.

This world of nature was conceived as somehow eternal, without beginning or end, as uncreated, as ruled by its Law or *Logos*, which mandates the kinds or patterns of change permeating it (see Frags. 30, 103). Behind all change, all strife, all conflict—which mark the world as we experience it—there is the "hidden harmony" of the *Logos* (see Frag. 54), the law of change which is itself changeless, the law of generation which is itself ungenerated, the law of decay which is itself immune to decay.

Heraclitus is best known for the focus he placed on the dynamic continuity of change we experience (see Frags. 8, 12, 51, 53, 65, 91, 126). This dynamicity does, indeed, present many examples of serial order although the primary notion of change is that, not of development, but of change of quality between contrary states.[3] Fire, the

fourth of the Greek elements, which Thales had apparently overlooked, was regarded as either a symbol of the dynamicity of nature or
as, in fact, its cause or true expression (see Frags. 30, 64, 67). But, when
we move from our perceptual experience to our understanding of it,
we find the pervasive nature of the *Logos,* wholly immanent as the
"hidden harmony" which keeps disparate multiplicity in functioning ordered unity. The uncreated and continuing cyclical process of
change constitutes the internal temporal nature of the world.

Hardly mentioned, time is always implicitly involved. Used to mark
out the stages of serial order, it seems to have been conceived
primarily in terms of the periodicities or pulsings of the regular cycles
of these series of change which we now term processes (see Frags. 31,
52, 67, 91). Used by reason in its understanding of these constituting
changes, rather than the governor of reason itself, we see here the dim
beginning of many philosophic themes as well as that of the consideration of time which, articulated quite differently by Plato and by
Aristotle, served to shape the subsequent discussion.

For each of them was to bring a notion of time, as Heraclitus had
already implicitly done, into a spatial world in order to comprehend
its internal order. In contrast, the biblical tradition had urged that the
spatial world, and the events within it, be seen in a frame of serial time
which provides its source of significance, movement, and meaning.[4]

However disparate these two conceptions of time may have been,
they share in one paradox: the rooting of time in the supratemporal or
the timeless. For the biblical God, who created time and was able to
enter into it and use it, was yet prior to its numbered days. And the
Heraclitean *Logos,* although continually expressed in the ordered
cycles marking the changes within the world, was yet their governor.
For neither, however, did their view of the world of time and change
imply any suggestion of ontological illusion or any diminution of its
reality or its being.

As the biblical and Greek traditions came together in the later days
of the Roman Empire to form the merged common core of the Western intellectual tradition, their similarities and disagreements
provided the root source of the dialectic marking subsequent discussion. This is particularly true of the disparate approaches they brought
to the interpretation of the time of human experience and the paradoxes which were engendered.

The prime paradox, of course, was the subsequent denigration of the import of time itself. The standard reading of the historic development of philosophic thought is to see a continuing subordination of time to eternity, of concrete temporal experience to non-temporal abstractions—concepts, logics, categories—conceived to be supratemporal or more real. Having accepted the distinction between 'reality' and 'appearance', already implicit in Heraclitus, the notion of 'appearance' was somehow identified, not with 'real appearing' but with 'illusion'; in contrast to the alleged timelessness of eternal truths, which were, of course, the thinker's real concern, time was relegated to illusion while thought continually aspired to reaching beyond it. It has, indeed, with some truth, been observed that "when philosophers who have dominated occidental philosophy thought of time at all, they put it on the level of appearance [illusion] and left it there with contempt." [5] This outcome is, indeed, ironic for a civilization, a culture, a tradition, whose historic roots place awareness of time and change at the foundation of their concerns.

If a prime function of philosophic thought is to interpret and comprehend the nature and meaning of human experience—not merely the content but also the experiencing itself—that thinking needs to be true to the experience within which it arises. Our experiencing is indeed sequential, marked by periodicity, development, and change. To denigrate time, then, is to denigrate the experience which we seek to understand. Arising as it does out of human temporal experience, philosophic thinking cannot come to terms with itself until it begins to take the time of its experience as at least the perduring context within which it is and it functions.

If we determine to take time seriously, we need to reconsider the ways in which time has been examined in the developing philosophic heritage that is now ours. We reread the old texts with the question of time as our guiding concern and seek to uncover something of the original thinking that was there expressed. Doing so often throws a standard text into a new perspective, casts a different light on the other work of the particular thinker, perhaps calls for a systematic reevaluation of his thought, and yields grounds for a new understanding of philosophic history and of the development over time of philosophic thought. With the question of time as our guide, we rediscover its central, if usually veiled, import in our intellectual tradition.

If the question of time is as fundamental as the two diverse roots of our heritage indicate, we can see why it has implications beyond itself and for most philosophic inquiries. We then need to retrace the prime landmarks which marked the development of the questions which concern us today. Because experiential time has so often been lost sight of, even when it is implicitly invoked, it is crucial not only to sketch out these prime stages of philosophical development but also to place a large emphasis on the original texts themselves, for they carry the burden of history and interpretation. Too often too many of us have relied on fourth- or fifth-hand reports instead of going back to the texts themselves to see for ourselves what their authors intended to say. Very often facing the original texts throws a new light on issues involved and permits us to see, from our own perspective, what might not have been so readily seen before.

The presentation is, then, primarily chronological and thematic. We see that each discernible group of thinkers managed to define together a core issue which framed the center of the dialogue bringing them together, and the questions they proposed to those who followed them. The gathering together of separate thinkers in a given period has been marked more by the common questions they felt were important to ask than by agreement on any set of answers they might have produced. This is not to suggest that subsequent discussions ignored earlier questions or topics of focus. Rather, intellectually responsible thinkers felt obligated to bring earlier questions into newer contexts so as to reexamine them in terms of inquiries regarded as even more encompassing or fundamental.

Informed and responsible discussion is not rootless; it generally becomes more complex as it develops; it does not cut itself off from its roots, but sees itself as obligated to explain what may have once been taken as 'obvious' and to integrate the insights of earlier thinkers into its own hopefully more mature perspective. Even then, when we seek to focus on questions of contemporary discussions—and the second half of this book is concerned with philosophers who have played important roles in the thinking of *this* twentieth century—we need to recognize the pluralism of approaches and the common root out of which they have sprung.

This is to say that in order to understand the contexts, areas of agreement and of concern today, we need to look back and take in the development out of our own beginnings. It was a wise man who had

suggested that those who are not ignorant of their past are not condemned to repeat it. If philosophic thought is to open new ground, it
must do so with conscious self-awareness of how its current starting
place was itself cleared. In such an endeavor, focus is crucial; following a generally humanistic orientation and without any implied denigration of science, the focus here is not on the doctrines of temporal
measurement produced or used by the physical sciences, but with the
meaning of everyday temporal experience as understood by reflective
philosophic consciousness.

If time is real in human experience, if there is any truth to James's
contention that time is somehow the meeting ground of human consciousness and the world that it is conscious of, temporal development
should certainly be manifested in the developing discussion concerning the nature and import of human time. To do this, we need to
retrieve our past which produced the questions we now ask and the
concerns we have now defined for ourselves, so that the possibilities
for future development may be more fully nourished.

1.

Genesis:

THE TIME OF CREATION

The opening passages of the Book of Genesis present the two biblical versions of the traditional story of the Creation. The selection comprises the first of these accounts in its entirety (1:1–2:3).*

In the beginning God created the heavens and the earth.

The earth was without form and void, and darkness was upon the face of the deep; and the Spirit of God was moving over the face of the waters.

And God said, "Let there be light"; and there was light. And God saw that the light was good; and God separated the light from the darkness. God called the light Day, and the darkness he called Night. And there was evening and there was morning, one day.

And God said, "Let there be a firmament in the midst of the waters, and let it separate the waters from the waters." And God made the firmament and separated the waters which were under the firmament from the waters which were above the firmament. And it was so. And God called the firmament Heaven. And there was evening and there was morning, a second day.

And God said, "Let the waters under the heavens be gathered together into one place, and let the dry land appear." And it was so. God called the dry land Earth, and the waters that were gathered together he called Seas. And God saw that it was good. And God said, "Let the earth put forth vegetation, plants yielding seed, and fruit trees bearing fruit in which is their seed, each according to its kind, upon the earth." And it was so. The earth brought forth vegetation,

* This selection, as well as other scriptural quotations in this volume, are taken from the Revised Standard Version of the Bible, copyright 1946, 1952, and (©) 1971, and used by permission, Division of Education and Ministry, National Council of the Churches of Christ in the United States of America.

plants yielding seed according to their own kinds, and trees bearing fruit in which is their seed, each according to its kind. And God saw that it was good. And there was evening and there was morning, a third day.

And God said, "Let there be lights in the firmament of the heavens to separate the day from the night; and let them be for signs and for seasons and for days and years, and let them be lights in the firmament of the heavens to give light upon the earth." And it was so. And God made the two great lights, the greater light to rule the day, and the lesser light to rule the night; he made the stars also. And God set them in the firmament of the heavens to give light upon the earth, to rule over the day and over the night, and to separate the light from the darkness. And God saw that it was good. And there was evening and there was morning, a fourth day.

And God said, "Let the waters bring forth swarms of living creatures, and let birds fly above the earth across the firmament of the heavens." So God created the great sea monsters and every living creature that moves, with which the waters swarm, according to their kinds, and every winged bird according to its kind. And God saw that it was good. And God blessed them, saying, "Be fruitful and multiply and fill the waters in the seas, and let birds multiply on the earth." And there was evening and there was morning, a fifth day.

And God said, "Let the earth bring forth living creatures according to their kinds: cattle and creeping things and beasts of the earth according to their kinds." And it was so. And God made the beasts of the earth according to their kinds and the cattle according to their kinds, and everything that creeps upon the ground according to its kind. And God saw that it was good.

Then God said, "Let us make man in our image, after our likeness; and let them have dominion over the fish of the sea, and over the birds of the air, and over the cattle, and over all the earth, and over every creeping thing that creeps upon the earth." So God created man in his own image, in the image of God he created him; male and female he created them. And God blessed them, and God said to them, "Be fruitful and multiply, and fill the earth and subdue it; and have dominion over the fish of the sea and over the birds of the air and over every living thing that moves upon the earth." And God said, "Behold, I have given you every plant yielding seed which is upon the face of all the earth, and every tree with seed in its fruit; you shall have them for food. And to every beast of the earth, and to every bird of the air, and to everything that creeps on the earth, everything that has the breath of life, I have given every green plant for food." And it was so. And God saw everything that he had made, and behold, it was very good. And there was evening and there was morning, a sixth day.

Thus the heavens and the earth were finished, and all the host of them. And on the seventh day God finished his work which he had done, and he rested on the seventh day from all his work which he had done. So God blessed the seventh day and hallowed it, because on it God rested from all his work which he had done in creation.

2.

Heraclitus:

THE LAW OF CHANGE

Heraclitus lived about 500 B.C. in the city of Ephesus, a Greek
settlement on the coast of Asia Minor. His only book, written in
an oracular style, aimed to present the knowledge of his time.
Approximately one hundred and twenty-five fragments are
known to have survived. The following selection is taken from
these, as collected by Diels, and translated by Kathleen
Freeman.*

1. The Law [1] *(of the universe)* is as here explained; but men are
always incapable of understanding it, both before they hear it, and
when they have heard it for the first time. For though all things come
into being in accordance with this Law, men seem as if they had never
met with it, when they meet with words *(theories)* and actions
(processes) such as I expound, separating each thing according to its
nature and explaining how it is made. As for the rest of mankind, they
are unaware of what they are doing after they wake, just as they forget
what they did while asleep.

2. Therefore one must follow (the universal Law, namely) that
which is common *(to all)*. But although the Law is universal, the
majority live as if they had understanding peculiar to themselves.

8. That which is in opposition is in concert, and from things that
differ comes the most beautiful harmony.

12. Anhalation *(vaporisation)*. Those who step into the same river
have different waters flowing ever upon them. (Souls also are vapor-
ised from what is wet).

* Kathleen Freeman, trans., *Ancilla to the Pre-Socratic Philosophers: A complete
translation of the Fragments in Diels*, Fragmente der Vorsokratiker (Cambridge:
Harvard University Press, 1957), pp. 24–34.

[1] Logos, the intelligible Law of the universe, and its reasoned statement by
Heracleitus.

20. When they are born, they are willing to live and accept their fate *(death);* and they leave behind children to become victims of fate.

23. They would not know the name of Right, if these things *(i.e. the opposite)* did not exist.

27. There await men after they are dead things which they do not expect or imagine.

30. This ordered universe *(cosmos),* which is the same for all, was not created by any one of the gods or of mankind, but it was ever and is and shall be ever-living Fire, kindled in measure and quenched in measure.

31. The changes of fire: first, sea; and of sea, half is earth and half fiery water-spout . . . Earth is liquified into sea, and retains its measure according to the same Law as existed before it became earth.

45. You could not in your going find the ends of the soul, though you travelled the whole way: so deep is its Law *(Logos).*

49a. In the same river, we both step and do not step, we are and we are not.

50. When you have listened, not to me but to the Law *(Logos),* it is wise to agree that all things are one.

51. They do not understand how that which differs with itself is in agreement: harmony consists of opposing tension, like that of the bow and the lyre.

52. Time is a child playing a game of draughts; the kingship is in the hands of a child.

53. War is both king of all and father of all, and it has revealed some as gods, others as men; some it has made slaves, others free.

54. The hidden harmony is stronger *(or,* 'better') than the visible.

56. Men are deceived over the recognition of visible things, in the same way as Homer, who was the wisest of all the Hellenes; for he too was deceived by boys killing lice, who said: 'What we saw and grasped, that we leave behind; but what we did not see and did not grasp, that we bring.'

64. The thunder-bolt *(i.e. Fire)* steers the universe.

65. Need and satiety.

67. God is day-night, winter-summer, war-peace, satiety-famine. But he changes like (fire) which when it mingles with the smoke of incense, is named according to each man's pleasure.

72. The Law *(Logos):* though men associate with it most closely, yet

they are separated from it, and those things which they encounter daily seem to them strange.

75. Those who sleep are workers and share in the activities going on in the universe.

80. One should know that war is general *(universal)* and jurisdiction is strife, and everything comes about by way of strife and necessity.

85. It is hard to fight against impulse; whatever it wishes, it buys at the expense of the soul.

86. *(Most of what is divine)* escapes recognition through unbelief.

87. A foolish man is apt to be in a flutter at every word *(or, 'theory': Logos)*.

88. And what is in us is the same thing: living and dead, awake and sleeping, as well as young and old; for the latter *(of each pair of opposites)* having changed becomes the former, and this again having changed becomes the latter.

89. To those who are awake, there is one ordered universe common *(to all)*, whereas in sleep each man turns away *(from this world)* to one of his own.

91. It is not possible to step twice into the same river. *(It is impossible to touch the same mortal substance twice, but through the rapidity of change)* they scatter and again combine *(or rather, not even 'again' or 'later', but the combination and separation are simultaneous)* and approach and separate.

103. Beginning and end are general in the circumference of the circle.

104. What intelligence or understanding have they? They believe the people's bards, and use as their teacher the populace, not knowing that 'the majority are bad, and the good are few'.

115. The soul has its own Law *(Logos)*, which increases itself *(i.e. grows according to its needs)*.

116. All men have the capacity of knowing themselves and acting with moderation.

119. Character for man is destiny.

126. Cold things grow hot, hot things grow cold, the wet dries, the parched is moistened.

126b. One thing increases in one way, another in another, in relation to what it lacks.

II.

Time and Motion

We all know the myth that Plato declined to take time seriously; indeed, its unquestioned acceptance has colored our reading of the development of philosophic thought. His two-realm hypothesis does urge that time is ontologically derived from a timeless eternity; but it is nevertheless true that Plato was deeply concerned with the crucial role of time in human experience. These two sides of Plato's thought were reflected by those who came after him. Succeeding thinkers carried forth his ontological stance while they wrestled with the import of time in the experiential world.

Plato grounded time in the work of the timeless Demiurge but insisted that time is the necessary principle of order in the world of nature. Aristotle grounded time in the unmoved (and, thereby, timeless) prime mover, Plotinus in the eternal One, and Augustine in the transcendent God. Each of them conceived of time as dependent upon a supervening timeless reality and sought out its meaning in some aspect of this transient world.

Each of them tied time to sequential order in nature, to visible movement and perception of change. Although claiming to ask the ontological question 'what is time?', each one discussed its essential nature in terms of its experiential connection with motion and argued about which of the two—time or motion—is ontologically prior; for, if the experience of time requires the perception of motion, and if motion itself requires time, the question of the nature of time raises the question of which is more fundamental or which is to be understood in terms of the other. If time is somehow tied to number, which is used for measuring, we then face the question of whether we measure time in terms of motion or motion in terms of time.

In facing either alternative, each of them seems to have taken space for granted. Essentially following Plato's lead, the (temporally?) prior being of space was presumed; time, as Plato saw it, was a created mode of ordering of spatial movement. Disagree with the creation thesis as Aristotle might, his argument that time is merely the measure of motion in space took spatiality as given and notably neglected to raise any question of its relation to time. Plotinus, who regarded time as primordial in the world of engendered beings, and Augustine, who regarded it as at least primordial in our experiencing itself, both seem to have presumed the being of space in which time was displayed. However the spatial backdrop be explained or ignored, the debate concerning the nature of time arose from the experiential fact that time and change (or motion) are apparently given together in human experience, that one is somehow requisite for the experience of the other.

In asking about the nature of time, the debate revolved around the two alternatives: Is time a kind of reality that is measured by motion, or is it rather a measure of that motion which is real? The substance of the argument was drawn from the world of experience and served to illuminate ways in which time functions within it. However the ontological question of the nature of time itself was to be resolved, it seems apparent that it was regarded as crucial to life, to intelligible activity, and somehow to the activity of the intellect as well.

The concept of time plays an important role in several of Plato's dialogues. In the *Phaedo,* for example, knowledge, identified with recollection, was seen as dependent upon the ability to recapture selected aspects of past experience and to bring them into the present as memory. In the *Theatetus,* a prime argument against a sense perception theory of knowledge presumes our ability to achieve insight into the future as well—for, judgments concerning wise actions "are all concerned with the future" which must be somehow discernible in the present.[1]

It was in the *Timaeus* that Plato most explicitly discussed time as such. This is the only dialogue devoted to cosmology and the natural sciences; it is, then, in terms of the visible world of nature which science seeks to comprehend that time was explicitly brought to the forefront of his discussion. In a radical departure from the heritage of Heraclitus and of Greek thought generally, Plato presented an account of the creation—albeit as a mythological account which claims to be rationally plausible but does not claim to be true. Here he set out a number of themes which defined the direction of subsequent discussion.

Time is not eternity; this was Plato's ontological thesis. Time is a derived order of being; it is the mark of rational order in spacebound change. Dependent on eternity for its being, it came into existence with the ordering of the created universe. It is, then, the prime characteristic of that ordered change which is termed 'becoming' or 'generation'. Time, distinguished from timeless eternity which is ultimately real, is the rational mode of mimicking eternal truth by unending rational sequential order; time is, in Plato's famous remark, the "moving image of eternity" (p. 43). Time is not "the mark of mere change but of the change that bridges the gap, as it were, between change and the immutability of the eternal nature." [2] Time, then, it is clearly implied, is a kind of reality.

In the opening statement, Timaeus sets forth the essential Platonic thesis of two realms. In contrast to the intelligible which is eternal, unchangeable, and tenseless, the world that can be seen and touched is one of changing things that first come into existence and then go out of existence. Such things are sequentially ordered in their being and must have a beginning and a prior cause of that beginning; the entire collection that we term the sensible world, therefore, must have a beginning and a cause of it. Hence, there must be a first cause, an intelligent, purposive "artificer" (p. 41) or "creator" (p. 41) who brings the sensible world into its ordered existence.

Like any craftsman, this essentially finite artist-god, usually referred to in the literature as the Demiurge, is limited by the materials with which he must work and the model he uses as a standard for the form he imposes on the material with which he is working. Conceived as a craftsman undertaking a deliberate creation, he is portrayed as looking to the unchanging world of pure rationality or pure being as the model for the world of change in order to invest it with the highest possible degree of rationality. The creative effort is to build order into a presumably already existent spatial "receptacle." Blind necessity, which characterizes space and the material which he does not create but with which he works, sets limits to the possible perfection of the ordered world of nature which he brings into existence. As Plato urged later in the dialogue, "Mind, the ruling power, persuaded necessity to bring the greater part of created things to perfection, and thus and after this manner in the beginning, when the influence of reason got the better of necessity, the universe was created." [3] The Demiurge, as a transcendent purposive mind, then, brought rationality into space by imposing the rule of a time-order. Time, which binds the direction of change, is the rational principle which brings meaningful change out of chaos; it makes the notion of causal sequence and development possible; it is the mark of rational order in the world we come to know.

In mythological terms, Plato saw the natural world as a dynamic interrelated organism. The world is not to be construed as 'dead matter' but as a living being, a god, with soul in its body and reason in its soul. The Idea for this visible world is intelligible, "a living creature" (p. 42). The world-soul permeates the visible body of the physical world which is a living organic being, all of whose changes are internal to it. Insofar as the Greek word for 'time' was also their word for 'life', it is not at all strange that Plato should have seen a time-ordered world as one that is alive. [4]

Once created, this living world is to be "eternal, so far as it might be" and the "image we call time" is also to be eternal and "moving according to number" (p. 44). Time and the heaven of the regularly moving bodies were created together; for the moving stars and especially the sun are necessary to time, to our knowledge of it, and of number which we learn from the sequences we observe in the heavens.

In contrast to space which is a preexistent necessity for the creation of time, time is a product of rational purpose and not of blind necessity. If space is a condition which reason uses for the production of visible temporal order, then time is inherent in the rational structure of the physical world. Time came into existence, not as a preexistent framework in which things happen, but as the rational ordering of the things in space in accord with the eternal model. Using the Greek concept of time as inseparable from regular circular motion, time's absolutely uniform flow was seen as an imperfect but continuing mirror of the self-sameness of eternity. As such it is regular, numerically measurable, and involved in concrete events. Time is then something of the living principle of the dynamic continuity of the visible world.

Plato appears to have been alone among the Greeks in suggesting the creation, thereby the beginning, of time. But this conjecture and the myth in which it was expressed have been most fruitful. Many of the themes—such as the identification of time and number—and many of the prime questions—such as the priority relationship of time and space—have been generated from his work.

Ontologically, time was seen as contrasted to and dependent upon eternity; time, then, is a second or derived order of reality. But as the principle of rational order in the organization of the dynamic content of space and as the dynamic imitation of a static eternity, Plato seems to have emphasized the similarity between time and eternity, a similarity that somehow bridges the radical opposition between them. Hence, it could be argued that time has some rational relation to the ultimately real.

As the principle of order, time would seem to be essentially sequential as it is tied to the mathematical sequential order of number. Time as measurable seems to provide for regularity of sequence. It is to be known by visual perception of the heavenly spheres. Indeed, Plato, later in the dialogue, urged that it is by "the sight of day and night, and the months and the revolutions of the years," the mark of timed motions, that men have "created number," obtained "a conception of time," the "power of enquiring about the nature of the universe; and from this source we have derived philosophy." [5] Although Plato generally contrasted perception to the intelligibility of pure reasoning, it would seem that he found its genetic, if not its logical, source in temporal apprehension. For time is not only tied to sense-perception; it is also tied to the intelligible by virtue of being tied to number; it might then seem that it is also the ground of human knowledge generally, even if human knowledge should be able to transcend its limits and ascend to eternity itself.

Literally unperceptible, time—in its nature and in our apprehension—is tied to motions which are perceptible, so that time can be understood and can function in the use of intellect. Literally unseen, time is requisite for the rational soul to discover order in the changing world that is seen. Though we need spatial movement to know and number time, time is a product, not of necessity, space, or matter, but of mind. Produced by mind and incarnate in soul, time is rooted in nature and discovered by men in the perceiving of motion in space. As the human mind uses the perception of motion in space to comprehend the rationality of the temporal order, it is, Plato argued, able to rise above perception and turn away in thought from the world of the "moving image" to its eternal standard of unchanging truth.

But with regard to time itself, how are we to understand Plato's startling thesis that time had a beginning? The content of the spatial world somehow existed 'before' time and temporal ordering.[6] But, then, there must have been a kind of time before the time of the world we inhabit; time would seem to exist 'in time'. And space would then seem to be independent of time for its existence and to be as eternal as being is. Perhaps Plato meant to take the beginning of time seriously—as Aristotle believed:[7] but this leads to contradiction or paradox. Perhaps Plato meant that the precedence of eternity is a logical or metaphysical priority as Plotinus seems to have understood his meaning; but this view suggests that time is temporally eternal and always *is;* how, then, except in terms of dynamism and change, could we distinguish an eternal time from eternity itself?

What, then, is time? Plato's text suggests that there is a kind of supervening duration of the world within which time, as we know it, made its appearance when ordered processes of change were rendered out of meaningless chaos. Time would then seem to be a kind of duration that is measurable in terms of change. Inextricably related in human experience, we are then forced to question whether time is something more than the changes by which we mark it. Is time itself measured by the order of change and motion, or is it nothing more than the measure itself, the measure of sequential relations in terms of which we understand the process of change? It is a focus on these questions that Aristotle made central to the ensuing discussion.

Aristotle did not start by seeking a view of the universe as a whole. His prime search seems to have been for the principles by which natural phenomena are to be understood. These constitute the order of nature which he conceived as a system of ordered motions that involve time. Taking motion and its temporal structure as given, he sought to understand their relationship.

But just how are they related? Aristotle's central thesis, which has dominated most philosophic and scientific discussion since, is that time is merely a mark of change or of the experience of change, that time is nothing more than the way we measure motion. Yet he did not develop this view without suggesting serious reservations and qualifications which themselves were to become questions for discussion.

Experience of change, all philosophers agree, has at least some kind of essential relationship to sense perception. But how does this bear on the nature of time? In the closing chapter of his work, *On the Senses,* Aristotle emphatically pointed out the central role of time in human sense experience. His specific argument arose out of what he regarded as the need for simultaneous perception of different qualities in objects together with the thesis that perception of succession requires the ability "to perceive every instant of time" (p. 47). In support of these positions he presented two arguments with implications obviously pointing beyond his own immediate discussion.

First, he emphatically tied time perception to self-awareness: the perception of any entity in a continuous time requires awareness of one's own existence. Therefore, no moment of time can be so small as to be imperceivable because that would involve ignorance of one's own existence as well as of one's perceptual experiencing. The continuity of the awareness of one's own being is necessary, he argued, for awareness of the moments constituting the time-continuum.

This line of argument immediately raises several questions. Are we, indeed, continually aware of our own existing selves? Need we be in order to have a continuity of experience or to experience the continuity of our own being? Do we need to be aware of every moment, no matter how minutely defined, in a temporal continuum, in order to experience its essential continuity? Indeed, is experiential time made up of truly separable moments? If, as Aristotle is to say, "nothing which it is possible to perceive is indivisible," [8] how are we to define the minuteness of these moments which he regarded as necessarily perceptible? And, if we grant his argument, how do we perceive time, aside from the objects that appear 'in' it? For we have no specific organ of time perception—only organs for perceiving surface aspects of things. Is time, then, really 'perceived', or is it a concomitant of the perception of things?

His second argument turned outwards and sought to establish the thesis that the perception of the instants of time is a necessary condition for the perception of any existent whole object. He did this by pursuing the hypothesis of the denial. But he used a phrase such as "a really whole time" which makes one wonder whether he was not, indeed, intimating that time is something more than merely the measure of motion (an intimation running through his essay on time which sought to establish that that is all it is).

Reverting to the original discussion which prompted these two arguments, he took up the question of the 'coinstantaneous' perception of different things: just as any perceived object appears as a unified whole, the perceiving soul must be one whole also. Aristotle here suggested not only some kind of parallelism between perceiver and perceived, but also the unified soul as the 'organ' of time perception. If awareness of self as well as of object depends upon the ability to perceive each instant of time, and if different objects appear 'across' those 'instants', then the soul as "the true perceiving subject" [9] must, in its unity be bridging the times of separate perceptions. It apparently does this by perceiving changes in its own states of awareness (see p. 51).

One of the most remarkable of issues which Aristotle has raised in this generally ignored discussion is that of the necessary relation between awareness of time and awareness of self. At first blush, his statement seems like a temporalized anticipation of the principle of the Cartesian cogito; but Descartes had neither built the notion of continuity into his principle nor did he suggest either a temporal form or a subjective basis for the temporal continuity of experience. For these reasons, Aristotle's identification of self-awareness and time-awareness seems much more of an anticipation of some contemporary thinkers to whom we turn in the later sections of this book.

In thinking through Aristotle's discussion of the relationship between time and experience, one should refer to his essay on *Memory and Recollection*, which he regarded as a sequel to *On the Senses;* he augmented this discussion of perceptual time by distinguishing memory, which he tied to sense perception, from recollection, which seems tied to conceptual reasoning.

Aristotle's prime motive seems to have been the attempt to understand the principles of the natural world. Both of his prime treatments of the concept of time are in works that are described as "physical." *On the Senses,* although a study of mental functioning and thereby a psychological work, was explicitly written from the vantage point of the "physical philosopher;" [10] Artistotle's famed and classic essay on the nature of time is in his *Physics.*

The thesis in the psychological work was that time perception is a necessary condition of awareness—of the self and of the objects of experience. In the *Physics,* despite the fact that he claims to tell us *what* time is, the focus is really on *how* it functions in our understanding of the objects in nature that we seek to understand in scientific terms. The discussion is within the context of natural science; all we have a right to expect, then, is a discussion of its character in scientific inquiry.

The essay is divided into five sections. The discussion is introduced by reviewing some of the special puzzles surrounding the nature of time: Does it exist? If so, how? We are quickly led to the centrality of the concept of the 'now'—reminding us of the psychological work which tied time perception to the present. Pointing to his essential argument, Aristotle associated time with motion while denying their identity.

The second section (sec. 11) is really the starting point. Time and movement are somehow intrinsically related. We always perceive them together: we do not perceive motion without time; we are unaware of time except in the presence of motion. But, if movement always has magnitude, then time must also: motion is perceived in terms of a sequence of 'before-and-after' and the interval between is measurable. Hence, we come upon the famed and crucial definition: "For time is just this—number of motion in respect of 'before' and 'after'" (p. 52). And, indeed, this is precisely what time is—for the student of movement (or change) in the physical world. Time, then, is measurable motion. Time is reckoned as a sequence of now-points and is continuous because it is an attribute of motion that is continuous.

In many ways, the heart of the essay is its third section (sec. 12) where the relationship between time and number is developed. Time is not the counting, but what is counted—in terms of before-and-after sequence. But, Aristotle noted, the relationship between time and movement is reciprocal "because they define each other." This reciprocity is even extended to the meaning of the phrase "to be in time;"—but it must be noted that the explanation is essentially temporal: 'to be in time' and 'to be in motion' are both defined in terms of the time-word 'when' (see p. 56). Despite the claim of reciprocity, the discussion proceeds in terms of time *qua* measure of motion; we are reminded that it is thereby also the measure of rest. Those things, then, that are "in time" are those things "whose existence is measured by it" in terms of motion and rest.

Such things are subject to "perishing and becoming." It is clear that Aristotle enunciated what has since become an old tradition of viewing time as destructive (rather than creative), as being "the cause . . . of decay" and consequently the burden we must bear. This thesis is reiterated in the fourth section (sec. 13), although a bit more cautiously: Time is "the condition of destruction," but not its cause. This fourth section discusses the concept of the 'now' as both unifying and dividing time; it also elucidates the meaning of some prime temporal expressions.

The final section (sec. 14) seems to have been written in a different key. At its outset we are told that "every change and everything that moves is in time"—but the meaning of this is difficult to comprehend if time is *merely* the measure of motion. In Sec. 11 we were told that the 'now' is not time but an attribute of it; here we are told that not only the 'now' but also the 'before-and-after' are *in* time. Since every movement or change involves these terms, it would seem to follow that *every* movement is also *in* time.

Aristotle then moves to the relationship of time and soul. He is obviously struggling with the seemingly simultaneous subjective and objective aspects of time already suggested in *On the Senses*. Different parts of the same long sentence—see 223a25 on pp. 60-61—suggest different positions: (a) If time is number it depends on soul; hence, "there would not be time unless there were soul." Motion would, indeed, exist, but time is apparently an attribute of the way in which a rational soul apprehends motion and is, therefore, in a real sense subjectively based. But (b), "before-and-after are attributes of movement and time is these *qua* numerable." But if so, aren't they still "numerable" even if no rational soul is around to number sequences of motion?

If we recall the psychological thesis that we actually perceive time, that it is from this perception that experience arises, it is difficult to escape the conclusion that Aristotle is working two different temporal perspectives against each other. He is not really concerned with the first because it can be regarded as merely speculative and beyond the context of his discussion. He can dismiss the relation of time to soul just because it is of no direct concern to the student of nature who does have a rational soul and, therefore, can be presumed to start with time even if it is soul-dependent. The scientist, as the ordinary man, is concerned to use the concept of time in order to understand change. For time is used as a conceptual instrument for understanding motion or change—and an aspect of any change is intelligible in terms of its quantifiable time, whatever else its time may also be.

Time brings into movement a direction of before-and-after sequence and a numbering process based upon it. Motion is understood in terms of successively discriminated points, or nows, along a line of continued sequence. These now-points have the order of number and each functions as a numerical unit. As time ties a before-and-after discriminated sequence into a movement, so the 'now' numbers the changing states of the particular thing. Our presumption is that perceived motion is intrinsically numerable, and it is this aspect of that motion we call time.

However, the continuing ambiguous relationship between time and motion is confounded when we are next told that "everything is measured by some one thing homogeneous with it . . . times by some definite time." And, again, "time is measured by motion as well as motion by time." The two are then essentially reciprocal and it is not clear why Aristotle persists in regarding motion as the fundament. The prime measure, anticipating the apex of the cosmology, is "regular circular motion;" hence, it might appear that time is measured *by* motion which is somehow homogeneous with it. With the introduction of the thesis of the primacy of circular motion and the consequent "circle of time," Aristotle brought his essay to an end.

One might well, however, refer to those sections of the *Physics* and the *Metaphysics* where Aristotle brought the main strands of his thinking together. For on the highest metaphysical level he continued the intrinsic relationship of time and motion by eternalizing both. Emphatically dissenting from Plato's view that time was created out of a timeless eternity, he conceived eternity as an endless series of moments; thus, he did not distinguish time from eternity as much as he eternalized time. But this unvarying continuity of temporal moments was tied to the unvarying continuity of the unmoved prime mover. Consequently, in a way, Aristotle came very close to Plato in one regard: he tied time to number, and based both on the "motion that is eternal" and the "infinite time" "during" which it transpires on the timeless perfection of the first cause of all that is.[11]

To the end, Aristotle consistently maintained the close connection of time and motion. As well he should—because they are inextricably connected in our experience. But no argument has been offered for his persistent attempt to reduce time to motion. From the fact that each is necessary for the experience of the other, it does not follow that time is nothing but the measure of motion; one could argue just as plausibly that motion is the measure of time, that we measure time lapses by motion, and, indeed, time-telling instruments claim to do just this.

Although Aristotle opened with the question 'what is time?', most of his discussion was concerned to ascertain its meaning in scientific and non-speculative experience. Perhaps the conflicts which arise stem from the confusion of the different questions being faced. And most curiously, the notion of spatiality, which is certainly crucial to the objectivity of physical objects and to the measuring of their changing states, is noticeable by its absence. We generally need a spatial as well as a time reference to say something about motion. But Aristotle ignored the spatial consideration except to note that 'place' is motionless (and thereby atemporal). Yet his final description of motion as of time is in spatial terms—circularity. Perhaps an unvoiced reason for the primacy he accords to motion is just that it is essentially a spatial activity. Time may be the measure of motion but the only prescription we have for so using it is to measure now-points along a spatial line.

He insisted on the priority of motion, subordinated time to it as a measurable attribute, yet often treated time almost as a reality parallel to it. And he could not refrain from lapsing into that ordinary expression which ordinary experience cannot avoid or explain, viz., our continual descriptions of things, events, and temporal sequences as being *in* time.

He suggested, and for reasons indicated, bypassed the notion that time may be mind-dependent. But the only probing of this suggestion he made is in terms of certain rudimentary linguistic analyses which did not tell us how time may relate to mind, but only how we—in Greek—talk about time. And the expressions he selected are, indeed, revealing. They all relate to our talk about succession of perceptual objects along the line of before-and-after sequence; none of them refers, despite his initial opening with inner experience, to the experience of temporal passage, of duration, of a now that is a living spread of time and not a mathematical point.

This is to suggest that Aristotle's working concept of time is strangely static; it lacks the dynamic aspect of time as we actually experience it and which his preoccupation with motion might have urged. The numbering of a before-and-after sequence, once done, does not change its truth value. To say that 'Plato was born some forty years before Aristotle' is, once true always true. If time is merely this unalterable chronology, it cannot itself explain why it is essentially asymmetrical unless motion is—but some motions, such as the movement of a pendulum, ideally need not be.

Although Aristotle urged that self-awareness and time-awareness are inextricably related, he did not pursue this thought to consider either the temporality of the self or the essentially dynamic quality of the self's experiencing. Really concerned with the sequence of events observed in nature, he found these sequences, in contrast to inner experience, to be essentially linear—even if the linearity be that of the circumference of a circle rather than that of an arrow. We may, then, understand his focus on the role of 'before-and-after' which is, indeed, time insofar as it may be said to provide the measure of motion in space. But Aristotle's many digressions and equivocations might be indicative of his own feeling that somehow time was also something more than just this.

Approximately five hundred years later, Plotinus, building on his own reading of Plato's *Timaeus*, entered the major attack on Aristotle's mode of treating time. Unlike Aristotle, Plotinus was not a practicing scientist; a notable influential metaphysician, his work has yet been described by Windelband as "the most definitive and thoroughly constructed system of science that antiquity produced." [12] He has exerted a deep influence on subsequent thought but, perhaps because his kind of metaphysics is currently out of fashion, his *Enneads* is unfortunately not often read today.

Born in Egypt, Plotinus came to have an honored position as a teacher in imperial Rome. He has usually been described as a religious writer and even as a mystic. But he was, first of all, a speculative metaphysician who built on his reading of Plato while borrowing freely from pre-Socratics, Aristotle, and some Asian thinkers. His new synthesis is usually referred to as neo-Platonism. Its essential thrust is the spiritualization of the universe which is seen as one integrated whole.

It is then, not surprising that Plotinus raised the question of the nature of time within the perspective of a world view.[13] Presuming Plato's thesis that eternity is more primordial than time, Plotinus viewed our understanding of time and eternity as in something of a dialectic relationship. Although time is not ultimately primordial, he argued, time is real; it cannot be reduced to number and is not merely a function of measurement; time is, indeed, the principle of order in the physical universe and the form of being in the world of engendered beings, the world to which we belong.

Given Plotinus' portrait of the structure of reality, it is clear that he had to insist on the correlate nature of time and of eternity, which is not time and more primordial than time. One may also anticipate that if time is to have any important level of reality—which he believed it does—it must be dependent on its ultimate source and thereby cannot be dependent on aspects of the material world which merely display it. Although Plotinus' theodicy explains the logic of his approach to the nature of time, one certainly need not accept it in order to derive an insight into the force of his understanding of time. In order to appreciate the thrust of his contribution to the discussion, it may be helpful to read his essay on time in what we might call secularized terms.

The thirteen sections of the essay divide into three groups. The first six sections ask us to look at eternity from the vantage of temporal beings: we can begin to understand the nature of time by looking for predicates which contrast to those ascribable to eternity. The second part (sections seven through ten) is concerned to examine in a 'cursory review of ancient opinions' what enlightenment they offer: we witness a sustained attack on all attempts to identify time with motion. Finally, beginning with section eleven, the essay seeks to tell us "what time really is."

In accord with Plotinus' attempt to spiritualize reality, he identified eternity, not with any kind of materiality, but with the life principle: it is "a life changelessly motionless," the entire system of interrelationships of all that is, was, or can be. That system is not itself changing; it is the unchanging, living framework within which all change occurs. It is not the Divine, but "the announcement" or sign of the Divine—which is always complete and, in contrast to the world of engendered beings, without need of any future. We are warned not to separate eternity from nature; eternity is an aspect of being which is always present and integral to nature.

Eternity may be likened to those aspects of nature which are not changeable—such as, perhaps, the laws of mathematics and physics, which are usually regarded as unchangeable even if our understandings of them change. Engendered or temporal beings depend upon futurity for their being: for futurity is the sign of lack of fulfillment, and consequently of novelty and change. In contrast, then, the eternal is changeless and does not need the 'more time' which futurity holds forth. Eternity is then perduring, changeless being and contains plurality within it. This is to say that eternity, in contrast to time, is without any sequence of 'before-and-after'. Time, therefore, cannot be eternal; time is an aspect of the created universe; one cannot use temporal terms, therefore, such as 'before' to refer to what was 'before the created universe came into being'; temporal terms refer only to what is *in* created time.

Then, what is time in which we exist? How do we, as temporal or engendered beings, participate in eternity as well? We must, Plotinus urged, "descend from eternity" and investigate the nature of time itself. Previous attempts to understand the nature of time did so in terms of its relationship with motion; time was identified as motion, as some kind of moving entity, as a phenomenon attendant upon motion such as measurement. Against all such identifications, Plotinus entered a strong attack.

Time cannot be motion because motion takes place *in* time and is, therefore, distinguishable from it. Time is continuous and, therefore, cannot be identified with the motion of the whole of the heavens—just because we measure its times in terms of time: time is not the motion of the entire cosmos or of parts of it. Search as we may into any motion, we find space and movement, but we do not perceive time. When we speak of the duration of things, we have variable movements which give us "not time, but times."

As we look at succession we do get number; the extent of motion is magnitude and is measurable. But the quantitative aspect of motion is not time but is found *in* time. And, again, if time is the measure of both motion and rest, it must be somehow different from both of them.

If time is not motion, but somehow related to it, Aristotle's central position might seem plausible: time is a number that measures motion. But what does this really mean? If time is an abstraction such as a number is—then why distinguish it from the number system? If time is a measure, with an extent of its own, it must have quantity and would be like a line parallel to motion. But if it shares in a moving or extendedness, how can it measure the moving and the extended? If they are parallel, why place the priority on the motion? Plotinus pointed us to the alternative: motion is really the measure of time. If time is continuous and unbroken, we must then find a continuous motion with distinguishable stages so that it is parallel to time and can be used to measure time.

Again, if the prime concern is the measured motion, is time the measured motion or the measuring magnitude? If time is the measured motion, then a measure outside of the motion is required for the measuring. But the measuring magnitude requires a measure. If time is the numerical value of the magnitude of motion, how can an abstract number undertake any act of measuring at all? And, if this last query be pushed aside, we have, at most, a particular quantity of time and not time itself. Whatever measures a succession of motions does so by time and in terms of time. Whatever time may be, it cannot be *merely* a numbering of motion. And, if we apply number to a portion of time, time cannot be merely number; it must have been before the number was applied to it.

Time must have existed before the measure and the measuring—for time, itself, does not need to be measured in order to have its being. Time cannot be described as a sequence attendant on motion because two questions then ensue: why the sequence? and, may not time be its cause rather than its result?

Attempts to understand time in terms of motion must all fail just because they constitute a superficial confusion of measure and measured. Essentially, Plotinus argued, we must see what time is in itself before we can speak intelligently about a certain amount of it.

What, then, is time?

Time is, first, not eternal; it is an engendered existent. It is an image of eternity, as produced by the Soul, which "clothed itself with time." Its energy brings forth "a constant progress of novelty" as succession is produced. Time is then "contained in differentiation of life;" it is defined as "the life of the Soul in movement as it passed from one stage of act or experience to another." Time is, then, a principle of the engendered world which manifests itself in the activity of being sequential; for only in terms of sequential order is the all-complete unity of the world able to have its essential unity expressed in the diversity that marks it. Time expresses the non-spatial or unextended oneness of the eternal in the unity of sequential succession of before-and-after, the immeasurable in an infinite futurity, the time-less concentration of the whole in a developing whole without final stage. In this essentially sequential way, time provides a moving approximation of the unified and unitary timeless order of reality.

The physical universe is in time as something of a frame for its being, its motion and its possible rest; but time is not exterior to it; time infuses it and all that is in it. The genesis of time is to be found in "the first stir of the Soul's tendency towards the production of the sensible universe," in the creative force, the *élan vital,* that forms and sustains the physical world. Time is the creative activity; what we usually call nature is its content.

But time, itself primordial in the created world, is not itself primordial. It depends upon the source of the created ordered world of time-bound things; it depends upon soul, the creative force which is present throughout the sensible universe, just as one's own individual soul is presumably present throughout one's body. The soul-act, on an individual or cosmic level, brings time with it as its activity; as the soul-activity is a continuity, so its time is also. Time, in its being, is then linked to soul as its source; it is endless, stretching into an infinite future because the soul's activity, in seeking a temporal counterpart to the timeless perfection of eternity, is endless. Complete fulfillment is unending and always needs look to its future; time can do no less.

Because time is not an object of perception as such, it is not directly measurable. We measure time by the movements of the heavenly bodies in space; from the two-fold alternation of night and day, we produce the concept of number which we use for this purpose. We number intervals within uniform motion and use this as a standard for measuring time. Continuous celestial movements, then, provide the measure of time but do not create it.

But this celestial movement is "performed in time." For the physical universe, time is primordial and is that in which everything that happens happens, "smoothly and under order." Order is its sign, not its source; for we know developmental order only because it signifies the rational relations of things in time. To define time, then, in terms of a sign by which we come to know it is a confession of a confusion of thought. To identify time as the measure of motion is to identify it with one of its non-essential uses. The nature of time is to serve as the ground of ordered change; we measure time by motion and should avoid confusing the measured with the measure.

Although Plotinus obviously built upon Plato's metaphor of time as the moving image of eternity, he abandoned Plato's identification of time and number. Plato had urged that time and eternity are in the same proportion as number and unity with number generated out of unity. Aristotle had carried this forward in his reduction of time to number. But Plotinus could not derive our notion of number from unity because he placed unity above even eternity, and he did not reduce time to number because he derived number from time. The concept of number, he argued, comes not from unity but from duality, the duality of the sequential alternation of day and night, from the first way in which we mark temporal passage.

Plato had suggested that the endlessness of time is tied to the endlessness of the number system; Aristotle had tied it to the endlessness of heavenly movement. Plotinus urged that time is tied neither to the abstraction of number nor to material entities external to soul or self.

Time is tied to concrete conscious existence, to the soul-act or creative activity which needs time for its being, for its essential activity, for its quest of fulfillment. Time is the form of activity of purposive life. The concrete self is inextricably involved with time in its activity of existing and seeking continued fulfillment. As the form of the activity in which the soul is engaged, its direction is toward the future which time presents for self-fulfillment. The self does, indeed, as part of its continuing activity, measure motion as an aspect of its continuing striving. But measuring requires a measurer, a measured, and a measure. The conscious self who sets out to do the measuring creates and uses the abstraction of number in application to observed motions in order to measure sequential time.

Plotinus has strongly attacked the thesis that time is to be understood as merely attendant upon motion. He reversed the priority and argued that motion is the measure of time which has a kind of reality of its own. His argument accords better with our ordinary experience—as evidenced by my use of the watch on my wrist—and with the way in which we express our experience in language—for we speak of things and events as happening "in time." But it still remains that the only way in which we can experience time or the passage of time and speak of different times is in terms of observed changes, of motion. Plotinus has, indeed, offered an explanation why this is so: dissenting from Aristotle, he argued that time itself cannot be perceived by us; consequently, we can only note its passage indirectly in terms of changes which are sequential and thereby serve as a sign of time-order. But when all is said and done, it is not clear just how Plotinus' insistence that time is more basic than motion alters their relationship in human experience. For we do number motion as we number time, and we measure each in terms of the other. In human experience, time and motion function together in terms of numerical judgments about sequential relationships.

The theoretical advantage of Plotinus' subordination of motion to time is one which he did not seem to notice. It offers us the beginning of an explanation as to why the sequence of before-and-after is irreversible. For, if time is real, independently of the motions 'in' it, time provides the form and possibility of change only in a 'forward' direction—from a prior state, to a present state to a state thereafter.

Plotinus did shift the Aristotelian focus on the past as the source of generation to the future as the source of continuing fulfillments. He shifted the emotive judgment that time is a burden to one that takes time as the arena of creativity. But these shifts are only suggested and left without development.

Despite the strong attack on Aristotle and the similarities that still remain, Plotinus has introduced a radically new departure into the discussion. He has taken up Aristotle's casual question which Aristotle casually brushed aside: the thought that somehow time is dependent upon the measurer, upon the soul or self doing the measuring. And this casual query he has built into a prime philosophical principle.

What emerges, if one excuses Plotinus' luxuriant metaphysical back-drop, is the thesis that time is objective—more primordial than any observable motion or sequence of change—and yet, paradoxically perhaps, grounded on soul as its 'subjective' act. Time, described as a unique kind of substance, is essentially activity or activity-related. Time is the primordial mark of the world as we experience it and act in it. Time is the primordial mark of our own selves. Time marks our experience and activity as essentially and pervasively temporal, for-ward-looking, ongoing, conceptually divisible yet essentially contin-uous with conscious being. In all of these ways, Plotinus initiated a tradition for understanding time and temporal experience which emerged in more modest metaphysical dress in the eighteenth cen-tury, nurtured prime philosophic movements of the nineteenth, and comes forcefully into our own day.

Approximately one hundred and fifty years after Plotinus, Augus-tine died in the North African city of Hippo while it was under Vandal attack. He was a prolific writer and was deeply influenced by Plotinus. As the first major philosopher of the Christian era, we see in Augus-tine the systematic effort to integrate the biblical and Greek cultural inheritances into one intellectual tradition. His work was to dominate Western thought until the thirteenth century. Although his influence then waned with the rise of a revived Aristotelianism, he can be seen as anticipating many elements of thought that entered into the rise of modern philosophy in the seventeenth and eighteenth centuries and into the existentialism of the twentieth.

His discussion of time starts from one problem posed in the opening pages of the Book of Genesis: if God "in the beginning" created time with the creation of the world, then what was God doing *before* the creation? Drawing the sharpest distinction between the timeless eter-nity of God and the essential temporality of man, Augustine faced questions raised by Plato, Aristotle, and Plotinus under the aegis of early Christian theology. But the radical turn in the way he faced their questions set the stage for very contemporary secular thought.

At the outset, he shifted the focus from the processes of change in the physical world to the human perspective which considers them. Apparently taking up Aristotle's casual question as to whether there could be time without soul, a measuring without a measurer, Augustine focused his inquiry on how we do indeed measure temporal passage. Dispensing with Plotinus' notion of a world-soul and focusing on individual soul instead, he pursued Plotinus' point that time is somehow linked to soul, the soul of the individual perceiver.

The human perspective, Augustine argued, is an essentially dynamic one in which past and future somehow arise out of the present perceiving moment. What is now seen as future will be seen as past. In contrast to Aristotle's discussion, which was wholly in terms of the ordered sequence of before-and-after stages in the development of external things, Augustine forcefully brought in a different analysis in terms of the shifting past-present-future of the experiencing subject. In this continual shifting, the three modes of experiential time are somehow tied together. In one sense the future follows from the past, but in another it seems that the past is "driven on" by the encroaching future. Yet past and future are literally non-existent and the only actuality is the point of perspective, the perceptual present.

This consideration led to Augustine's famed and anguished cry, "What then is time? I know what it is if no one asks me what it is; but if I want to explain it to someone who has asked me, I find that I do not know" (p. 82). Time, as future and as present, depends upon change; but insofar as the future, which is non-existent, passes through the actual present, into the no-longer-existent past, time somehow seems to depend upon non-existence. Facing a number of questions which arise from the essentially dynamic nature of temporal experience, Augustine noted that the measuring of time, in terms of long and short, is only possible in present perception. *In the present,* memory presents the presence of things past, expectation the presence of things future, while sight reports the presence of things present. We measure time in the present, but how can we do so if the present, as Aristotle's 'now', has no extension?

The expressions which we use to talk about time themselves consume time. We cannot really know what they mean until we know what time itself is. Language, then, does *not* give us a key to the nature of time, even if time is always referred to when we speak and write.

Time has been identified with the movements of the heavenly bodies. But these movements take time in which to transpire. Their durations are measured in terms of time. They can, then, be but signs of time but not time itself.

Yet time, as the measure of temporal extension, must be some kind of extension which bodily movements manifest to us. What we measure is, in one sense, not time but motion; in another, we use motion to measure time. But we can only measure what is appearing in our present experience. Time, then, seems to be a kind of extension that enters and permeates our entire range of experience.

What we really seem to be measuring are not the past or future changes about which we talk and which we cannot presently see, but our memories and expectations which are present to us in actual conscious awareness. If I can, indeed, measure changes from future to past within the content of my own mind, if I can compare the extension of different durations from within my present experience, then, whatever else time may be, it is at least somehow an extension of the perceiving mind itself. I can, indeed, know the time of my experiencing self—but hardly more than that, just because I find myself "spilled and scattered among times whose order I do not know" (p. 96). Time, itself, must be some concomitant of "created being" (p. 96). As such, its being is dependent upon eternity in the face of which man's knowledge stops.

All that we can know is experiential time; by looking inward we find that our experience is constituted in dynamic temporal terms. This temporal experience is formulated in the mind's comparing of what is present to it as memory, as perception, as expectation. What is given from memory we call the past; what is given in immediate perception we call the present; what is given in terms of expectation we call the future. And it is to this future that we look as the source of hope, salvation, and deliverance.

Radical as Augustine's turn was—from time as the mark of change in nature to time as the mode of human perception, from temporal description in terms of unalterable before-and-after sequence to the moving experiential perspective of past-present-future—Augustine remained within the context of the classical discussion which he substantively brought to a close. For, whether in terms of an external or internal point of reference, each of these four classical writers approached time questions by asking *what* time is and seeking the answer in terms of its relationship to the measuring of observed change.

Yet similar as Augustine's voiced concern was to that of his

predecessors, his introspective shift had far-reaching implications—
for his own discussion, for the new questions it suggests, and for its
impact on later philosophic thought.

In common with Plato and Plotinus, Augustine grounded the concept of time in that of eternity. For these three, time is a mark of
created being, of a derivative order of reality; nevertheless, this is not
to say that it is *un*real. All three linked time to mind; in contrast,
Augustine linked it, not to transcendent, but to individual mind or
soul.

Augustine, in common with Aristotle, pursued an empirical examination—but it is a different kind of empiricism. Aristotle's empiricism had looked out onto the world of nature and was concerned with
the objects of experience. Augustine's looked into the experiencing
mind and asked about the modes of its experiencing. Although they
both spoke in terms of the measure of motion, their concepts of
measure were not identical. Aristotle conceived of measure in quantitative terms; in contrast, Augustine saw measuring as an essentially
vital activity which is essentially not quantitative although it may *use*
quantitative terms in comparing durations. Aristotle's focus on the
'now'—as an indivisible moment which is not part of time but a
boundary of before-and-after discrimination—has been transformed
into the present attention of the soul which is, in a sense, part of time:
the present not only divides the future from the past but extends into
both of them in order to take them into its perspective. Merely as a
boundary, the present is conceivable as unextended. But in order to
measure time, it must take in both memory and expectation—and thus
cannot be reduced to an indivisible momentary point.

In Augustine's treatment, time is not an aspect of motion but an
almost absolute standard of even 'flow' by which motion can be
measured. This suggests that if all external motions were speeded up,
it would not, in contrast to Aristotle, mean that time is also; hence, we
would be able to note the acceleration of change. Time, again, is not
pure motion. For, Augustine had argued that even if the heavenly
motion by which we generally measure time would cease (or cease to
be observable), there would still be intervals to be measured, and any
measuring device, such as the potter's wheel, will do.

In contrast to Plotinus, Augustine declared the autonomy of motion
and of time. Plotinus had reversed the Aristotelian identification and

argued that time is the cause of motion. Augustine carried this forward by arguing that we measure motion by means of time only by comparing intervals of time; therefore, we could not measure motion unless we were able to measure time itself. Using time as measure requires that time not be dependent upon what is being measured. It also requires that time be permanent. Time has, therefore, a metaphysical independence of measure and the introspective search has been merely for that aspect of time which we use in the measure of motion.

But to do this in terms of memory and expectation raises at least two questions. We generally find that when we perceive much activity in a short duration, that time seems short but when we recall it in memory it seems to have been long; likewise, waiting for something to happen, the expectation, can seem very long, but once it has happened, the duration appears to have been short. In these conflicts of temporal perspective, which are we to take as authoritative—or should we really be quantifying them at all? Augustine has, in effect, defined time in terms of memory and expectation. But how can we define memory and expectation without already presuming the modes of past and future time?

What is time? As Augustine thinks about it, it is somehow at once composed of indivisible nows, yet extended; made up of discrete perceptual presents, yet essentially continuous; internal in mind, but referent to an external reality in space.

All three predecessors regarded time as external to the individual's perception. Plato and Plotinus had identified time with transcendent soul but little had been said about the individual soul's activity in the perception—and this has been the focus of Augustine's concern. But this suggests that the individual soul is somehow an autonomous being whose mind is, in some sense, independent of the 'external' world. As a being, essentially different from the external things whose motions are measured, it somehow is the permanent, at least in its attention, amidst the changes that it discerns. But how can there be any sense of objective time unless there is some kind of preestablished harmony between the individual soul and the things it observes? How can individual souls correlate their temporal experiences unless there is some kind of harmony between them or a common membership in some kind of over-soul? And, if the soul is the permanent amidst the

changes it considers, is the soul itself "in" (pp. 91, 95) the time it measures? It perceives only the present but it also, somehow, transcends the
present in providing the past and the future. The individual soul must
transcend the separate passing moments; it must provide the continuity of change. It is, therefore, not legitimate in those terms, to speak
of the atomicity of separate perceptual moments because we can only
become aware of them in the continuity of present with future and
with past. But, if the soul must somehow transcend the moments of its
continuing experience in order to discern them, can we truly say that
the self is "in" the time it reckons?

Augustine's perplexity seems to argue that although time is subjective insofar as it is measurable, it is objective insofar as awareness
of time is dependent upon external motion. Does time, then, exist
except for the individual mind? And what is the relation between time
and space?

Augustine had used the essentially spatial analogy of extension to
define time. In the end he followed his predecessors in explicitly
grounding time upon space. For Plato, the creation of temporal order
was an imposition of transcendent mind on the receptacle, on a
pre-existent spatial container. For Aristotle, time was the measure of
motion in space and even his transcendent unmoved mover, his analogue of the priority of the eternal, was described in spatial terms as
being at the edge of the universe. Augustine's final description of time
as some kind of 'extension' of mind is a spatial kind of description
which perhaps derives from the fact that he saw time as some kind of
concomitant of the divine creation of the "heavens and the earth."

This ontological priority of space before time, a priority presumed
if not explicitly voiced, marks a common feature of these classical
discussions (from which only Plotinus may conceivably be exempted).
It is to be challenged in the period marking the rise of modern
philosophy, a movement which will take up Augustine's methodological revolution and seek the ground of human knowledge, not
so much in the world that is experienced as in the experiencing itself.

3.

Plato:

THE CREATION OF TIME

The *Timaeus* of Plato (427?–347 B.C.) is his only dialogue on cosmology and the natural sciences and the only one in which he explicitly discussed the ontology of time as such. The selection (in the standard Jowett translation) is taken from the discourse of Timaeus in which he offered a mythological account of creation as a "likely" account.

Tim. First then, in my judgment, we must make a distinction and ask, What is that which always is and has no becoming; and what is that which is always becoming and never is? That which is apprehended by intelligence and reason is always in the same state; but that which is conceived by opinion with the help of sensation and without reason, is always in a process of becoming and perishing and never really is. Now everything that becomes or is created must of necessity be created by some cause, for without a cause nothing can be created. The work of the creator, whenever he looks to the unchangeable and fashions the form and nature of his work after an unchangeable pattern, must necessarily be made fair and perfect; but when he looks to the created only and uses a created pattern, it is not fair or perfect. Was the heaven then or the world, whether called by this or by any other more appropriate name—assuming the name, I am asking a question which has to be asked at the beginning of an enquiry about anything—was the world, I say, always in existence and without beginning? or created, and had it a beginning? Created, I reply, being visible and tangible and having a body, and therefore sensible; and all sensible things are apprehended by opinion and sense and are in a process of creation and created. Now that which is created must, as we affirm, of necessity be created by a cause. But the father and maker of all this universe is past finding out; and even if we found him, to tell of him to all men would be impossible. And there is still a question to be asked about him: Which of the patterns had the artificer in view when he made the world,—the pattern of the unchangeable, or of that which is created? If the world be indeed fair and the artificer good, it is manifest that he must have looked to that which is eternal; but if what cannot be said without blasphemy is true, then to the created pattern. Every one will see that he must have looked to the eternal; for the world is the fairest of creations and he is the best of causes. And having been created in this way, the world has been framed in the likeness of that which is apprehended by reason and mind and is unchangeable, and must therefore of necessity, if this is admitted, be a copy of something. Now it is all-important that the beginning of everything should be according to nature. And in speaking of the copy and the original we may assume that words are akin to the matter which they describe; when they relate to the lasting and permanent and intelligible, they ought to be lasting and unalterable, and, as far as their nature allows, irrefutable and immovable—nothing less. But when they express only the copy or likeness and not the eternal things themselves, they need only to be likely and analogous to the real words. As being is to becoming, so is truth to belief. If then, Socrates, amid the many opinions about the gods and the generation of the universe, we are not able to give notions which are altogether and in every respect exact and consistent with one another, do not be surprised. Enough, if we adduce probabilities as likely as any others; for we must remember that I who am the speaker, and you who are the judges, are only mortal men, and we ought to accept the tale which is probable and enquire no further.

Soc. Excellent, Timaeus; and we will do precisely as you bid us. The prelude is charming, and is already accepted by us—may we beg of you to proceed to the strain?

Tim. Let me tell you then why the creator * made this world of generation. He was good, and the good can never have any jealousy of anything. And being free from jealousy, he desired that all things should be as like himself as they could be. This is in the truest sense the origin of creation and of the world as we shall do well in believing on the testimony of wise men: God desired that all things should be

29

30

* [Often referred to in the literature as the Demiurge.—C. M. S.]

good and nothing bad, so far as this was attainable. Wherefore also finding the whole visible sphere not at rest, but moving in an irregular and disorderly fashion, out of disorder he brought order, considering that this was in every way better than the other. Now the deeds of the best could never be or have been other than the fairest; and the creator, reflecting on the things which are by nature visible, found that no unintelligent creature taken as a whole was fairer than the intelligent taken as a whole; and that intelligence could not be present in anything which was devoid of soul. For which reason, when he was framing the universe, he put intelligence in soul, and soul in body that he might be the creator of a work which was by nature fairest and best. Wherefore, using the language of probability, we may say that the world became a living creature truly endowed with soul and intelligence by the providence of God.

This being supposed, let us proceed to the next stage. In the likeness of what animal did the Creator make the world? It would be an unworthy thing to liken it to any nature which exists as a part only; for nothing can be beautiful which is like any imperfect thing; but let us suppose the world to be the very image of that whole of which all other animals both individually and in their tribes are portions. For the original of the universe contains in itself all intelligible beings, just as this world comprehends us and all other visible creatures. For the Deity, intending to make this world like the fairest and most perfect of intelligible beings, framed one visible animal comprehending within 31   itself all other animals of a kindred nature. Are we right in saying that there is one world, or that they are many and infinite? There must be one only, if the created copy is to accord with the original. For that which includes all other intelligible creatures cannot have a second or companion; in that case there would be need of another living being which would include both, and of which they would be parts, and the likeness would be more truly said to resemble not them, but that other which included them. In order then that the world might be solitary, like the perfect animal, the creator made not two worlds or an infinite number of them; but there is and ever will be one only begotten and created heaven. . . .

Now the creation took up the whole of each of the four elements; for the Creator compounded the world out of all the fire and all the water and all the air and all the earth, leaving no part of any of them nor any power of them outside. His intention was, in the first place, that the animal should be as far as possible a perfect whole and of perfect parts: secondly, that it should be one, leaving no remnants out 33 of which another such world might be created: and also that it should be free from old age and unaffected by disease. Considering that if heat and cold and other powerful forces which unite bodies surround and attack them from without when they are unprepared, they decompose them, and by bringing diseases and old age upon them, make them waste away—for this cause and on these grounds he made the world one whole, having every part entire, and being therefore perfect and not liable to old age and disease. And he gave to the world the figure which was suitable and also natural. . . .

Now God did not make the soul after the body, although we are speaking of them in this order; for having brought them together he would never have allowed that the elder should be ruled by the younger; but this is a random manner of speaking which we have, because somehow we ourselves too are very much under the dominion of chance. Whereas he made the soul in origin and excellence prior to and older than the body, to be the ruler and mistress, of whom 35 the body was to be the subject. . . .

Now when the Creator had framed the soul according to his will, he formed within her the corporeal universe and brought the two together, and united them centre to centre. The soul, interfused everywhere from the centre to the circumference of heaven, of which also she is the external envelopment, herself turning in herself, began a divine beginning of never-ceasing and rational life enduring 37 throughout all time. The body of heaven is visible, but the soul is invisible, and partakes of reason and harmony, and being made by the best of intellectual and everlasting natures, is the best of things created. . . .

When the father and creator saw the creature which he had made moving and living, the created image of the eternal gods, he rejoiced, and in his joy determined to make the copy still more like the original; and as this was eternal, he sought to make the universe eternal, so far as might be. Now the nature of the ideal being was everlasting, but to bestow this attribute in its fulness upon a creature was impossible. Wherefore he resolved to have a moving image of eternity, and when he set in order the heaven, he made this image eternal but moving according to number, while eternity itself rests in unity; and this image we call time. For there were no days and nights and months and years before the heaven was created, but when he constructed the heaven he created them also. They are all parts of time, and the past and future are created species of time, which we unconsciously but wrongly transfer to the eternal essence; for we say that he 'was,' he 'is,'

38   he 'will be,' but the truth is that 'is' alone is properly attributed to him, and that 'was' and 'will be' are only to be spoken of becoming in time, for they are motions, but that which is immovably the same cannot become older or younger by time, nor ever did or has become, or hereafter will be, older or younger, nor is subject at all to any of those states which affect moving and sensible things and of which generation is the cause. These are the forms of time, which imitates eternity and revolves according to a law of number. Moreover, when we say that what has become *is* become and what becomes *is* becoming, and that what will become *is* about to become and that the non-existent *is* non-existent,—all these are inaccurate modes of expression. But perhaps this whole subject will be more suitably discussed on some other occasion.

Time, then, and the heaven came into being at the same instant in order that, having been created together, if ever there was to be a dissolution of them, they might be dissolved together. It was framed after the pattern of the eternal nature, that it might resemble this as far as was possible; for the pattern exists from eternity, and the created heaven has been, and is, and will be, in all time. Such was the mind and thought of God in the creation of time. The sun and moon and five other stars, which are called the planets, were created by him in order to distinguish and preserve the numbers of time, and when he had made their several bodies, he placed them in the orbits in which the circle of the other was revolving—in seven orbits seven stars. . . .

39   That there might be some visible measure of their relative swiftness and slowness as they proceeded in their eight courses, God lighted a fire, which we now call the sun, in the second from the earth of these orbits that it might give light to the whole of heaven, and that the animals, as many as nature intended, might participate in number, learning arithmetic from the revolution of the same and the like. Thus, then, and for this reason the night and the day were created, being the period of the one most intelligent revolution. And the

month is accomplished when the moon has completed her orbit and overtaken the sun, and the year when the sun has completed his own orbit. Mankind, with hardly an exception, have not remarked the periods of the other stars, and they have no name for them, and do not measure them against one another by the help of number, and hence they can scarcely be said to know that their wanderings, being infinite in number and admirable for their variety, make up time. And yet there is no difficulty in seeing that the perfect number of time fulfils the perfect year when all the eight revolutions, having their relative degrees of swiftness, are accomplished together and attain their completion at the same time, measured by the rotation of the same and equally moving. After this manner, and for these reasons, came into being such of the stars as in their heavenly progress received reversals of motion, to the end that the created heaven might imitate the eternal nature, and be as like as possible to the perfect and intelligible animal.

4.

Aristotle:

PERCEIVING TIME AND SELF

Aristotle (384–322 B.C.) seems to have provided the first systematic attempt to correlate different areas of knowledge and
inquiry into a philosophic synthesis. This selection is taken from
the concluding seventh chapter of *On the Senses,* his study of the
specific senses, in the standard Ross translation. It is one of his
psychological studies which he regarded as a sequel to *De Anima*
and as prefatory to his study of *Memory and Recollection.*

448a   If, then, the sensibles denominated co-ordinates though in different
provinces of sense (e.g. I call Sweet and White co-ordinates though in
different provinces) stand yet more aloof, and differ more, from one
another than do any sensibles in the same province; while Sweet
differs from White even more than Black does from White, it is still
less conceivable that one should discern then [viz. sensibles in different sensory provinces whether co-ordinates or not] coinstantaneously
than sensibles which are in the same province. Therefore, if coinstantaneous perception of the latter be impossible, that of the former
is *a fortiori* impossible.

Some of the writers who treat of concords assert that the sounds
combined in these do not reach us simultaneously, but only appear to
do so, their real successiveness being unnoticed whenever the time it
involves is [so small as to be] imperceptible. Is this true or not? One
might perhaps, following this up, go so far as to say that even the
current opinion that one sees and hears coinstantaneously is due
merely to the fact that the intervals of time [between the really successive perceptions of sight and hearing] escape observation. But this
can scarcely be true, nor is it conceivable that any portion of time
should be [absolutely] imperceptible, or that any should be absolutely

unnoticeable; the truth being that it is possible [1] to perceive every instant of time. [This is so]; because if it is inconceivable that a person should, while perceiving himself or aught else in a continuous time, be at any instant unaware of his own existence; while, obviously, the assumption, that there is in the time-continuum a time so small as to be absolutely imperceptible, carries the implication that a person would, during such time, be unaware of his own existence, as well as of his seeing and perceiving; [this assumption must be false].

Again, if there is any magnitude, whether time or thing, absolutely imperceptible owing to its smallness, it follows that there would not be either a thing which one perceives, or a time in which one perceives it, unless in the sense that in some part of the given time he sees some part of the given thing. For [let there be a line $\alpha\beta$, divided into two 448b parts at γ, and let this line represent a whole object and a corresponding whole time. Now,] if one sees the whole line, and perceives it during a time which forms one and the same continuum, only in the sense that he does so in some portion of this time, let us suppose the part $\gamma\beta$, representing a time in which by supposition he was perceiving nothing, cut off from the whole. Well, then, he perceives *in* a certain part [viz. in the remainder] of the time, or perceives *a part* [viz. the remainder] of the line, after the fashion in which one sees the whole earth by seeing some given part of it, or walks in a year by walking in some given part of the year. But [by hypothesis] in the part $\beta\gamma$ he perceives nothing: therefore, in fact, he is said to perceive the whole object and during the whole time simply because he perceives [some part of the object] in some part of the time $\alpha\beta$. But [2] the same argument holds also in the case of $\alpha\gamma$ [the remainder, regarded in its turn as a whole]; for it will be found [on this theory of vacant times and imperceptible magnitudes] that one always perceives only in some part of a given whole time, and perceives only some part of a

1. To demonstrate this directly, Aristotle might have again employed his distinction between actuality and potentiality. But he chooses here the method of *reductio ad absurdum.*—Trans.
2. Since it is not really possible in any concrete case to divide a whole object and the time of its perception, as we have divided the line, secluding, as if known, the part not perceived and the time in which no perception takes place.—Trans.

whole magnitude, and that it is impossible to perceive any [really] whole [object in a really whole time; a conclusion which is absurd, as it would logically annihilate the perception of both Objects and Time].

Therefore we must conclude that all magnitudes are perceptible, but their actual dimensions do not present themselves immediately in their presentation as objects. One sees the sun, or a four-cubit rod at a distance, as a magnitude, but their exact dimensions are not given in their visual presentation: nay, at times an object of sight appears indivisible, but [vision, like other special senses, is fallible respecting 'common sensibles', e.g. magnitude, and] nothing that one sees is really indivisible. The reason of this has been previously explained. It is clear then, from the above arguments, that no portion of time is imperceptible.

5.

Aristotle:

TIME AS MEASURE

This selection comprises the whole of Aristotle's famous essay on the nature of time. It appeared as the third part of Book IV of the *Physics* (in the standard Ross translation), the first and second parts being concerned to elucidate the concepts of 'place' and 'void'. It should, then, be borne in mind that, because of its context, the essay was primarily concerned with the concept of time as it functions in the study of nature.

10.

Next for discussion after the subjects mentioned is Time. The best plan will be to begin by working out the difficulties connected with it, making use of the current arguments. First, does it belong to the class of things that exist or to that of things that do not exist? Then secondly, what is its nature? To start, then: the following considerations would make one suspect that it either does not exist at all or barely, and in an obscure way. One part of it has been and is not, while the other is going to be and is not yet. Yet time—both infinite time and any time you like to take—is made up of these. One would naturally suppose 218ª that what is made up of things which do not exist could have no share in reality.

Further, if a divisible thing is to exist, it is necessary that, when it exists, all or some of its parts must exist. But of time some parts have been, while others have to be, and no part of it *is*, though it is divisible. For what is 'now' is not a part: a part is a measure of the whole, which must be made up of parts. Time, on the other hand, is not held to be made up of 'nows'.

Again, the 'now' which seem to bound the past and the future—does it always remain one and the same or is it always other and other? It is hard to say.

(1) If it is always different and different, and if none of the *parts* in time which are other and other are simultaneous (unless the one contains and the other is contained, as the shorter time is by the longer), and if the 'now' which is not, but formerly was, must have ceased-to-be at some time, the *'nows'* too cannot be simultaneous with one another, but the prior 'now' must always have ceased-to-be. But the prior 'now' cannot have ceased-to-be in[1] itself (since it thenexisted); yet it cannot have ceased-to-be in another 'now'. For we may lay it down that one 'now' cannot be next to another, any more than point to point. If then it did not cease-to-be in the next 'now' but in another, it would exist simultaneously with the innumerable 'nows' between the two—which is impossible.

Yes, but (2) neither is it possible for the 'now' to remain always the same. No determinate divisible thing has a single termination, whether it is continuously extended in one or in more than one dimension: but the 'now' is a termination, and it is possible to cut off a determinate time. Further, if coincidence in time (i.e. being neither prior nor posterior) means to be 'in one and the same "now" ', then, if both waht is before and what is after are in the same 'now', things which happened ten thousand years ago would be simultaneous with what has happened to-day, and nothing would be before or after anything else.

This may serve as a statement of the difficulties about the attributes of time.

As to what time is or what is its nature, the traditional accounts give us as little light as the preliminary problems which we have worked through.

Some assert that it is (1) the movement of the whole, others that it is (2) the sphere itself.[2]

(1) Yet part, too, of the revolution is a time, but it certainly is not a revolution: for what is taken is part of a revolution, not a revolution. Besides, if there were more heavens than one, the movement of any of

218ᵇ

---

1. The argument would be clearer if we could say 'during' itself. If the existent perished 'in' itself, it would never exist without perishing.—Trans.
2. Aristotle is probably referring to Plato and the Pythagoreans respectively.—Trans.

them equally would be time, so that there would be many times at the same time.

(2) Those who said that time is the sphere of the whole thought so, no doubt, on the ground that all things are in te and all things are in the sphere of the whole. The view is too naive for it to be worth while to consider the impossibilities implied in it.

But as time is most usually supposed to be (3) motion and a kind of change, we must consider this view.

Now (a) the change or movement of each thing is only *in* the thing which changes or *where* the thing itself which moves or changes may chance to be. But time it present equally everywhere and with all things.

Again, (b) change is always faster or slower, whereas time is not: for 'fast' and 'slow' are defined by time—'fast' is what moves much in a short time, 'slow' what moves little in a long time; but time is not defined by time, by being either a certain amount or a certain kind of it.

Clearly then it is not movement. (We need not distinguish at present between 'movement' and 'change'.)

11.

But neither does time exist without change; for when the state of our own minds does not change at all, or we have not noticed its changing, we do not realize that time has elapsed, any more than those who are fabled to sleep among the heroes in Sardinia do when they are awakened; for they connect the earlier 'now' with the later and make them one, cutting out the interval because of their failure to notice it. So, just as, if the 'now' were not different but one and the same, there would not have been time, so too when its difference escapes our notice the interval does not seem to be time. If, then, the non-realization of the existence of time happens to us when we do not distinguish any change, but the soul seems to stay in one indivisible state, and when we perceive and distinguish we say time has elapsed, evidently time is not independent of movement and change. It is evident, then, that time is neither movement nor independent of   219a movement.

We must take this as our starting-point and try to discover—since we wish to know what time is—what exactly it has to do with movement.

Now we perceive movement and time together: for even when it is dark and we are not being affected through the body, if any movement takes place in the mind we at once suppose that some time also has elapsed; and not only that but also, when some time is thought to have passed, some movement also along with it seems to have taken place. Hence time is either movement or something that belongs to movement. Since then it is not movement, it must be the other.

But what is moved is moved from something to something, and all magnitude is continuous. Therefore the movement goes with the magnitude. Because the magnitude is continuous, the movement too must be continuous, and if the movement, then the time; for the time that has passed is always thought to be in proportion to the movement.

The distinction of 'before' and 'after' holds primarily, then, in place; and there in virtue of relative position. Since then 'before' and 'after' hold in magnitude, they must hold also in movement, these corresponding to those. But also in time the distinction of 'before' and 'after' must hold, for time and movement always correspond with each other. The 'before' and 'after' in motion identical in substratum with motion yet differs from it in definition, and is not identical with motion.

But we apprehend time only when we have marked motion, marking it by 'before' and 'after'; and it is only when we have perceived 'before' and 'after' in motion that we say that time has elapsed. Now we mark them by judging that A and B are different, and that some third thing is intermediate to them. When we think of the extremes as different from the middle and the mind pronounces that the 'nows' are two, one before and one after, it is then that we say that there is time, and this that we say is time. For what is bounded by the 'now' is thought to be time—we may assume this.

When, therefore, we perceive the 'now' as one, and neither as before and after in a motion nor as an identity but in relation to a 'before' and an 'after', no time is thought to have elapsed, because there has been no motion either. On the other hand, when we do 219ᵇ perceive a 'before' and an 'after', then we say that there is time. For time is just this—number of motion in respect of 'before' and 'after'.

Hence time is not movement, but only movement in so far as it admits of enumeration. A proof of this: we discriminate the more or the less by number, but more or less movement by time. Time then is a kind of number. (Number, we must note, is used in two senses—both of what is counted or the countable and also of that with which we count. Time obviously is what is counted, not that with which we count: these are different kinds of thing.)

Just as motion is a perpetual succession, so also is time. But every simultaneous time is self-identical; for the 'now' as a subject is an identity, but it accepts different attributes.[3] The 'now' measures time, in so far as time involves the 'before and after'.

The 'now' in one sense is the same, in another it is not the same. In so far as it is in succession, it is different (which is just what its being now was supposed to mean), but its substratum is an identity: for motion, as was said, [219a 11], goes with magnitude,[4] and time, as we maintain, with motion. Similarly, then, there corresponds to the point[5] the body which is carried along, and by which we are aware of the motion and of the 'before and after' involved in it. This is an identical *substratum* (whether a point or a stone or something else of the kind), but it has different *attributes*—as the sophists assume that Coriscus' being in the Lyceum is a different thing from Coriscus' being in the market-place.[6] And the body which is carried along is different, in so far as it is at one time here and at another there. But the 'now' corresponds to the body that is carried along, as time corresponds to the motion. For it is by means of the body that is carried along that we become aware of the 'before and after' in the motion, and if we regard these as countable we get the 'now.' Hence in these also the 'now' as substratum remains the same (for it is what is before and after in movement), but what is predicated of it is different; for it is in so far as the 'before and after' is numerable that we get the 'now'.

3. E. g., if you come in when I go out, the time of your coming in is in fact the time of my going out, though for it to be the one and to be the other are different things.—Trans.
4. i. e. with the path traversed.—Trans.
5. i. e. in the path.—Trans.
6. sc, to prove that Coriscus is different from himself. I. e., they assume that a difference in the attribute means a difference in the substratum.—Trans.

This is what is more knowable: for, similarly, motion is known because of that which is moved, locomotion because of that which is carried. For what is carried is a real thing, the movement is not. Thus what is called 'now' in one sense is always the same; in another it is not the same: for this is true also of what is carried.

Clearly, too, if there were no time, there would be no 'now', and vice versa. Just as the moving body and its locomotion involve each other mutually, so too do the number of the moving body and the number of its locomotion. For the number of the locomotion is time, while the 'now' corresponds to the moving body, and is like the unit of number.

Time, then, also is both made continuous by the 'now' and divided at it. For here too there is a correspondence with the locomotion and the moving body. For the motion or locomotion is made one by the thing which is moved, because *it* is one—not because it is one in its own nature (for there might be pauses in the movement of such a thing)—but because it is one in definition[7]: for this determines the movement as 'before' and 'after'. Here, too, there is a correspondence with the point; for the point also both connects and terminates the length—it is the beginning of one and the end of another. But when you take it in this way, using the one point as two, a pause is necessary, if the same point is to be the beginning and the end. The 'now' on the other hand, since the body carried is moving, is always different.

Hence time is not number in the sense in which there is 'number' of the same point because it is beginning and end, but rather as the extremities of a line form a number, and not as the parts of the line do so, both for the reason given (for we can use the middle point as two, so that on that analogy time might stand still), and further because obviously the 'now' is no *part* of time nor the section any part of the movement, any more than the points are parts of the line—for it is two *lines* that are *parts* of one line.

In so far then as the 'now' is a boundary, it is not time, but an attribute of it; in so far as it numbers, it is number; for boundaries belong only to that which they bound, but number (e.g. ten) is the number of these horses, and belongs also elsewhere.

It is clear, then, that time is 'number of movement in respect of the

220ª

7. i.e. as *moved.*—Trans.

before and after', and is continuous since it is an attribute of what is continuous.

12.

The smallest number, in the strict sense of the word 'number', is two. But of number as concrete, somtimes there is a minimum, sometimes not: e.g. of a 'line', the smallest in respect of *multiplicity* is two (or, if you like, one), but in respect of *size* there is no minimum; for every line is divided *ad infinitum.* Hence it is so with time. In respect of number of the minimum is one (or two); in point of extent there is no minimum.

It is clear, too, that time is not described as fast or slow, but as many or few[8] and as long or short. For as continuous it is long or short and as a number many or few, but it is not fast or slow—any more than any number with which we number is fast or slow.

Further, there is the same time everywhere at once, but not the same time before and after, for while the present change is one, the change which has happened and that which will happen are different. Time is not number with which we count, but the number of things which are counted, and this according as it occurs before or after is always different, for the 'nows' are different. And the number of hundred horses and a hundred men is the same, but the things numbered are different—the horses from the men. Further, as a movement can be one and the same again and again, so too can time, e.g. a year or a spring or an autumn.

Not only do we measure the movement by the time, but also the time by the movement, because they define each other. The time marks the movement, since it is its number, and the movement the time. We describe the time as much or little, measuring it by the movement, just as we know the number by what is numbered, e.g. the number of the horses by one horse as the unit. For we know how many horses there are by the use of the number; and again by using the one horse as unit we know the number of the horses itself. So it is with the time and the movement; for we measure the movement by the time

8. e.g. 'many years.'—Trans.

and vice versa. It is natural that this should happen; for the movement goes with the distance and the time with the movement, because they are quanta and continuous and divisible. The movement has these attributes because the distance is of this nature, and the time has them because of the movement. And we measure both the distance by the movement and the movement by the distance; for we say that the road is long, if the journey is long, and that this is long, if the road is long—the time, too, if the movement, and the movement, if the time.

221ª   Time is a measure of motion and of being moved, and it measures the motion by determining a motion which will measure exactly the whole motion, as the cubit does the length by determining an amount which will measure out the whole. Further 'to be in time' means, for movement, that both it and its essence are measured by time (for simultaneously it measures both the movement and its essence, and this is what is being in time means for it, that its essence should be measured).

Clearly then 'to be in time' has the same meaning for other things also, namely, that their being should be measured by time. 'To be in time' is one of two things: (1) to exist when time exists, (2) as we say of some things that they are 'in number'. The latter means either what is a part or mode of number—in general, something which belongs to number—or that things have a number.

Now, since time is number, the 'now' and the 'before' and the like are in time, just as 'unit' and 'odd' and 'even' are in number, i.e. in the sense that the one set belongs to number, the other to time. But things are in time as they are in number. If this is so, they are contained by time as things in place are contained by place.

Plainly, too, to be in time does not mean to coexist with time, any more than to be in motion or in place means to coexist with motion or place. For if 'to be in something' is to mean this, then all things will be in anything, and the heaven will be in a grain; for when the grain is, then also is the heaven. But this is a merely incidental conjunction, whereas the other is necessarily involved: that which is in time necessarily involves that there is time when *it* is, and that which is in motion that there is motion when *it* is.

Since what is 'in time' is so in the same sense as what is in number is so, a time greater than everything in time can be found. So it is necessary that all things in time should be contained by time, just like other things also which are 'in anything', e.g. the things 'in place' by place.

A thing, then, will be affected by time, just as we are accustomed to say that time wastes things away, and that all things grow old through time, and that there is oblivion owning to the lapse of time, but we do not say the same of getting to know or of becoming young or fair. For time is by its nature the cause rather of decay, since it is the number of   221ᵇ change, and change removes what is.

Hence, plainly, things which are always are not, as such, in time, for they are not contained by time, nor is their being measured by time. A proof of this is that none of them is *affected* by time, which indicates that they are not in time.

Since time is the measure of motion, it will be the measure of rest too—indirectly. For all rest is in time. For it does not follow that what is in time is moved, though what is in motion is necessarily moved. For time is not motion, but 'number of motion': and what is at rest, also, can be in the number of motion. Not everything that is not in motion can be said to be 'at rest'—but only that which can be moved, though it actually is not moved, as was said above [202ᵃ4].

'To be in number' means that there is a number of the thing, and that its being is measured by the number in which it is. Hence if a thing is 'in time' it will be measured by time. But time will measure what is moved and what is at rest, the one *qua* moved, the one *qua* at rest; for it will measure their motion and rest respectively.

Hence what is moved will not be measurable by the time simply in so far as it has quantity, but in so far as its *motion* has quantity. Thus none of the things which are neither moved nor at rest are in time: for 'to be in time' is 'to be measured by time', while time is the measure of motion and rest.

Plainly, then, neither will everything that does not exist be in time, i.e. those non-existent things that cannot exist, as the diagonal cannot be commensurate with the side.

Generally, if time is directly the measure of motion and indirectly of other things, it is clear that a thing whose existence is measured by it will have its existence in rest or motion. Those things therefore which are subject to perishing and becoming—generally, those which at one time exist, at another do not—are necessarily in time: for there is a greater time which will extend both beyond their existence and

beyond the time which measures their existence. Of things which do
not exist but are contained by time some were, e.g. Homer once was,
some will be, e.g. a future event; this depends on the direction in
which time contains them; if on both, they have both modes of
existence. As to such things as it does not contain in any way, they
neither were nor are nor will be. These are those non-existents whose
opposites always are, as the incommensurability of the diagonal always is—and this will not be in time. Nor will the commensurability,
therefore; hence this eternally is not, because it is contrary to what
eternally is. A thing whose contrary is not eternal can be and not be,
and it is of such things that there is coming to be and passing away.

13.

The 'now' is the link of time, as has been said [220a5] (for it connects
past and future time), and it is a limit of time (for it is the beginning of
the one and the end of the other). But this is not obvious as it is with
the point, which is fixed. It divides potentially, and in so far as it is
dividing the 'now' is always different, but in so far as it connects it is
always the same, as it is with mathematical lines. For the intellect it is
not always one and the same point, since it is other and other when
one divides the line; but in so far as it is one, it is the same in every
respect.

So the 'now' also is in one way a potential dividing of time, in
another the termination of both parts, and their unity. And the
dividing and the uniting are the same thing and in the same reference,
but in essence they are not the same.

So one kind of 'now' is described in this way: another is when the
time is *near* this kind of 'now'. 'He will come now' because he will
come to-day; 'he has come now' because he came to-day. But the
things in the *Iliad* had not happened 'now', nor is the flood 'now'—not
that the time from now to them is not continuous, but because they are
not near.

'At some time' means a time determined in relation to the first of the
two types of 'now', e.g. 'at some time' Troy was taken, and 'at some
time' there will be a flood; for it must be determined with reference to
the 'now'. There *will* thus be a determinate time from this 'now' to that

and there *was* such in reference to the past event. But if there be no time which is not 'sometime', every time will be determined.

Will time then fail? Surely not, if motion always exists. Is time then always different or does the same time recur? Clearly time is, in the same way as motion is. For if one and the same motion sometimes recurs, it will be one and the same time, and if not, not.

Since the 'now' is an end and a beginning of time, not of the same 222b time however, but the end of that which is past and the beginning of that which is to come, it follows that, as the circle has its convexity and its concavity, in a sense, in the same thing, so time is always at a beginning and at an end. And for this reason it seems to be always different; for the 'now' is not the beginning and the end of the same thing; if it were, it would be at the same time and in the same respect two opposites. And time will not fail; for it is always at a beginning.

'Presently' or 'just' refers to the part of future time which is near the indivisible present 'now' ('When do you walk?' 'Presently', because the time in which he is going to do so is near), and to the part of past time which is not far from the 'now' ('When do you walk?' 'I have just been walking'). But to say that Troy has just been taken—we do not say that, because it is too far from the 'now'. 'Lately', too, refers to the part of past time which is near the present 'now'. 'When did you go?' 'Lately', if the time is near the existing now. 'Long ago' refers to the distant past.

'Suddenly' refers to what has departed from its former condition in a time imperceptible because of its smallness; but it is the nature of *all* change to alter things from their former condition. In time all things come into being and pass away; for which reason some called it the wisest of all things, but the Pythagorean Paron[9] called it the most stupid, because in it we also forget; and his was the truer view. It is clear then that it must be in itself, as we said before [221G1] the condition of destruction rather than of coming into being (for change, in itself, makes things depart from their former condition), and only incidentally of coming into being, and of being. A sufficient evidence of this is that nothing comes into being without itself moving somehow and acting, but a thing can be destroyed even if it does not

9. Nothing further is known of Paron.—Trans.

move at all. And this is what, as a rule, we chiefly mean by the thing's being destroyed by time. Still, time does not work even this change; even this sort of change takes place *incidentally* in time.

We have stated, then, that time exists and what it is, and in how many senses we speak of the 'now', and what 'at some time', 'lately', 'presently' or 'just', 'long ago', and 'suddenly' mean.

14.

These distinctions having been drawn, it is evident that every change and everything that moves is in time; for the distinction of faster and slower exists in reference to all change, since it is found in every instance. In the phrase 'moving faster' I refer to that which changes before another into the condition in question, when it moves over the same interval and with a regular movement; e.g. in the case of locomotion, if both things move along the circumference of a circle, or both along a straight line; and similarly in all other cases. But what is *before* is in time; for we say 'before' and 'after' with reference to the distance from the 'now', and the 'now' is, the boundary of the past and the future; so that since 'nows' are in time, the before and the after will be in time too; for in that in which the 'now' is the distance from the 'now' will also be. But 'before' is used contrariwise with reference to past and to future time; for in the past we call 'before' what is farther from the 'now', and 'after' what is nearer, but in the future we call the nearer 'before' and the farther 'after'. So that since the 'before' is in time, and every movement involves a 'before', evidently every change and every movement is in time.

223ª

It is also worth considering how time can be related to the soul; and why time is thought to be in everything, both in earth and in sea and in heaven. Is it because it is an attribute, or state, of movement (since it is the number of movement) and all these things are movable (for they are all in place), and time and movement are together, both in respect of potentiality and in respect of actuality?

Whether if soul did not exist time would exist or not, is a question that may fairly be asked; for if there cannot be some one to count there cannot be anything that can be counted, so that evidently there cannot be number; for number is either what has been, or what can be, counted. But if nothing but soul, or in soul reason, is qualified to count, there would not be time unless there were soul, but only that of which time is an attribute, i.e. if *movement* can exist without soul, and the before and after are attributes of movement, and time is these *qua* numerable.

One might also raise the question what sort of movement time is the number of. Must we not say 'of *any* kind'? For things both come into being in time and pass away, and grow, and are altered in time, and are moved locally; thus it is of each movement *qua* movement that time is the number. And so it is simply the number of continuous movement, not of any particular kind of it.

But other things as well may have been moved now, and there would be a number of each of the two movements. Is there another time, then, and will there be two equal times at once? Surely not. For a time that is both equal and simultaneous is one and the same time, and even those that are not simultaneous are one in kind; for if there were dogs, and horses, and seven of each, it would be the same number. So, too, movements that have simultaneous limits have the same time, yet the one may in fact be fast and the other not, and one may be locomotion and the other alteration; still the time of the two changes is the same if their number also is equal and simultaneous; and for this reason, while the movements are different and separate, the time is everywhere the same, because the number of equal and simultaneous movements is everywhere one and the same.

Now there is such a thing as locomotion, and in locomotion there is included circular movement, and everything is measured by some one thing homogeneous with it, units by a unit, horses by a horse, and similarly times by some definite time, and, as we said [220ᵇ28] time is measured by motion as well as motion by time (this being so because by a motion definite in time the quantity both of the motion and of the time is measured): if, then, what is first is the measure of eveything homogeneous with it, regular circular motion is above all else the measure, because the number of this is the best known. Now neither alteration nor increase nor coming into being can be regular, but locomotion can be. This also is why time is thought to be the movement of the sphere, viz. because the other movements are measured by this, and time by this movement.

This also explains the common saying that human affairs form a circle, and that there is a circle in all other things that have a natural movement and coming into being and passing away. This is because all other things are discriminated by time, and end and begin as though conforming to a cycle; for even time itself is thought to be a circle. And this opinion again is held because time is the measure of this kind of locomotion and is itself measured by such. So that to say that the things that come into being form a circle is to say that there is a circle of time; and this is to say that it is measured by the circular movement; for apart from the measure nothing else to be measured is observed; the whole is just a plurality of measures.

224ª

It is said rightly, too, that the number of the sheep and of the dogs is the same *number* if the two numbers are equal, but not the same *decad* or the same *ten;* just as the equilateral and the scalene are not the same *triangle,* yet they are the same *figure,* because they are both triangles. For things are called the same so-and-so if they do not differ by a differentia of that thing, but not if they do; e.g. triangle differs from triangle by a differentia of triangle, therefore they are different triangles; but they do not differ by a differentia of figure, but are in one and the same division of it. For a figure of one kind is a circle and a figure of another kind a triangle, and a triangle of one kind is equilateral and a triangle of another kind scalene. They are the same figure, then, and that, triangle, but not the same triangle. Therefore the number of two groups also is the same number (for their number does not differ by a differentia of number), but it is not the same decad; for the things of which it is asserted differ; one group are dogs, and the other horses.

We have now discussed time—both time itself and the matters appropriate to the consideration of it.

6.

Plotinus:

TIME AND ENGENDERED BEING

The sole book by Plotinus (A.D. 205–270) was entitled *The En-neads*. Actually a collection of his fifty-four essays arranged in systematic order by his pupil, Porphyry, (who earned his own place in the history of logic), the book is divided into six groups of nine essays each—hence, its name. The selection is taken from the thirteen sections comprising the Seventh Tractate (entitled "Time and Eternity") of the Third Ennead as translated by Stephen MacKenna.*

1.

Eternity and Time; two entirely separate things, we explain "the one having its being in the everlasting Kind, the other in the realm of Process, in our own Universe"; and, by continually using the words and assigning every phenomenon to the one or the other category, we come to think that, both by instinct and by the more detailed attack of thought, we hold an adequate experience of them in our minds without more ado.

When, perhaps, we make the effort to clarify our ideas and close into the heart of the matter we are at once unsettled: our doubts throw us back upon ancient explanations; we choose among the various theories, or among the various interpretations of some one theory, and so we come to rest, satisfied, if only we can counter a question with an approved answer, and glad to be absolved from further enquiry.

Now, we must believe that some of the venerable philosophers of old discovered the truth; but it is important to examine which of them

---

*The text is taken from the edition published by Charles T. Branford Co., Boston, n.d.

really hit the mark and by what guiding principle we can outselves attain to certitude.

What, then, does Eternity really mean to those who (thus casually) describe it as something different from Time? We begin with Eternity, since when the standing Exemplar is known, its representation in image—which Time is understood to be—will be clearly apprehended—though it is of course equally true, admitting this relationship of Time as image to Eternity the original, that if we chose to begin by identifying Time we could thence proceed upwards by Recognition (the Platonic Anamnesis) and become aware of the Kind which it images.

3.

. . . We know it [Eternity] as a Life changelessly motionless and ever holding the Universal content (time, space and phenomena) in actual presence; not this now and now that other, but always all; not existing now in one mode and now in another, but a consummation without part or interval. All its content is in immediate concentration as at one point; nothing in it ever knows development: all remains identical within itself, knowing nothing of change, for ever in a Now since nothing of it has passed away or will come into being, but what it is now, that it is ever.

Eternity, therefore—while not the Substratum (not the essential foundation of the Divine or Intellectual Principle)—may be considered as the radiation of this substratum: it exists as the announcement of the Identity in the Divine, of that state—of being thus and not otherwise—which characterises what has no futurity but eternally is.

What future, in fact, could bring to that Being anything which it now does not possess; and could it come to be anything which it is not once for all?

There exists no source or ground from which anything could make its way into that standing present; any imagined entrant will prove to be not alien but already integral. And as it can never come to be anything at present outside it, so, necessarily, it cannot include any past; what can there be that once was in it and now is gone? Futurity, similarly, is banned; nothing could be yet to come to it. Thus no ground is left for its existence but that it be what it is.

That which neither has been nor will be, but simply possesses being; that which enjoys stable existence as neither in process of change nor having ever changed—that is Eternity. Thus we come to the definition: the Life—instantaneously entire, complete, at no point broken into period or part—which belongs to the Authentic Existent by its very existence, this is the thing we were probing for—this is Eternity.

4.

We must, however, avoid thinking of it as an accidental from outside grafted upon that Nature: it is native to it, integral to it.

It is discerned as present essentially in that Nature like everything esle that we can predicate There—all immanent, springing from that Essence and inherent to that Essence. For whatsoever has primal Being must be immanent to the Firsts and be a First—Eternity equally with The Good that is among them and of them and equally with the truth that is among them.

In one aspect, no doubt, Eternity resides in a partial phase of the All-Being; but in another aspect it is inherent in the All taken as a totality, since that Authentic All is not a thing patched up out of external parts, but is authentically an all because its parts are engendered by itself. It is like the truthfulness in the Supreme which is not an agreement with some outside fact or being but is inherent in each member about which it is the truth. To an authentic All it is not enough that it be everything that exists: it must possess all-ness in the full sense that nothing whatever is absent from it. Then nothing is in store for it: if anything were to come, that thing must have been lacking to it, and it was, therefore, not All. And what, of a Nature contrary to its own, could enter into it when it is (the Supreme and therefore) immune? Since nothing can accrue to it, it cannot seek change or be changed or ever have made its way into Being.

Engendered things are in continuous process of acquisition; eliminate futurity, therefore, and at once they lost their being; if the non-engendered are made amenable to futurity they are thrown down from the seat of their existence, for, clearly, existence is not theirs by their nature if it appears only as a being about to be, a becoming, an advancing from stage to stage.

The essential existence of generated things seems to lie in their existing from the time of their generation to the ultimate of time after which they cease to be: but such an existence is compact of futurity, and the annulment of that futurity means the stopping of the life and therefore of the essential existence. . . .

5.

. . . Eternity, thus, is of the order of the supremely great; it proves on investigation to be identical with God: it may fitly be described as God made manifest, as God declaring what He is, as existence without jolt or change, and therefore as also the firmly living.

And it should be no shock that we find plurality in it; each of the Beings of the Supreme is multiple by virtue of unlimited force; for to be limitless implies failing at no point, and Eternity is pre-eminently the limitless since (having no past or future) it spends nothing of its own substance.

Thus a close enough definition of Eternity would be that it is a life limitless in the full sense of being all the life there is and a life which, knowing nothing of past or future to shatter its completeness, possesses itself intact for ever. To the notion of a Life (a Living-Principle) all-comprehensive add that it never spends itself, and we have the statement of a Life instantaneously infinite.

6.

. . . Things and Beings in the Time order—even when to all appearance complete, as a body is when fit to harbour a soul—are still bound to sequence; they are deficient to the extent of that thing, Time, which they need: let them have it, present to them and running side by side with them, and they are by that very fact incomplete; completeness is attributed to them only by an accident of language.

But the conception of Eternity demands something which is in its nature complete without sequence; it is not satisfied by something measured out to any remoter time or even by something limitless, but, in its limitless reach, still having the progression of futurity: it requires something immediately possessed of the due fullness of Being, something whose Being does not depend upon any quantity (such as instalments of time) but subsists before all quantity.

Itself having no quantity, it can have no contact with anything quantitative since its Life cannot be made a thing of fragments, in contradiction to the partlessness which is its character; it must be without parts in the Life as in the essence.

The phrase "He was good" [1] (used by Plato of the Demiurge) refers to the Idea of the All; and its very indefiniteness signifies the utter absence of relation to Time: so that even this Universe has had no temporal beginning; and if we speak of something "before" it, that is only in the sense of the Cause from which it takes its Eternal Existence. Plato used the word merely for the convenience of exposition, and immediately corrects it as inappropriate to the order vested with the Eternity he conceives and affirms.

7.

Now comes the question whether, in all this discussion, we are not merely helping to make out a case for some other order of Beings and talking of matters alien to ourselves.

But how could that be? What understanding can there be failing some point of contact? And what contact could there be with the utterly alien?

We must then have, ourselves, some part or share in Eternity.

Still, how is this possible to us who exist in Time?

The whole question turns on the distinction between being in Time and being in Eternity, and this will be best realised by probing to the Nature of Time. We must, therefore, descend from Eternity to the investigation of Time, to the realm of Time: till now we have been taking the upward way; we must now take the downward—not to the lowest levels but within the degree in which Time itself is a descent from Eternity.

If the venerable sages of former days had not treated of time, our method would be to begin by linking to (the idea of) Eternity (the idea of) its Next (its inevitable downward or outgoing subsequent in the same order), then setting forth the probable nature of such a Next and

1. [See Plato, *Timaeus*, 29 (p. 41).—C. M. S.]

proceeding to show how the conception thus formed tallies with our own doctrine.

But, as things are, our best beginning is to range over the most noteworthy of the ancient opinions and see whether any of them accord with ours.

Existing explanations of time seem to fall into three classes:—

Time is variously identified with what we know as Movement, with a moved object, and with some phenomenon of Movement: obviously it cannot be Rest or a resting object or any phenomenon of rest, since, in its characteristic idea, it is concerned with change.

Of those that explain it as Movement, some identify it with Absolute Movement (or with the total of Movement), others with that of the All. Those that make it a moved object would identify it with the orb of the All. Those that conceive it as some phenomenon, or some period, of Movement treat it, severally, either as a standard of measure or as something inevitably accompanying Movement, abstract or definite.

8.

Movement Time cannot be—whether a definite act of moving is meant or a united total made up of all such acts—since movement, in either sense, takes place in Time. And, of course, if there is any movement not in Time, the identification with Time becomes all the less tenable.

In a word, Movement must be distinct from the medium in which it takes place.

And, with all that has been said or is still said, one consideration is decisive: Movement can come to rest, can be intermittent; Time is continuous.

We will be told that the Movement of the All is continuous (and so may be identical with Time).

But, if the reference is to the circuit of the heavenly system (it is not strictly continuous, or equable, since) the time taken in the return path is not that of the outgoing movement; the one is twice as long as the other: this Movement of the All proceeds, therefore, by two different degrees; the rate of the entire journey is not that of the first half.

Further, the fact that we hear of the Movement of the outermost sphere being the swiftest confirms our theory. Obviously, it is the swiftest of movements by taking the lesser time to traverse the greater space—the very greatest—all other moving things are slower by taking a longer time to traverse a mere segment of the same extension: in other words, Time is not this movement.

And, if Time is not even the movement of the Kosmic Sphere much less is it the sphere itself though that has been identified with Time on the ground of its being in motion.

Is it, then, some phenomenon or connection of Movement?

Let us, tentatively, suppose it to be extent, or duration, of Movement.

Now, to begin with, Movement, even continuous, has no unchanging extent (as Time the equable has), since, even in space, it may be faster or slower; there must, therefore, be some unit of standard outside it, by which these differences are measurable, and this outside standard would more properly be called Time. And failing such a measure, which extent would be Time, that of the fast or of the slow—or rather which of them all, since these speed-differences are limitless?

Is it the extent of the subordinate Movement (= movement of things of earth)?

Again, this gives us no unit since the movement is infinitely variable: we would have, thus, not Time but Times.

The extent of the Movement of the All, then?

The Celestial Circuit may, no doubt, be thought of in terms of quantity. It answers to measure—in two ways. First there is space; the movement is commensurate with the area it passes through, and this area is its extent. But this gives us, still, space only, not Time. Secondly, the circuit, considered apart from distance traversed, has the extent of its continuity, of its tendency not to stop but to proceed indefinitely: but this is merely amplitude of Movement; search it, tell its vastness, and, still, Time has no more appeared, no more enters into the matter, than when one certifies a high pitch of heat; all we have discovered is Motion in ceaseless succession, like water flowing ceaselessly, motion and extent of motion.

Succession or repetition gives us Number—dyad, triad, etc.—and the extent traversed is a matter of Magnitude; thus we have Quantity of Movement—in the form of number, dyad, triad, decade, or in the form of extent apprehended in what we may call the amount of the Movement: but, the idea of Time we have not. That definite Quantity is (not Time but) merely something occurring within Time, for, otherwise Time is not everywhere but is something belonging to Movement which thus would be its substratum or basic-stuff: once more, then, we would be making Time identical with Movement; for the extent of Movement is not something outside it but is simply its continuousness, and we need not halt upon the difference between the momentary and the continuous, which is simply one of manner and degree. The extended movement and its extent are not Time; they are in Time. Those that explain Time as extent of Movement must mean not the extent of the movement itself but something which determines its extension, something with which the movement keeps pace in its course. But what this something is, we are not told; yet it is, clearly, Time, that in which all Movement proceeds. This is what our discussion has aimed at from the first: "What, essentially, is Time?" It comes to this: we ask "What is Time?" and we are answered, "Time is the extension of Movement in Time!"

On the one hand Time is said to be an extension apart from and outside that of Movement; and we are left to guess what this extension may be: on the other hand, it is represented as the extension of Movement; and this leaves the difficulty what to make of the extension of Rest—though one thing may continue as long in repose as another in motion, so that we are obliged to think of one thing Time that covers both Rest and Movements, and, therefore, stands distinct from either.

What then is this thing of extension? To what order of beings does it belong?

It obviously is not spatial, for place, too, is something outside it.

9.

"A Number, a Measure, belonging to Movement?" [2]

This, at least, is plausible since Movement is a continuous thing; but let us consider.

2. [See Aristotle, *Physics*, 219b (pp. 52-53).—C. M. S.]

To begin with, we have the doubt which met us when we probed its identification with extent of Movement: is Time the measure of any and every Movement?

Have we any means of calculating disconnected and lawless Movement? What number or measure would apply? What would be the principle of such a Measure?

One Measure for movement slow and fast, for any and every movement: then that number and measure would be like the decade, by which we reckon horses and cows, or like some common standard for liquids and solids. If Time is this Kind of Measure, we learn, no doubt, of what objects it is a Measure—of Movements—but we are no nearer understanding what it is in itself.

Or: we may take the decade and think of it, apart from the horses or cows, as a pure number; this gives us a measure which, even though not actually applied, has a definite nature. Is Time, perhaps, a Measure in this sense?

No: to tell us no more of Time in itself than that it is such a number is merely to bring us back to the decade we have already rejected, or to some similar collective figure.

If, on the other hand, Time is (not such an abstraction but) a Measure possessing a continuous extent of its own, it must have quantity, like a foot-rule; it must have magnitude; it will, clearly, be in the nature of a line traversing the path of Movement. But, itself thus sharing in the movement, how can it be a Measure of Movement? Why should the one of the two be the measure rather than the other? Besides an accompanying measure is more plausibly considered as a measure of the particular movement it accompanies than of Movement in general. Further, this entire discussion assumes continuous movement, since the accompanying principle, Time, is itself unbroken (but a full explanation implies justification of Time in repose).

The fact is that we are not to think of a measure outside and apart, but of a combined thing, a measured Movement, and we are to discover what measures it.

Given a Movement measured, are we to suppose the measure to be a magnitude?

If so, which of these two would be Time, the measured movement or the measuring magnitude? For Time (as measure) must be either the movement measured by magnitude, or the measuring magnitude itself or something using the magnitude like a yard-stick to appraise the movement. In all three cases, as we have indicated, the application is scarcely plausible except where continuous movement is assumed; unless the Movement proceeds smoothly, and even unintermittently and as embracing the entire content of the moving object, great difficulties arise in the identification of Time with any kind of measure.

Let us, then, suppose Time to be this "measured Movement," measured by quantity. Now the Movement if it is to be measured requires a measure outside itself; this was the only reason for raising the question of the accompanying measure. In exactly the same way the measuring magnitude, in turn, will require a measure, because only when the standard shows such and such an extension can the degree of movement be appraised. Time then will be, not the magnitude accompanying the Movement, but that numerical value by which the magnitude accompanying the Movement is estimated. But that number can be only the abstract figure which represents the magnitude, and it is difficult to see how an abstract figure can perform the act of measuring.

And, supposing that we discover a way in which it can, we still have not Time, the measure, but a particular quantity of Time, not at all the same thing: Time means something very different from any definite period: before all question as to quantity is the question as to the thing of which a certain quantity is present.

Time, we are told, is the number outside Movement and measuring it, like the tens applied to the reckoning of the horses and cows but not inherent in them: we are not told what this Number is; yet, applied or not, it must, like that decade, have some nature of its own.

Or "it is that which accompanies a Movement and measures it by its successive stages"; but we are still left asking what this thing recording the stages may be.

In any case, once a thing—whether by point or standard or any other means—measures succession, it must measure according to time: this number appraising movement degree by degree must, therefore, if it is to serve as a measure at all, be something dependent upon time and in contact with it: for, either, degree is spatial, merely—the beginning and end of the Stadium, for example—or in the only alternative, it is a pure matter of Time: the succession of early and late is stage of Time, Time ending upon a certain Now or Time beginning from a Now.

Time, therefore, is something other than the mere number measuring Movement, whether Movement in general or any particular tract of Movement.

Further:—Why should the mere presence of a number give us Time—a number measuring or measured; for the same number may be either—if Time is not given us by the fact of Movement itself, the movement which inevitably contains it itself a succession of stages? To make the number essential to Time is like saying that magnitude has not its full quantity unless we can estimate that quantity.

Again, if Time is, admittedly, endless, how can number apply to it?

Are we to take some portion of Time and find its numerical statement? That simply means that Time existed before number was applied to it.

We may, therefore, very well think that it existed before the Soul or Mind that estimates it—if, indeed, it is not to be thought to take its origin from the Soul—for no measurement by anything is necessary to its existence; measured or not, it has the full extent of its being.

And suppose it to be true that Soul is the appraiser, using Magnitude as the measuring standard, how does this help us to the conception of Time?

10.

Time, again, has been described as some sort of a sequence upon Movement, but we learn nothing from this, nothing is said, until we know what it is that produces this sequential thing; probably the cause and not the result would turn out to be Time.

And, admitting such a thing, there would still remain the question whether it came into being before the movement, with it, or after it; and, whether we say before or with or after, we are speaking of order in Time: and thus our definition is "Time is a sequence upon movement *in Time!*"

Enough. Our main purpose is to show what Time is, not to refute false definition. To traverse point by point the many opinions of our many predecessors would mean a history rather than an identification; we have treated the various theories as fully as is possible in a cursory review: and, notice, that which makes Time the Measure of the All-Movement is refuted by our entire discussion and, especially, by the observations upon the Measurement of Movement in general, for all the argument—except, of course, that from irregularity—applies to the All as much as to particular Movement.

We are, thus, at the stage where we are to state what Time really is.

11.

To this end we must go back to the state we affirmed of Eternity, unwavering Life, undivided totality, limitless, knowing no divagation, at rest in unity and intent upon it. Time was not yet: or at least it did not exist for the Eternal Beings, though its being was implicit in the Idea and Principle of progressive derivation.

But from the Divine Beings thus at rest within themselves, how did this Time first emerge?

We can scarcely call upon the Muses to recount its origin since they were not in existence then—perhaps not even if they had been. The engendered thing, Time, itself, can best tell us how it rose and became manifest; something thus its story would run:

Time at first—in reality before that "first" was produced by desire of succession—Time lay, self-concentrated, at rest within the Authentic Existent: it was not yet Time; it was merged in the Authentic and motionless with it. But there was an active principle there, one set on governing itself and realising itself (= the All-Soul), and it chose to aim at something more than its present: it stirred from its rest, and Time stirred with it. And we (i.e. human Souls as summed in the principle of developing Life, the All-Soul?) we, stirring to a ceaseless succession, to a next, to the discrimination of identity and the establishment of ever new difference, traversed a portion of the outgoing path and produced an image of Eternity, produced Time.

For the Soul contained an unquiet faculty, always desirous of translating elsewhere what it saw in the Authentic Realm, and it could not bear to retain within itself all the dense fullness of its possession.

A Seed is at rest; the nature-principle within, uncoiling outwards, makes way towards what seems to it a large life; but by that partition it loses; it was a unity self-gathered, and now, in going forth from itself, it fritters its unity away; it advances into a weaker greatness. It is so with this faculty of the Soul, when it produces the Kosmos known to sense—the mimic of the Divine Sphere, moving not in the very movement of the Divine but in its similitude, in an effort to reproduce that of the Divine. To bring this Kosmos into being, the Soul first laid aside its eternity and clothed itself with Time; this world of its fashioning it then gave over to be a servant to Time, making it at every point a thing of Time, setting all its progressions within the bournes of Time. For the Kosmos moves only in Soul—the only Space within the range of the All open to it to move in—and therefore its Movement has always been in the Time which inheres in Soul.

Putting forth its energy in act after act, in a constant progress of novelty, the Soul produces succession as well as act; taking up new purposes added to the old it brings thus into being what had not existed in that former period when its purpose was still dormant and its life was not as it since became: the life is changed and that change carries with it a change of Time. Time, then, is contained in differentiation of Life; the ceaseless forward movement of Life brings with it unending Time; and Life as it achieves its stages constitutes past Time.

Would it, then, be sound to define Time as the Life of the Soul in movement as it passes from one stage of act or experience to another?

Yes; for Eternity, we have said, is Life in repose, unchanging, self-identical, always endlessly complete; and there is to be an image of Eternity—Time—such an image as this lower All presents of the Higher Sphere. Therefore over against that higher life there must be another life, known by the same name as the more veritable life of the Soul; over against that movement of the Intellectual Soul there must be the movement of some partial phase; over against that identity, unchangeableness and stability there must be that which is not constant in the one hold but puts forth multitudinous acts; over against that oneness without extent or interval there must be an image of oneness, a unity of link and succession; over against the immediately infinite and all-comprehending, that which tends, yes, to infinity but by tending to a perpetual futurity; over against the Whole in concentration, there must be that which is to be a Whole by stages never final. The lesser must always be working towards the increase of its Being, this will be its imitation of what is immediately complete, self-

realised, endless without stage: only thus can its Being reproduce that
of the Higher.

Time, however, is not to be conceived as outside of Soul; Eternity is
not outside of the Authentic Existent: nor is it to be taken as a
sequence or succession to Soul, any more than Eternity is to the
Divine. It is a thing seen upon Soul, inherent, coeval to it, as Eternity
to the Intellectual Realm.

12.

We are brought thus to the conception of a Natural-Principle—
Time—a certain expanse (a quantitative phase) of the Life of the Soul,
a principle moving forward by smooth and uniform changes following silently upon each other—a Principle, then, whose Act is (not one
like that of the Supreme but) sequent.

But let us conceive this power of the Soul to turn back and withdraw
from the life-course which it now maintains, from the continuous and
unending activity of an ever-existent soul not self-contained or self-
intent but concerned about doing and engendering: imagine it no
longer accomplishing any Act, setting a pause to this work it has
inaugurated; let this outgoing phase of the Soul become once more,
equally with the rest, turned to the Supreme, to Eternal Being, to the
tranquilly stable.

What would then exist but Eternity?

All would remain in unity; how could there be any diversity of
things? What Earlier or Later would there be, what long-lasting or
short-lasting? What ground would lie ready to the Soul's operation
but the Supreme in which it has its Being? Or, indeed, what operative
tendency could it have even to That since a prior separation is the
necessary condition of tendency?

The very sphere of the Universe would not exist; for it cannot
antedate Time: it, too, has its Being and its Movement in Time; and if
it ceased to move, the Soul-Act (which is the essence of Time) continuing, we could measure the period of its Repose by that standard
outside it.

If, then, the Soul withdrew, sinking itself again into its primal unity,
Time would disappear: the origin of Time, clearly, is to be traced to
the first stir of the Soul's tendency towards the production of the

sensible universe with the consecutive act ensuing. This is how "Time"—as we read—"came into Being simultaneously" with this All: the Soul begot at once the Universe and Time; in that activity of the Soul this Universe sprang into being; the activity is Time, the Universe is a content of Time. No doubt it will be urged that we read also of "the orbit of the Stars being Times": but do not forget what follows; "the stars exist," we are told, "for the display and delimitation of Time," and "that there may be a manifest Measure." [3] No indication of Time could be derived from (observation of) the Soul; no portion of it can be seen or handled, so it could not be measured in itself, especially when there was as yet no knowledge of counting; therefore the Soul brings into being night and day; in their difference is given Duality—from which, we read, arises the concept of Number.

We observe the tract between a sunrise and its return and, as the movement is uniform, we thus obtain a Time-interval upon which to support ourselves, and we use this as a standard. We have thus a measure of Time. Time itself is not a measure. How would it set to work? And what kind of thing is there of which it could say, "I find the extent of this equal to such and such a stretch of my own extent?" What is this "I"? Obviously something by which measurement is known. Time, then, serves towards measurement but is not itself the Measure: the Movement of the All will be measured according to Time, but Time will not, of its own Nature, be a Measure of Movement: primarily a Kind to itself, it will incidentally exhibit the magnitudes of that movement.

And the reiterated observation of Movement—the same extent found to be traversed in such and such a period—will lead to the conception of a definite quantity of Time past.

This brings us to the fact that, in a certain sense, the Movement, the orbit of the universe, may legitimately be said to measure Time—in so far as that is possible at all—since any definite stretch of that circuit occupies a certain quantity of Time, and this is the only grasp we have of Time, our only understanding of it: what that circuit measures—by indication, that is—will be Time, manifested by the Movement but not brought into being by it.

This means that the measure of the Spheric Movement has itself

3. [See Plato, *Timaeus*, 38-39 (p. 44).—C. M. S.]

been measured by a definite stretch of that Movement and therefore is something different; as measure, it is one thing and, as the measured, it is another; (its being measure or) its being measured cannot be of its essence.

We are no nearer knowledge than if we said that the foot-rule measures Magnitude while we left the concept of Magnitude undefined; or, again, we might as well define Movement—whose limitlessness puts it out of our reach—as the thing measured by Space; the definition would be parallel since we can mark off a certain space which the Movement has traversed and say the one is equivalent to the other.

13.

The Spheral Circuit, then, performed in Time, indicates it: but when we come to Time itself there is no question of its being "within" something else: it must be primary, a thing "within itself." It is that in which all the rest happens, in which all movement and rest exist smoothly and under order; something following a definite order is necessary to exhibit it and to make it a subject of knowledge—though not to produce it—it is known by order whether in rest or in motion; in motion especially, for Movement better moves Time into our ken than rest can, and it is easier to estimate distance traversed than repose maintained.

This last fact has led to Time being called a measure of Movement when it should have been described as something measured by Movement and then defined in its essential nature; it is an error to define it by a mere accidental concomitant and so to reverse the actual order of things. Possibly, however, this reversal was not intended by the authors of the explanation: but, at any rate, we do not understand them; they plainly apply the term Measure to what is in reality the measured and leave us unable to grasp their meaning: our perplexity may be due to the fact that their writings—addressed to disciples acquainted with their teaching—do not explain what this thing, measure, or measured object, is in itself.

Plato does not make the essence of Time consist in its being either a measure or a thing measured by something else.

Upon the point of the means by which it is known, he remarks that the Circuit advances an infinitesimal distance for every infinitesimal segment of Time so that from that observation it is possible to estimate what the Time is, how much it amounts to: but when his purpose is to explain its essential nature he tells us that it sprang into Being simultaneously with the Heavenly system, a reproduction of Eternity, its image in motion, Time necessarily unresting as the Life with which it must keep pace: and "coeval with the Heavens" because it is this same Life (of the Divine Soul) which brings the Heavens also into being; Time and the Heavens are the work of the one Life.

Suppose that Life, then, to revert—an impossibility—to perfect unity: Time, whose existence is in that Life, and the Heavens, no longer maintained by that Life, would end at once.

It is the height of absurdity to fasten on the succession of earlier and later occurring in the life and movement of this sphere of ours, to declare that it must be some definite thing and to call it Time, while denying the reality of the more truly existent Movement, that of the Soul, which has also its earlier and later: it cannot be reasonable to recognise succession in the case of the Soulless Movement—and so to associate Time with that—while ignoring succession and the reality of Time in the Movement from which the other takes its imitative existence; to ignore, that is, the very Movement in which succession first appears, a self-actuated movement which, engendering its own every operation, is the source of all that follows upon itself, to all which, it is the cause of existence, at once, and of every consequent.

But:—we treat the Kosmic Movement as overarched by that of the Soul and bring it under Time; yet we do not set under Time that Soul-Movement itself with all its endless progression: what is our explanation of this paradox?

Simply, that the Soul-Movement has for its Prior (not Time but) Eternity which knows neither its progression nor its extension. The descent towards Time begins with this Soul-Movement; it made Time and harbours Time as a concomitant to its Act.

And this is how Time is omnipresent: that Soul is absent from no fragment of the Kosmos just as our Soul is absent from no particle of ourselves. As for those who pronounce Time a thing of no substantial existence, of no reality, they clearly belie God Himself whenever they say "He was" or "He will be": for the existence indicated by the "was and will be" can have only such reality as belongs to that in which it is said to be situated:—but this school demands another style of argument.

Meanwhile we have a supplementary observation to make.

Take a man walking and observe the advance he has made; that advance gives you the quantity of movement he is employing: and when you know that quantity—represented by the ground traversed by his feet, for, of course, we are supposing the bodily movement to correspond with the pace he has set within himself—you know also the movement that exists in the man himself before the feet move.

You must relate the body, carried forward during a given period of Time, to a certain quantity of Movement causing the progress and to the Time it takes, and that again to the Movement, equal in extension, within the man's soul.

But the Movement within the Soul—to what are you to (relate) refer that?

Let your choice fall where it may, from this point there is nothing but the unextended: and this is the primarily existent, the container to all else, having itself no container, brooking none.

And, as with Man's Soul, so with the Soul of the All.

"Is Time, then, within ourselves as well?"

Time is in every Soul of the order of the All-Soul, present in like form in all; for all the Souls are the one Soul.

And this is why Time can never be broken apart, any more than Eternity which, similarly, under diverse manifestations, has its Being as an integral constituent of all the eternal Existences.

7.

Augustine:

EXPERIENTIAL TIME

The eleventh book of *The Confessions of St. Augustine**
(354–430) is one of the most celebrated and searching discussions
of the nature of time. Taking the opening passages of the Book of
Genesis as prompting his opening query, Augustine focused on
the way time functions in human experience.

10

Certainly it is a mark of being full of the old error when people say
to us: "What was God doing before He made heaven and earth?"
Their argument is as follows: If God was unoccupied with the making
of anything, why did He not go on forever remaining in the same state
as that in which He had always been? For if there came into existence
in God a new motion and a new will to make something which He had
never made before, how can one call it a true eternity, when a will
arises which was not previously in existence? For the will of God is not
a creature; it is prior to all creation, because nothing could be created
unless the will of the creator had come first. Therefore, God's will
belongs to the very substance of God. But if something which was
previously not there arose in God's substance, then one could not
truly call that substance eternal. If, on the other hand, God's will that
there should be a creation had been in existence from eternity, why is
creation also not from eternity?

* Rex Warner, trans., *The Confessions of St. Augustine* (New York: The New
American Library of World Literature, Inc., Mentor-Omega Books, 1963). By
permission.

11

Those who say this do not yet understand you, O Wisdom of God, light of minds. They do not yet understand how these things are made which are made by you and in you, and they are trying to taste eternity while their mind is still fluttering about in the past and future movements of things, and so is still unstable.

Can we not hold the mind and fix it firm so that it may stand still for a moment and for a moment lay hold upon the splendor of eternity which stands forever, and compare it with the times that never stand, and see that no comparison is possible? Then it would see that a long time is long only because of the numbers of movements passing by in succession, which cannot have a simultaneous extension; but that in eternity nothing passes by; everything is present, whereas time cannot be present all at once. It would be seen too that all time past is driven on by time future, and all the future follows from the past, and that both past and future are created by and proceed from that which is perpetually present. Who can so hold the mind of man that it may stand and see how eternity, which stands still and is neither past nor future, dictates the times that are past and the times that are to come? Could my hand have strength enough, or could the hand of my mouth by speech achieve so great a thing?

14

Therefore, since you made time itself, one cannot say that there was any time in which you had not made anything. And no times are coeternal with you, because you are permanent, whereas, if they were permanent, they would not be times. What then is time? Who can find a quick and easy answer to that question? Whoever in his mind can grasp the subject well enough to be able to make a statement on it? Yet in our ordinary conversation we use the word "time" more often and more familiarly than any other. And certainly we understand what we mean by it, just as we understand what others mean by it when we hear the word from them.

What then is time? I know what it is if no one asks me what it is; but if I want to explain it to someone who has asked me, I find that I do not know. Nevertheless, I can confidently assert that I know this: that if nothing passed away there would be no past time, and if nothing were coming there would be no future time, and if nothing were now there would be no present time.

But in what sense can we say that those two times, the past and the future, exist, when the past no longer is and the future is not yet? Yet if the present were always present and did not go by into the past, if would not be time at all, but eternity. If, therefore, the present (if it is to be time at all) only comes into existence because it is in transition toward the past, how can we say that even the present *is?* For the cause of its being is that it shall cease to be. So that it appears that we cannot truly say that time exists except in the sense that it is tending toward nonexistence.

15

Nevertheless, we say "a long time" and "a short time," though we only use these expressions about the past and the future. A hundred years ago, for instance, we say is a long time past, and a hundred years from now a long time ahead; ten days ago a short time past, ten days from now a short time ahead. But how can we say that something which does not exist at all is either long or short? For the past no longer is, and the future has not yet come to be. Should we then not say: "It is long?" Should we instead say of the past: "It was long" and of the future: "It will be long?"

My Lord, my light, is it not so that here also man is mocked by your truth? For as to this time in the past that was long, when was it long? When it was already past or when it was still present? It could be long only while it was in existence to be long. But once past, it had ceased to exist. So, not being in existence at all, it could not be long.

Therefore we must not say: "the time past was long"; for we shall not find anything in it which could be long, since from the moment when it becomes past, it ceases to exist. Let us say instead: "that particular present time was long," because while it was present, it was long. For it had not yet passed away so as not to exist, and consequently there was something which could be long; though once it had passed away, it ceased to be long by also ceasing to exist.

Let us see, therefore, soul of man, whether present time can be long.

For to you, soul of man, it has been granted to feel spaces of time and to measure them. Now what answer will you give me?

Is a present time of a hundred years a long time? But first let me see whether a hundred years can be present. If we are in the first of these years, it is present, but the ninety-nine other years are still to come and therefore do not yet exist. And if we are in the second year, one year is already past, one is present, and the rest are to come. In the same way, whatever year we may care to choose in the hundred-year series as being present, all the years before it will be past and all the years after it will be future. Therefore a hundred years cannot be present.

But let us at least see whether this one year itself which we have selected is present. Here too, if we are in the first month, the other months are still to come, and if we are in the second month, the first is already in the past and the others do not yet exist. For a year is twelve months, of which just that one month in which we are is present, all the rest being either past or future. Yet even the month in which we are is not present; only one day of it is. If that day is the first, the other days are still to come; if the last, all the other days are in the past; if any intermediate day, it is between days past and days to come.

So now we see that this "present time," which we discovered to be the only time that could be called "long," has contracted to the space of scarcely one day. But let us look into this one day too, because not even a single day is wholly present. It (including the hours of day and night) is made up of twenty-four hours; the first of these has all the other hours still to come; the last of these has them all in the past; any intermediate hour has those before it in the past, those after it in the future. And that very hour itself is made of fleeting moments; whatever part of these has fled away is in the past, whatever remains is in the future. If anything can be meant by a point of time so small that it cannot be divided into even the most minute particles of moments, that is the only time that can be called "present." And such a time must fly so rapidly from future to past that it has no duration and no extension. For if it does have any extension, it can be divided into past and future; whereas the present does not take up any space.

Where, then, is the time that we can call "long"? Is it in the future? But we cannot say of the future: "It is long," because it is not yet in existence to be long. We have to say: "It will be long." But when will it be long? If we are still imagining it in the future, it will not be long, because there will still not be anything in existence to be long. If, on the other hand, we say that it will be long at the moment when out of the future (which does not yet exist) it will begin to take on being and become present, so that there can be something in existence to be long, then the present cries aloud in the words which we have just heard that it cannot be long.

16

And yet, Lord, we do perceive definite periods of time, and we compare them with each other and say that some are longer and others shorter. We even measure how much longer or shorter this time is than that, and we say that it is twice or three times as long, or equivalent—one time being of the same length as another. But when we measure time by our perception of it, it is time passing that we are measuring. For it is impossible to measure the past, which is no longer in existence, or the future which is not yet in existence, unless perhaps one is going to be rash enough to maintain that it is possible to measure something which does not exist. When, therefore, time is passing, it can be perceived and measured, but when it has passed, it cannot, because it does not exist.

17

I am asking questions, Father, not making statements. My God, govern me and direct me. We learned at school and we teach at school that there are three times—past, present, and future. Am I now to be told that this is not so, that there is only the present, since the other two times do not exist? Or can we say that they do exist, but that there is some secret place from which time emerges when from the future the present comes into existence, and again some secret place into which it withdraws when the past comes out of the present? For where can those who have prophesied the future see the future, if the future is not yet in existence? For that which does not exist cannot be seen. And those who tell us about the past certainly could not tell us the truth unless they saw it in their mind's eye, and if the past were nonexistent, it would be quite impossible for it to be perceived. Therefore both the future and the past exist.

18

Allow me, Lord, to push my questions further. My hope, let me maintain this line of inquiry without being distracted from it.

If the future and the past exist, I want to know where they are. And if I still lack the strength to know this, nevertheless one thing I do know, which is that, wherever they are, they are not there as future and as past, but as present. For if there too they are future, they are not yet there, and if there too they are past, they are no longer there. Thus, wherever they are, and whatever they are, they cannot be anything except present. Although with regard to the past, when this is reported correctly what is brought out from the memory is not the events themselves (these are already past) but words conceived from the images of those events, which, in passing through the senses, have left as it were their footprints stamped upon the mind. My boyhood, for instance, which no longer exists, exists in time past, which no longer exists. But when I recollect the image of my boyhood and tell others about it, I am looking at this image in time present, because it still exists in my memory. Whether a similar cause operates with regard to predictions of the future—namely, that images of things which do not yet exist are felt in advance as already existing—this, I confess, my God, I do not know. But I do know this, that we often premeditate our future actions, and this premeditation is present while the action which we are premeditating, being in the future, does not yet exist. But when we have embarked on it and begun to do what we were premeditating, then that action will exist, because then instead of being in the future, it will be in the present.

Whatever, then, may be the mode of this secret foreknowledge of the future, nothing can be seen which does not exist. And what is already in existence is not future, but present. Therefore, when we speak of seeing the future, what is seen is not the actual future itself (which, being future, does not yet exist), but the causes, or perhaps the signs of that future—causes and signs which are already in existence. And so to those who see them they are not future, but present, and from them future events are conceived in the mind and predicted. These concepts, again, are already in existence, and by those who make the predictions they are contemplated as being present in the mind.

Let me take one example out of the great number of possible examples. I am looking at the dawn sky and I foretell that the sun is going to rise. What I am looking at is present; what I foretell is future. What is future is not the sun, which is already in existence, but its rising, which has not yet taken place. Yet unless I could imagine in my mind this rising (as I do now in speaking of it), I should not be able to predict it. But the glow which I see in the sky is not the sunrise, although it comes before the sunrise; nor is the image in my mind the sunrise. Both those two are perceived in the present, so that the sunrise, which is in the future, can be foretold. The future, therefore, is not yet, and if it is not yet, it does not exist, and if it does not exist, it is quite impossible for it to be seen. But it can be predicted from the present which is already in existence and which can be seen.

20

It is now, however, perfectly clear that neither the future nor the past are in existence, and that it is incorrect to say that there are three times—past, present and future. Though one might perhaps say: "There are three times—a present of things past, a present of things present, and a present of things future." For these three do exist in the mind, and I do not see them anywhere else: the present time of things past is memory; the present time of things present is sight; the present time of things future is expectation. If we are allowed to use words in this way, then I see that there are three times and I admit that there are. Let us go further and say: "There are three times—past, present, and future." It is an incorrect use of language, but it is customary. Let us follow the custom. See, I do not mind, I do not object, I find no fault, provided that we understand what is said—namely, that neither what is to come nor what is past is now in existence. It is not often that we use language correctly; usually we use it incorrectly, though we understand each other's meaning.

21

I said just now that we measure time as it passes, and in such a way that we are able to say that one period of time is twice as great as another, or of the same length, and so on of any other parts of time which are measurable. For this reason, as I said, we measure time as it is passing, and if I am asked how I know this, I should reply that I know it because we do measure time, and we cannot measure what does not exist, and the past and the future do not exist. But how do we measure the present, since it has no extent? It is measured while it is passing; when it has passed by it is not measured, for then there will be nothing there to measure.

But where does time come from, by what way does it pass, and where is it going to when we are measuring it? It can only come from the future, it can only pass by way of the present, and it can only go into the past. Therefore it comes from something which is not yet in existence, it passes through something which has no extension, and it goes in the direction of something which has ceased to exist.

But how can we measure time except in some sort of extension? When we say single and double and triple and all the other expressions of this sort which we use about time, we must be speaking of extensions or spaces of time. In what kind of extension, then, do we measure time as it is passing by? In the future, from which it comes? But we cannot measure something which is not yet in existence. In the present, through which it passes? But we cannot measure something which has no extension. In the past, toward which it is going? But we cannot measure something which no longer exists.

22

... We are constantly using the word "time," both in the singular and the plural. We say: "How long did he speak?" "How long was he in doing that?" "For how long a time have I not seen that?" "This syllable is double the length of that short one." We use these expressions ourselves and we hear other people using them; we both understand and make ourselves understood; they are the commonest of expressions and the clearest, and yet on the other hand they are extraordinarily obscure and no one has yet discovered what they mean.

23

I once heard a learned man say that what constitutes time is the motions of the sun and moon and stars. I did not agree. For one might equally well say that the motions of all bodies constitute time. Suppose that the lights of the heavenly bodies constitute time. Suppose that the lights of the heavenly bodies were to cease and a potter's wheel were to be turning around: would there be no time by which we could measure its rotations and say that these rotations were of equal duration, or, if it turned sometimes faster and sometimes slower, that in the one case the period taken by the turn was shorter and in the other case longer? And when we were saying these things, would not we too be speaking in time? In our words would there not be syllables that are long and others that are short, simply because some take a longer and some a shorter time to pronounce? God, grant us men to see in a small thing principles which are common to things both small and great. There are stars and lights in the heavens to be *for signs, and for seasons, and for years and for days;* there certainly are; yet, just as I should not say that one turn of that little wooden wheel constituted a day, so that learned man should not say that it does not constitute any time at all.

What I want to know is the force and nature of time, by which we measure the motions of bodies and say, for example, that this motion is double the length of that one. And I put forward the following question: now what we mean by "a day" is not simply the time when the sun is above the earth (this merely distinguishes day from night), but the whole circuit of the sun from the east back to the east again (as, when we say "so many days have passed" we include the nights with the days and do not reckon up their spaces separately). Since, then, a day is constituted by the motion of the sun and its circuit from east to east, what I want to know is this: is it the motion itself which makes the day? Or is it the time taken by that motion? Or is it both?

If it were the movement of the sun that makes the day, then it would still be a day even if the sun completed its course in a time so small as to last only one hour. If it were the time now taken by the sun to complete its circuit that makes a day, then it would not be a day if there were only the space of one hour between one sunrise and the next; to make a day the sun would have to complete its circuit twenty-four times. If what makes a day is both the motion and the time taken, we could not call it "a day," if the sun stood still and a time passed which was equivalent to the time normally taken by the sun to go on its way from one dawn to the next.

So I shall not now inquire what it is which we call "a day." My question is: "what is time?" The time by which we measure the circuit of the sun and by which we should be able to say that the sun had gone around in half its normal time, if it went around in a time equal to twelve hours. The time by which we are able to compare the two periods and say that one is twice as long as the other, and this proportion of one to two would still hold good though the sun were, for its part, always doing its circuit from east to east; yet on some occasions it would be doing it in single, on others in double time.

Let no one tell me, then that the motions of the heavenly bodies constitute time. When, at a man's prayer, the sun stood still so that he might fight and win a battle, the sun did indeed stand still but time went on; it went on for the space necessary for that battle to be fought and brought to its conclusion.

I see, therefore, that time is an extension of some sort. But do I see this? Or do I only seem to see it? You, light, you, truth, will show me.

24

Is it your will that I should agree if someone tells me that time is the motion of a body? It is not your will. That no body can move except in time is something which I understand; it is what you say to me. But that the motion of a body actually *is* time is something which I do not understand; you do not say this to me. For when a body is in motion, I measure the length of the motion in time from the moment when it begins to move until the moment when it ceases to move. And if I did not observe the moment when the movement began and if the movement continues to go on so that I cannot observe the moment when it ends, then I am incapable of measuring it—except in the sense of measuring from the moment when I began to look until the moment when I stopped looking. If I look at it for a long time, all I can say is: "It was a long time." I cannot say how long a time; because we can only say "how long" by means of comparison—as, "this is as long as that," "this is twice as long," and so on. But when we can observe the points in space from which the body in motion comes and to which it goes (or the parts of a body, if it is revolving on its axis), then we can say exactly how much time has been taken to complete the movement of the body (or of its part) from one place to another place.

It is clear, then, that the motion of a body is one thing and the means by which we measure the duration of that motion is another thing. Is it not obvious which of the two deserves the name of "time"? A body may sometimes be in motion, at varying speeds, and may sometimes be standing still; but by means of time we measure not only its motion but its rest. We say: "It was at rest for the same time as it was in motion," or "It was at rest twice or three times as long as it was in motion," or any other proportion which we have either exactly measured or else guessed ("more or less" as we say).

Time, therefore, is not the motion of a body.

25

And I confess to you, Lord, that I still do not know what time is, and then again I confess to you, Lord, that I do know that I am saying these things in time, that I have been speaking of time for a long time and that this "long time" is only long because of the passage of time. But how do I know this, when I do not know what time is? Or by "not knowing" do I perhaps mean simply that I do not know how to express something which is in fact known to me? A bad state indeed to be in, not even to know what it is that I do not know! See, my God, I am in your presence and I do not lie. As I speak, so is my heart. *Thou shalt light my candle; Thou, O Lord my God, wilt enlighten my darkness.*

26

My soul speaks with truth when it confesses to you that I do measure time. Is it the case then, Lord my God, that I perform the act of measuring but do not know what I am measuring? I measure the motion of a body in time. Is it the case that I do not measure time itself? But could I measure the motion of a body—how long it lasts, how long it takes to go from one place to another—if I were not measuring the time in which the motion takes place? How, then, do I measure time itself? Do we measure a longer time by means of a shorter time, as, for instance, we measure a rood in terms of cubits? In this way, certainly, we seem to measure the quantity of syllables—the long by the short—and we say that a long syllable is double the length of a short. So we measure the length of poems by the lengths of the lines, and the lenghts of the lines by the lengths of the feet, and the lengths of the feet by the lengths of the syllables, and the lengths of the long syllables by the lengths of the short ones. I do not mean measuring poems by pages; that is a spatial and not a temporal measurement. I mean the measurement of words as they are pronounced and pass away, and we say: "That is a long poem, for it is made up of so many lines; the lines are long, for they are composed of so many feet; the feet are long, for they extend into so many syllables; that syllable is long, for it is the double of a short one."

Yet all this still does not give us a fixed measure of time. It may happen that a short line, if recited slowly, may take up more time than a longer line, if spoken hurriedly. The same holds good of a poem, or a foot or a syllable. Adn so it seems to me that time can only be a kind of extension; but I do not know what it is an extension of. Could it not be, I wonder, an extension of the mind itself? What is it, I beseech you my God, that I measure when I say, either in an indefinite way: "This time is longer than that" or, with precision: "This is double that"? That I am measuring time, I know. But I am not measuring the future, because it is not yet in existence; I am not measuring the present, because the present has no extension of space; I am not measuring the past, because it no longer exists. What then am I measuring? Is it time passing, but not past? That was what I said previously.

27

Press on, my mind! Go forward with all your strength! *God is our helper. He made us and not we ourselves.* Go forward toward the place where truth begins to dawn.

Let us consider the case of a bodily voice. The voice begins to sound, it sounds, it continues to sound, and then it stops sounding.

Now there is silence; the voice is past and is no longer a voice. Before it began to sound, it was in the future and could not be measured because it did not yet exist, and now it cannot be measured because it no longer exists. Therefore, it could only be measured while it was actually sounding, because only then was there something in existence which could be measured. But even then it was not static; it was going, and going away into the past. Was it this that made it the more measurable? For while it was in the process of going away it was extended through a certain space of time which made measurement possible; for the present occupies no space.

We grant, therefore, that then it was able to be measured. Now consider the case of another voice. This voice begins to sound; it still goes on sounding; it sounds at the same pitch continuously with no variation. Let us measure it while it is sounding; for when it has stopped sounding it will be in the past, and this will be nothing which can be measured. Obviously, then, we must measure it and say how long it is. But it is still sounding and measurement is only possible from the beginning, when it started to sound, to the end, when it ceased sounding. What we measure is the space between a beginning and an end. Therefore, a voice that has never ceased to sound cannot be measured; we cannot say how long or how short it is; it cannot be described as equal to another or single or double or anything of that sort. But when it has ceased to sound, it will no longer exist. How then shall we be able to measure it? And yet we do measure time. But the times we measure are not those which do not yet exist, not those which no longer exist, not those which are without duration, not those which are without beginning and end. Therefore, what we measure is neither the future nor the past nor the present nor what is passing. Yet nevertheless we do measure time.

"*Deus creator omnium*"—this line is composed of eight syllables, short and long alternately. Thus the four short syllables (the first, third, fifth, and seventh) are single in relation to the four long ones (the second, fourth, sixth, and eighth). Each long syllable has double the time of each short syllable—I pronounce them and I say that it is so, and, by the plain evidence of our senses, so it is. So far as sense can make things plain I measure a long syllable by a short, and I feel by means of my senses that it has twice the length. But when two syllables sound one after the other—the first short, the next long—how shall I

keep hold of the short one? How, in my measurement, shall I apply it
to the long one, so as to find that the long one has twice its length? The
long one has not even begun to sound unless the short one has ceased
to sound. And how can I measure the long syllable as something
present? I cannot begin to measure it until it is finished. And when it is
finished it has passed away.

What, then, is it that I measure? Where is that short syllable by
which I measure? Where is that long syllable which I measure? Both
have sounded, have fled away, have gone into the past, and no longer
exist; and yet I do measure; I reply in all sincerity (my reply being
based on the confidence one may have in a practiced sense) that one
syllable is, so far as space of time is concerned, twice the length of the
other. And I cannot make this judgment except when both the syllables have gone into the past and are finished. Therefore, what I am
measuring is not the syllables themselves (they no longer exist) but
something in my memory which remains there fixed.

It is in you, my mind, that I measure time. Do not interrupt me, or
rather, do not allow yourself to be interrupted by the thronging of
your impressions. It is in you, I say, that I measure time. As things pass
by they leave an impression in you; this impression remains after the
things have gone into the past, and it is this impression which I
measure in the present, not the things which, in their passage, caused
the impression. It is this impression which I measure when I measure
time. Therefore, either this itself is time or else I do not measure time
at all.

Now what happens when we measure periods of silence and say
that this period of silence occupied the same amount of time as that
period of speech? Is it not the case that we extend our thoughts up to
what would have been the length of a speech if that speech were
audible, and in this way are able to reach a conclusion about the
intervals of silence in a given space of time? Without making any use
of voice or tongue, we can go over in our mind poems, verses,
speeches, and we can form our conclusions about the measurements
of their movements and about the spaces of time taken up by one in
relation to another just as well as if we were actually reading these
passages aloud. If a man decides to utter a rather long sound and
makes up his mind how long it is going to be, he has passed through
that space of time in silence; then, committing it to memory, he begins

to utter the sound and it goes on sounding until it reaches the limit which he set for it. Or it would be truer to say "it did sound" and "it will sound"; for the part of it which at any moment is completed *has* sounded, and the part of it which remains to be uttered *will* sound, and so it goes on, as the act of will, which is in the present, transfers the future into the past, the past growing as the future diminishes, until the future is consumed and it is all past.

28

But how can the future which does not yet exist, be diminished or consumed? How can the past, which no longer exists, grow? Only because, in the mind, which performs all this, there are three things done. The mind looks forward to things, it looks at things, and it looks back on things. What it looks forward to passes on through what it looks at into what it looks back on. No one, of course, can deny that the future does not yet exist. But nevertheless there is in the mind already the expectation of the future. No one can deny that the past no longer exists. But nevertheless there is still in the mind the memory of the past. No one can deny that the present time has no extension, since it passes in a flash. But nevertheless our attention (our "looking at") is something constant and enduring, and through it what is to be proceeds into what has been. Thus it is not the future that is long, for the future does not exist; a long future is a long expectation of the future. Nor is the past long, since it does not exist; a long past is a long memory of the past.

Suppose I am about to recite a psalm which I know. Before I begin, my expectation (or "looking forward") is extended over the whole psalm. But once I have begun, whatever I pluck off from it and let fall into the past enters the province of my memory (or "looking back at"). So the life of this action of mine is extended in two directions—toward my memory, as regards what I have recited, and toward my expectation, as regards what I am about to recite. But all the time my attention (my "looking at") is present and through it what was future passes on its way to become past. And as I proceed further and further with my recitation, so the expectation grows shorter and the memory grows longer, until all the expectation is finished at the point when the whole of this action is over and has passed into the memory. And what is true of the whole psalm is also true of every part of the psalm and of every syllable in it. The same holds good for any longer action, of which the psalm may be only a part. It is true also of the whole of a man's life, of which all of his actions are parts. And it is true of the whole history of humanity, of which the lives of all men are parts.

29

But because *Thy loving kindness is better than all lives*, see, my life is a kind of distraction and dispersal. . . . But now *are my years spent in mourning*, and you, my comfort, my Lord, my Father, are eternal. But I have been spilled and scattered among times whose order I do not know; my thoughts, the innermost bowels of my soul, are torn apart with the crowding tumults of variety, and so it will be until all together I can flow into you, purified and molten by the fire of your love.

30

And I shall stand and become set in you, in my mold, in your truth. And I shall not endure the questions of men who, victims of a disease which is its own punishment, want to drink more than their stomachs can hold and who ask: "What was God doing before He made heaven and earth?" or: "Why did the idea of making something occur to Him, when previously He had never made anything?" Grant them, Lord, to think carefully what they are saying and to realize that when there is no time, one cannot use the word "never." To say, therefore, that God "never" made anything can only mean that God did not make anything "in any time." Let them see, then, that without created being, time cannot exist and let them cease to *speak* that *vanity*. I pray that they too may be *stretched out to those things which are before*, so that they may understand that before all times you are the eternal creator of all times, that no times are coeternal with you, nor is any other creature, even if there were a creature before all times.

III.

Time and Understanding

Beginning with the work of René Descartes, in the early part of the seventeenth century, modern philosophy arose as an attempt to achieve a new understanding of nature and of the human capacity to expand its realms of knowledge. The Cartesian principle of the cogito and the accompanying metaphysical justifications of its clear and distinct ideas were to enhance our ability to understand the world of nature. Concerned with justifying the application of mathematical truths to the study of natural phenomena, Descartes' task was to vindicate their use by the human mind.

Turning away from the perceptual world to his own thinking, he used an introspective method to establish the non-empirical validity of mathematical reasoning. The source of the ideas we use to examine nature was to be found in the thinking mind, not in the world it thinks about. Once the thinking mind's authority to utilize mathematical reasoning was confirmed by its interior examination, it was to use these clear and distinct ideas, which it validated, for a mechanistic understanding of the world of nature.

And this world of nature was conceived primarily in spatial terms. Descartes' fundamental concept was that of substance. Material substance, the thingness of physical entities, was described in terms of spatial extension. Change and motion, then, were conceived primarily in geometric terms as change of place or position, not in terms of process. As a result, the concept of time played a minor role in the Cartesian attempt to understand the world of nature in the idiom of mathematical mechanics.

Whatever the role assigned to time, one might have expected that Descartes would have developed the concept of a continuity of time, which seems implied by his doctrine of the permanence of substance. But the concept of time (only briefly alluded to in the *Meditations)*, which he used and defended, was conceived atomistically as a series of separate independent moments—analogous to separate points which are only connected into a line by imagination. He saw that this view precludes continuity of being; it means, he noted, that "my life may be divided into an infinite number of parts, none of which is in any way dependent on the other." Concerned about the continuity of the self, he worried how we can possibly get from one sovereign moment to the next and saw these independent moments as necessitating the repeated creative efforts of God "in each moment." [1] Indeed, this argument, and the atomistic view of temporal points on which it is based, was even advanced as an axiom:

> The present time has no causal dependence on the time immediately preceding it. Hence in order to secure the continued existence of a thing, no less a cause is required than that needed to produce it at the first.[2]

Again, when we speak "of the time or duration of something which endures," he has told us, "you will not deny that the single moments of this time can be separated from their neighbors . . ." [3]

Each 'present moment' is a discrete now that is completely autonomous, somehow infinitely divisible and yet irreducibly real. Without infinitely recurring divine intervention, Descartes had seen that this sovereign autonomy of discrete temporal now-moments, which are taken as objectively real, forecloses the possibility of that continuity, of self and of things, so central to our experience. But, if this experiential continuity is not authentic, we should ask whether it is an illusion ensuing from the perfect efficacy of God's recurrent creativity. Whatever the philosophic motive for insisting on discrete and independent real temporal points, its force is to throw into question his necessary thesis that God is not a deceiver, a thesis he recognized as fundamental to his philosophic enterprise. This thesis of temporal atomicity also raises the question of reconciling his doctrine of the permanence of substance, the possibility of causal explanation of the continuity of motion and, for that matter, of the continuity of the thinking of the scientist whose work he had set out to justify.

In his defense of temporal atomicity, Descartes had apparently used 'time' and 'duration' as interchangeable. But in his last major philosophic work, *The Principles of Philosophy* (1644), he made a sharp and influential distinction between them, a distinction which parallels that between the characteristics we truthfully see in things and those which are merely in our thoughts about them. This double distinction is set out in the only Principle concerned with time. Captioned, "Particular attributes are in things, others are in thinking. And what duration and time are," it reads:

> Some [attributes] are in things themselves, of which they are said to be attributes or modes; others surely are only in our thinking. Thus, when we distinguish time from duration taken generally and say that it [time] is the number of motion, it [time] is only a mode of thinking; for certainly we neither understand a different duration in motion than in things not moved: which is manifest from the fact that if two bodies are moved for an hour, the one slowly and the other quickly, we would not count more time in the one than in the other, although there is much more motion. However, in order to measure the duration of all things, we compare their duration with the duration of those greatest and most uniform motions from which arise the years and the days; and this we call the duration of time. This therefore adds nothing to duration taken generally except a mode of thinking.[4]

The statment, noteworthy for its lack of temporal predicates, leaves duration undefined but yet regards it as an absolute and real dimension in which motions or events transpire and in which things last or perish. As such, duration is an inherent aspect or condition of existent things. Time, in contrast, is the human perspective; it is the way *we* measure and compare 'chunks' of duration; we do this by measuring the extent of uniform motions, computing their durations in terms of a recurrent metric such as a day or year—and use this measure of motion in order to measure the duration itself. If we may presume Descartes' essentially spatial concept of motion, it then appears that it is duration that is objectively uniform and real while time is a human mode of quantifying it in terms of spatial displacement.

It would then seem that Descartes, by pursuing an Augustinian method, has amended the Aristotelian formula, 'time is the measure of motion' to read 'time is the measure of duration by means of numbering spatial motion'. Taking a uniform motion (e.g., the rotation of the earth around the sun), we number its movement—as a geometer marks off the circumference of a circle—into days, months, years, and take this numbering of motion as a standard metric by which to measure the duration of all other things and events.

We can now look back on Descartes' earlier statements and see that his thesis of the atomicity of now-movements was patterned after the image of a sequence of numbers—each of which can be regarded as a singular (even if collective) term, each of which is infinitely divisible, and each of which is not dependent on or 'caused by' any other single number. Although Descartes had inveighed against dealing with time in terms of abstractions, the perplexities which arise from his theory of discrete moments seem to do so because he modelled his vision of concrete time on the abstraction of a number series.

But Descartes had, in his distinction between duration and time, provided a framework which claims to touch on physical reality and permits us to understand and describe sequential relations of things with some degree of precision. Whatever the later philosophic implications to ensue from Descartes' view that spatial measurement is to be used to mediate between real duration and human modes of sequential description, its impact on the development of scientific thought appears to have been decisive. Some forty years later it seems to have been adopted by Sir Isaac Newton and offered, with minor terminological modifications, as a first principle of his new physics. Newton's famous definition reads:

> Absolute, true and mathematical time, of itself, and from its own nature, flows equably without relation to anything external, and by another name is called duration; relative, apparent and common time, is some sensible and external (whether accurate or unequable) measure of duration by means of motion, which is commonly used instead of true time, such as an hour, a day, a month, a year.[5]

As absolute time (duration) and absolute space became the two fundamental quantifiable coordinates of the new Newtonian physics, so they entered into modern thought.

The new philosophy followed Descartes' lead by focusing its attention on the nature of that mental activity that produced the new science. Because mental activity is more nearly described as temporal than spatial, and because the philosophic focus was on the process of understanding rather than on the world that was to be understood, the justification of the Descartes–Newton concept of time became crucially important to philosophic thought. As philosophers sought to understand the nature and functioning of man's thinking in its understanding of nature, they sought to understand the role of time in man's thinking that seeks to come to grips with nature.

Turning away from Descartes' metaphysical justification of the cogito to the activity of the cogito itself, a prime question slowly emerged: how is the thinking mind to understand the time that seems to be its prime dimension. Following the Cartesian introspective method, Locke, Leibniz, and Kant set out to face the questions posed by the theory of time in the new scientific outlook, questions about the functioning of mental activity in seeking to understand the phenomena which appear within its scope.

John Locke's *Essay Concerning the Human Understanding* is generally regarded as the fount of sense-data empiricism. Much in this celebrated book, however, goes beyond this usual categorization. In the chapter entitled "Idea of Duration and Its Simple Modes," Locke has provided an almost phenomenological description of the manner in which our temporal concepts arise. Turning away from the scientific preoccupation to quantify the sequential relationships of the objects of experience, Locke faced the question of just how our conception of time arises in the human understanding.

Taking the notions of time, duration and eternity, his essential argument here is that they can be traced back to the two "sources of all knowledge, viz. sensation and reflection" (p. 122). He took the concept of duration—seen as the lapse between two successive ideas in consciousness—as primary, and argued that our ideas of time and eternity are both derived from it. After drawing the distinction between the permanence of space and of duration, which he defined in terms of succession, Locke turned to the derivation of the concept of time from that of duration (secs. 3–15).

The notion of the succession of ideas in consciousness becomes fundamental to his argument. It is, he insisted, from this experience of the succession of ideas that we derive the notion of continuity—in ourselves and in the things that we observe. The concept of duration, then, arises from "reflection on the train of the ideas" which we observe within ourselves (see sec. 4). This succession of ideas is fundamental and the perception of external motions is dependent upon it. The succession of ideas is, then, the measure of all other successions (sec. 12) and would thus seem to have its own constancy (sec. 9). In contrast to duration, which he defined in terms of succession, the notion of 'instant' was defined in terms of the instantaneous event whose beginning and end are unperceivable in our experience of it. The succession of ideas is, then, the fundamental experience from which all time concepts emerge. This sequence, Locke argued, is periodic in nature and thus provides the regularity that is requisite for any measure. By comparing the intervals between separate ideas we derive the notions of duration and of ordered sequence; this regularity in "the constant train of *ideas* in our minds" gives us that measured duration which we call time (sec. 16).

The long tradition that had linked time and motion is then called into question—and on two separate grounds. First, Locke argued that we do not know if external motions are the cause of succession of our ideas. Our ideas are a series of periodic appearances, and it is only by comparing them that we can arrive at any concept of motion. Evidence for this argument is the fact that we can experience the 'constant train of ideas', and hence duration, even where there is no external motion. Time *qua* measure is thus dependent upon the periodicity of appearances in our minds, not on motion external to our minds. But, second, time cannot be the sole measure of motion; we need space, as well as time, to measure external motions. It is only the regularity we see in certain motions, not motion in general, that apparently had suggested the traditional view (see secs. 16 and 22).

Consequently, our "minutes, days, and years are, then, no more necessary to time or duration, than inches, feet, yards and miles, marked out in matter, are to extension" (p. 130). These names for periods of duration are merely our ways of distinguishing sequential intervals from the appearances which arise as ideas of consciousness. But, "*any* regular periodical appearances" (p. 132) will do: the periodicity of the train of ideas enables us to use any periodic sequence we choose as a measure of the duration of other sequences. What we measure in time need not be contemporaneous; so, we can, in imaginative reflection, extend our measure forwards and backwards into an indefinitely endless series that we call eternity.

Locke then turned to discuss the relationship between duration and space. In that discussion (not included in the selection) it becomes perfectly clear that he had taken up Newton's definitions of time and of space and sought out their source in the functioning of human experience. Locke took duration as "absolute time" and thereby as the "common measure of all existence whatsoever, wherein all things, while they exist, equally partake" 6 Time is described as "part" 7 of duration, and space and duration are described as co-existent (thus making it possible to measure specific motions): "every part of space being in every part of duration, and every part of duration in every part of [space]" 8 Time, then, is the measure of duration by means of periodic changes; when we apply this to the observed world we combine it with the measure of space in order to obtain a measure of motion. But the presumption of any temporal measurement is the constancy of duration as "constant, equal, uniform" (p. 129), an assumption which we certainly cannot prove.

One prime criticism of Locke's discussion was voiced by Leibniz, who called into question Locke's fundamental premise that our ideas follow each other in a regular sequence (see sec. 16). These changing ideas, Leibniz had suggested, do not really give us time, but do "furnish us with the occasion for thinking of time" which we measure in terms of observed motion in space and from which we derive the notion of duration.9 Locke had, indeed, internalized the notion of absolute time or duration, assumed that its even flow is also that of our ideas, and then presumed a complete parallelism between the sequential order in our internal ideas and in the world of nature in which things are. Especially in view of his own wise dismissal of any presumption of a causal explanation for this parallelism, there is no apparent ground for this claim.

Locke had pointed out the spatial component for motion, presumed in all classical discussion. It is well to make this explicit for unless we are discussing purely psychic changes, whenever we use time in the measuring of motion, we are measuring motion in space and take account of the quantity of space traversed. But, in making this important point, it seems clear that Locke's real concern was not the internal constitution of the time experience which he had claimed to make foundational, but the time of things in nature which it is science's business to calculate. What he has then done is to seek an internal ground for this measuring of external things. He has claimed to do this by a reduction of sequential consideration to sense *and* reflection. But it is not clear just how sense perception contributes to the development of his three temporal concepts (duration, time, eternity) as Locke developed them—seemingly out of reflection alone.

In his discussion, his working concept of the 'present' has not been the instantaneous now-moment, which he seems to have taken as primary (see sec. 10), as much as a spread over the 'interval'; for, if we truly can apprehend the periodicity of the sequence of ideas by introspection alone, as Locke had claimed, we must not only be able to hold in the present perspective the moments which make up the interval between one idea and the other; we must also be able to hold the several intervals in mind at once in order to compare them. It would be difficult to understand just how we are to arrive at even an approximate comparison if it were not possible for us to transcend somehow an atomicity of now-moments—or an atomicity of ideas.

And it is this latter that seems to be another premise (in accord with his whole philosophy) which Locke needed for his argument. If one would urge that the 'train of ideas' is modeled, not after the pendulum of the clock but after the flow of a stream, it is hard to see how this periodicity could be primary and would not itself have to be derived from something external to it.[10] Locke seems rather to have transformed the assumed Cartesian atomicity of temporal moments into the atomicity of discrete and separable ideas and then attempted to trace the continuity of moments back from it. For his notion of duration appears, in the end, to be an irreducible line of separable moments alongside of which things transpire. Time, then, becomes a method of counting these moments and apportioning to each event, the number of moment-points alongside of which it takes place. It is, then, not clear that, despite his disavowal and attack, he really succeeded in departing from the inherited notion that time is the measure of motion—for the time that emerges is a spatially conceived measure of a spatially conceived duration, arising from a sequence of motions whose duration points it is used to measure.

In many crucial ways, Locke worked out from what Descartes and

Newton had given him. He took up their distinction of (absolute) duration and (relative) time; he sought to ground these in the subjectivity of the cogito's understanding. For he had seen that man's scientific thinking about the dynamics of the world depends in the end upon man's capacity to think and on the nature of that thinking. Locke has, thereby, effectively urged that the thinking of the cogito is, itself, inherently temporal and that its intellectual activity, its thinking about the content of its ideas, necessitates comparison of stages in a temporal sequence. Whatever the evaluation of the way in which he formulated his discussion, his premises, or the specific conclusions he claimed to have established, he effectively moved the quest for the understanding of nature's temporal sequences to a quest for the temporal understanding of the understanding itself, to a quest for the temporal constitution of human experience.

Leibniz, Locke's junior by fourteen years, was often described as the most erudite man of his time. His work, scattered among innumerable notes, papers, memoranda, and letters to eminent persons, displays a sustained attempt to provide a coherent metaphysical foundation for both science and morals. His interests and activities covered a wide range of the knowledge and affairs of his era; the impact of his influence, still felt in several disciplines, has not been limited by the battles between philosophic schools.

He can be seen as having taken the Cartesian primacy of the cogito forth into that new system of speculative metaphysics which he worked out under the title of a 'monadology'. In so doing, he carried the Cartesian thrust forward by radically transforming its static mechanism into a dynamic teleological system. Rejecting the Descartes–Newton framework of absolute space and time, within which Locke tried to work, together with the companion rejection of the Cartesian preoccupation with spatiality, his work constituted a philosophic revolution of first magnitude.

Sharing with Newton the distinction of having invented calculus, Leibniz was impatient with the primacy of geometric reasoning in Cartesian thinking and with its emphasis on spatiality as the prime characteristic of existent things. He was preoccupied with the essentially dynamic character of the things that are experienced as well as of the experiencing self. Spatial extension could certainly characterize a static thing but in itself could not account for any changing state or any motion; spatiality is also inappropriate as a prime characterization of mental activity. He consequently found it necessary to reject the apparent Cartesian identification of substance or 'thingness' with the essentially static notion of spatial extension or displacement.

Leibniz conceived activity and individuality to be the first marks of any existent entity. If all cognition can arise only through individual perceptions, each of which has its own orientation or point of view onto the universe of activity it observes, then space cannot be the primary dimension and neither space nor time can be absolute.

In defiance of a long tradition hailing back to Aristotle, Leibniz reasoned that space and time, as such, are *not* perceived. What is actually perceived is the variety of objects in a changing order of relationships to each other and to the perceiver as well. What we do perceive is this order of changing things, and it is this ordering that we understand in terms of time and of space.

Space, then, is but "the order of co-existing things" and time is but the "order of non-contemporaneous things" (p. 136). Necessary to any understanding of events, changing things, or motions, he reasoned, are these two *temporally defined* coordinates: that order of simultaneous existences which we term 'space', and that order of sequential existences which we call 'time'. Space and time, then, are both to be understood in terms of the *temporal* predicates 'at the same time' or simultaneity, and 'at succeeding times' or sequentiality. In the face of virtually all preceding thought, the primacy of the spatial backdrop of all changing is effectively subordinated to temporal order.

Duration, then, in a further step from the tradition, is no longer regarded as that absolute 'flow' which time measures but the quantification of time, or temporal order, itself. For, if time is the prime 'dimension' of existent things but is nothing itself, then, when we measure time, we are not measuring a self-subsistent entity—but merely the sequential order of the changing relations of things which we arbitrarily mark off in terms of abstractions called 'moments'.

Time, Leibniz argued, is not some *thing* which is independent of existent things; it is not a frame or a container in which they subsist. Time and space cannot be absolute; they can have no reality independent of the entities that are related in terms of them. To maintain that space and time have any such independent reality would mitigate against the thesis of a rational universe—in which there must be a reason sufficient to explain why a particular event took place just when and where it did rather than at some other time or place. If time and space were absolutes—and here we may note the primarily theological level of much of the discussion of that time—there would be no rational explanation of why the world was created just *when* and *where* it was instead of at some other time in some other place. If time and space are not pre-existent containers for things, but are the system of relationships of those things to each other, no such problem arises. For space and time are then nothing in themselves; they are each but the *internal* and inherent order of aspects of the created world.

Leibniz offered a non-theological explanation as well. All knowledge and awareness, he noted, arises within the perceiving individual self. Each individual mind is in a continual change of state in its own consciousness—as Locke's notion of the 'train of ideas' indicated. Each individual is, then, characterized by its own sequential development. Its time is the sequence of its perceptual (and appetitive) life. It understands its time, then, in terms of the sequence of the changes that mark its being, which is a continuity of becoming. The objects of its perceptions are other individual entities, of high or low status, in the order of being, in the plenum of real individual beings or monads constituting the order of the universe. Time and space, then, are not independently real, but 'well-founded phenomena'; the correlation of the temporally marked perceptions of this multiplicity of individuals is possible just because they "will always have a harmony among themselves, because they always represent the same universe." [11]

This comprehension of reality, in terms of the outlooks of the individual perceivers, places a premium on the essentially temporal character of the individual. For, Leibniz argued, individual identity is constituted in terms of temporal continuity; this continuity is not merely subjective; it is the continuity of existence and is rooted in the essentially rational structure of the universe. If the continuity of individual identity is real, then its time must be a continuity also. If the universe is composed of perceiving entities—entities which, each in its own way, are continually reflecting its environment or context of being—then the universe as a whole, as every discernible aspect of it, is to be understood in terms of the continuity of internal temporal order.

If the existence of an individual is a continuity of becoming, if our perceptions arise out of our capacity to have them, then "all our phenomena, that is to say all that can ever happen to us, are only consequences of our being." [12] One's own nature is the ground of what is possible for him, and any future actuality, whether contingent or necessary, is already within the realm of those possibilities he represents. If so, then the continuity of one's own life, as of the time that marks its development, at any abstracted moment, should reveal something of its past and its future. "The present is great with the future; the future could be read in the past; the distant is expressed in the near. One could learn the beauty of the universe in each soul if one could unravel all that is rolled up in it, but that develops perceptibly only with time." [13] The *flow* of temporal order is, then, to be understood in the flow of one's life. Time, like life, is not to be understood in terms of atomically separated moments or ideas. Time is not to be understood as a lifeless container. Time is the system of relationships that bind the members of the universe to each other in a mutuality of development.

Time, then, to face the ancients' query, can be used to measure motions by taking the 'borders' of any given change as moments, as non-existent but useful abstractions from the continuity of becoming. How wide or narrow the intervals between these arbitrarily selected moments may be is just as dependent on the kind of measuring in which we are engaged as is our choice of meters or miles to measure the order of spatial relation of two simultaneously existing entities.

But this is to suggest that the notion of an objective time is an ideal thought, an abstract creation of the imaginative intellect—which may find such a conception useful but cannot claim that the thought has any real reference to an existent entity. Time is an existential phenomenon that is "well-founded" in the sequential ordering of the continuity of one's life experiencing. It marks a continuity of changing which can be marked off in the flow of one's being, in arbitrarily chosen markings called 'moments' which are abstracted from the flow. Such 'moments' are really indiscernible and are thereby unreal. We can then understand why the past flows into the present without our being able to mark any sharp borderline between past and present, why the present similarly flows on into the future. The continuity of order in nature is understood by us through the continuity of our experiencing of it in community with other beings who share this common experience, while bringing their own individualities to its appreciation.

The many strands of inquiry which arose from Descartes' attempt to validate the human understanding were brought together in the Critical philosophy of Immanuel Kant. Following out diverse leads developed by Locke and Leibniz, Kant turned his attention to the 'dissection' of the cogito, of the cognitive consciousness itself, in order to determine the ways in which it structures its knowledge, in order to discern the limitations which its cognitive capabilities impose on its possible attainments. Within the limits of cognitive competence, Kant believed that he had established the objective validity of cognitive thought—just as those same limits simultaneously mark out the limits beyond which cognitive claims could not possibly be validated. Central to his task was his success in establishing the "empirical reality" of both time and space, while dismissing the legitimacy of any claims concerning the ultimate nature of either beyond the bounds of possible human experience.

Raised on the predominant Wolffian philosophy of the German Enlightenment, Kant's prime departure from the common outlook of his colleagues was his defense of Newtonian views, developed by his interest in the new physics. In 1765, Leibniz' one major work, *The New Essays on the Human Understanding,* was finally unearthed and published. This critique of Locke, Leibniz's only sustained and systematic philosophic work, seems to have impelled Kant to think through the philosophic foundations of his own thinking. Developing Locke's notion of the 'train of ideas' and a number of Leibnizian distinctions (e.g., the phenomenal–noumenal distinction) while criticizing some of their formulations, Kant sought to ground the validity of the new physics while still providing a foundation for morality. Working within the general framework and orientation of Leibniz's monadological metaphysics, Kant developed his new Critical metaphysic of human experience. Concerned to elicit, from an interior analysis, the ways in which the human mind 'reflects' the world that appears to it, he effectively restricted Leibniz's principle of sufficient reason to the grounds of our knowledge of things that appear, and as they appear, to us—instead of permitting speculative application to things as they really might be beyond the ways in which we are able to experience them. He thus restricted all possible knowledge of the world of nature to the capability of this structured 'reflecting' process which produces it.

Crucial to Kant's new philosophy of experience was his new answer to the old question: 'What, then, are space and time?'. Disavowing the Newtonian view of absolute space and time on Leibnizian grounds and also because they could not possibly be experienced, he, in effect restated the Leibnizian view but with a radical restriction on its possible extension or area of applicability: time and space are indeed necessary relational predicates of things—*as we are able to perceive them*, as they can enter into the consciousness of human experience. Time and space are, then, the two "forms" in which all human perceptions occur. By thus 'reducing' time and space to the experiential form of the perceiving subject, Kant claimed to account for their pervasiveness in our experience and, thereby, for the universality and objective validity of our temporal and spatial predicates. Insofar as all external things are perceived by us in spatial terms, space is the form of what Kant, in the First *Critique,* was to term "outer sense." Insofar as *all* of our awarenesses—of outer things or of our own internal thoughts—are acts of consciousness in sequential order, the process of thinking, regardless of its object, is essentially temporal and time is the form of what Kant came to call "inner sense" in which all consciousness transpires.

Experiential time and space are then "empirically real"; they are real *within* our experience, but just what they might be beyond our possible experiences we are clearly unable to know. We do, however, need to use the ideas of one comprehensive 'absolute' space and one all-inclusive 'absolute' time-order so that we may understand the particular space and time relations of our particular experiences. These two ideas—of one universal space and one universal time—are then "transcendentally ideal," i.e., they are ideas which are necessarily presupposed in understanding our finite range of possible particular experiences; but they are ideas which can refer to no conceivable or directly verifiable experience we might have.

The first presentation of the outlook of the new Critical philosophy was offered by Kant in his "Inaugural Dissertation." Published in 1770, this short essay provides a lucid preview of the essential thrust of the *Critique of Pure Reason* which was to appear eleven years later. The central portion of this essay provides a clear and readily readable account of the new Kantian thesis concerning the nature of space and time and of the relationship between them.

The complete priority of time before space, already suggested in Leibniz's definitions of them, is made explicit in Kant's reasoning: some of our thoughts are about external objects; these are apprehended as spatial objects and are thus perceived as being in spatial form; but *all* of our thoughts—including those about spatial objects but also about our own thoughts and feelings—are experienced in temporal sequence and are thus inherently in temporal form. Time, then, is the prime and universal form of all sensibility; it is, to use the term by which Kant's *Anschauung* (outlook) is usually translated, the prime form of "pure intuition," the prime form of our capacity to have data to think about regardless of what that data may be. For Kant's argument has been that we are not able to 'receive' data except in some kind of ordered form, that without temporal order, any coherent experience would be impossible.

The Kantian shift—from the nature of the world as it may be in itself to the nature of the human knower of the world—suggests a significant reversal of Platonism. In contrast to Plato, who seemed to presume the existence of space which was *then* ordered in terms of time, Kant's argument, within the confines of human experientiality, is the reverse. From the viewpoint of the human experienc*er,* the dimension of time is fundamental, primary, and pervasive. In order that an X may be an object of our awareness, it may, indeed, be seen as spatial, but it must be seen as temporal—for it is only in the form of time that we become aware of it in consciousness. As Kant capsuled it, time "contains the universal form of phenomena" (sec. 14.7) and makes any experience, and our understanding of it, possible. And, again, "time is an absolutely primary, formal principle of the sensible world" (sec. 14.7). But this is to say that for Kant, as for Plato, time is *the* prime principle of order—for Plato of the physical universe itself, for Kant as it may be experienced by man. The Kantian shift of the space-or-time priority results from the shift of perspective from the world-as-it-is-in-itself to the knower to whom the world appears in terms of his finite capacity to apprehend it.

By thus internalizing the traditional thesis—that time is the principle of order—Kant had begun to see that this necessitates a revolutionary change in the understanding of the human understanding itself. However we may judge the notion of eternal truth or idea, Kant reasoned, ideas must be temporalized, must be brought within the form of time in order to have cognitive relevance. Even in using the hallowed logical principle of contradiction, "Reason itself cannot . . . dispense with the support of this concept of time, so primitive and original is it" (sec. 14.6). One essential thrust of the Critical philosophy is the thesis that possible human knowledge is essentially time-bound, that pure reason cannot of itself attain knowledge; that to do so, it must enter into and be molded by the nature of temporal experience. Human knowledge, then, as distinct from speculation, cannot, because of its inherent nature, attain any competence or attainment over any material outside of presentations in temporal form.

Foreshadowed in the "Dissertation," the First *Critique* presented what in many ways is but a working out of this essential thesis and some of its implications. Kant's detailed elaboration of his argument for the fundamentality of time in human cognition was developed in some four crucial stages.

First, (in the "Aesthetic"), he presented a complex restatement and development of the argument of the "Dissertation"—that all experiential content must be framed in the forms of space and time, that the form of time is more encompassing, and that what is conceived as non-temporal cannot be a possible object of human experiencing. Second, (in the First Edition "Preview" to the "Deduction"), he argued that our conscious awareness of objects is not merely the presence of sense-data but already a synthesis of that data with concepts structured in a specific categorial manner.[14] In presenting this argument, he added the crucial point that even the most rudimentary awareness already involves a synthesis of the present presentation with memory-recall—which is to say that the most rudimentary experiential notion of the present already involves a temporal synthesis with an aspect of the past.[15]

Third, in what may be the most crucial part of the *Critique* from a temporalist perspective, he draws the essential conclusion from the first two stages: any concept, abstract or concrete, that is involved in the understanding of any perceptual object, is temporally derived and structured. This is to say that the pure categories, in terms of which we conceptually comprehend what we see, are not ideal concepts (in a Platonic sense) but are abstractions from the essential temporal structure of experience. This third stage was presented in a very short and somewhat cryptic chapter entitled, "The Schematism of the Pure Concepts of Understanding." If we bear in mind that 'schema' is a model or diagramatic representation and not a photograph, that 'schemata' is its plural form, then the word, 'schematism' clearly means that system of schemata or conceptual models in terms of which we conceptualize our experiences. Any empirical concept, such as, for example, 'tree', is a schema; we recognize pines, oaks, and birches as 'embodiments' of the schema 'tree' although no one image could represent them all—their shapes, colors and changing states are too different. The 'schematism of the pure concepts or categories' then means that operationally active systematic procedure whereby the functioning understanding develops pure or non-experiential concepts or categories in terms of which it is able to intellectually understand the content of its time-bound sense experiences.

It was only after Kant had presented his new thesis embodied in the "Schematism," that he was able to develop and elucidate those actual principles which he saw as operative in the formation of actual human knowledge; it was there, for example, that the fuller implications of the complete priority accorded to time became evident—as he traced our working notions of such honored metaphysical concepts as substance, permanence, and matter back to three prime concepts of time: duration, succession, and co-existence.

For the present concern, the crucial point is the Kantian argument that all cognitive concepts arise from the manner in which we are able to organize the temporal field presented to us by our awareness of sense-perceptions, by our capability to have sense-experiences. Prefigured at the end of the "Dissertation," [16] the chapter on the "Schematism" constitutes a philosophic revolution of first magnitude.

Here he introduced the problem which arises from the fact, taken as established, that all awarenesses in consciousness (i.e., "inner sense") are in the form of time by virtue of the sequential nature of ideas. This means that concepts invoked in the interpretation and understanding of these presentations must be able to enter temporal form as well. So Kant proceeded to an analysis of a schema-in-general, of a model for the cognitive activity of the intellect. He offered a brief analysis of the basic conceptual schemata or diagramatic models, presupposed by the mind in its essential activity of interpreting the meaning of the data it receives. He then offered an explanation of the categories (already presented earlier in the book) [17] which, he here insists, are necessarily rooted in the schematic system. The twelve fundamental categories of thinking (roughly akin to Aristotle's "predicables") are divided into four groups of three each. Kant's essential argument here is that they each arise from one of the four ways in which we conceptualize the experience of time.

Thus, the quantitative categories—of unity, plurality, and totality—arise from the concept of number which, in turn, arises from the experience of a time-series, a series of abstracted 'moments' of time; the qualitative categories—of reality, negation, and limitation or finite reality (e.g., any specific object)—arise out of the experience of the content of any temporal representation; the relational categories—of quality inhering in a substance, of causal relations, or of simultaneous interaction—arise from the orders of events in time; finally, the modal categories—of possibility, existence, and necessity—arise from the ways in which *we* understand the entirety of what we experience with the internal complexities we see it as presenting to us through the whole scope of possible experience.

All of this is very difficult material; it is difficult to grasp even after going through Kant's complete text. The point, however, which is essential for this present overview of the development of a philosophic understanding of time in human experience, can be put rather simply. Kant's essential argument here is to challenge the entire tradition in its postulation of atemporal or supratemporal concepts as the source of cognitive truth. Against all rationalisms (including the rationalism present in traditional empiricisms), Kant argued that the human understanding is only able to obtain knowledge of the world in which it finds itself by the use of concepts which are temporally structured and which are thereby qualified to enter into the field of time in which all our experience transpires.

It is from the four ways in which we are able to structure time—time-series, time-content, time-order, and time-scope—that all cognitive predicates arise. It is the unity of this four-fold time experience that provides the unity of consciousness; it is these fundamental time-structuring rubrics, models, or schemata that enable us to develop empirical concepts, abstract categories, and the operating principles by which actual knowledge is built up. But, if these "schemata of sensibility first realise the categories, they at the same time restrict them" to the field of possible time-bound sense experience. These temporal notions or schemata are, then "the true and sole conditions under which these [categorial] concepts obtain relation to objects and so possess *significance*" (p. 156).

Time, then, Kant's revolution insisted, is not only the essential form and inherent limitation of any possible sense experience or awareness we might have. It is also the essential structure of any concept we can legitimately use to understand the objects of which we are aware in consciousness. These temporal concepts or schemata, which are the ways in which we organize temporal experience, comprise, then, the fundamental condition for any coherent intelligible experiences we may have. The capacity to have and organize the experience of time— which frames any experiential content we may have—is then the capacity to have any intelligible experiences at all.

If, as Kant was to say later, "experience is nothing but a continual synthesis of perceptions," [18] then experience is grounded on this capacity to unite disparate temporal segments into a continuity of temporal 'flow'; cognitive experience then essentially involves understanding the dynamic nature of temporal passage by means of temporally framed conceptual structuring.

In the fullest sense, then, time for Kant as for Plato, is *the* principle of order—for Plato in the world in itself, for Kant in our understanding of it as we are able to perceive it. But, in sharpest contrast to Plato, this does not mean that the temporal is derived from a timeless order and is to be somehow understood by timeless concepts. The essential thrust of Kant's argument here is that timeless concepts are not cognitive and that the only ways in which we can make legitimate cognitive claims about the world as we perceive it is to use concepts that are temporally structured, which, indeed, arise out of the way in which time can be understood. Such temporally structured concepts do ground the validity of empirical ideas about the world which appears to us as temporal. They are competent to handle what is *in* time because they are concerned with time and do not claim competence beyond time. Time is the principle of order and, one can even say, of intelligibility, in our understanding of the world as it appears to us within a temporal frame.

Kant also continued the ancient tradition that time and number are essentially related. The complete universality of application of mathematical principles to the entirety of the world of possible experience, Kant traced to his grounding of them in the forms of time and space while he continued also the ancient tradition of relating time and motion. "Geometry is based upon the pure intuition of space. Arithmetic accomplishes its concept of number by the successive addition of units in time; and pure mechanics especially cannot attain its concepts of motion without employing the representation of time." [19]

In view of the historic import and far-reaching implications of Kant's newly proclaimed primacy of time for the understanding, it is especially remarkable that he did not examine the notion of time itself. Using it repeatedly, he never seems to have really defined it. In the "Dissertation," he introduced time in terms of succession and simultaneity. In the First *Critique,* he seems to have understood it primarily in terms of succession, referring explicitly to simultaneity only in the section entitled "Analogies of Experience." He was, of course, quite aware of the difficulties involved with a comprehensive and precise definition of time. In one of the last pre-Critical essays, he had referred to some of the difficulties involved in attaining such a definition.[20] He then set out to attain a working definition by examining its meanings in different contexts of thought. But never having developed a comprehensive definition of the concept of time, he found himself using shifting predicates as he proceeded from one discussion to another, e.g., from the "Aesthetic" to various stages of the "Analytic" in the First *Critique,* from questions of theoretical understanding to those concerned with moral practical reason.

By and large, Kant's working notion of time seems to have been Locke's "train of ideas"; conceived along a linear analogy, the basic notion seems to have been that of the succession of discernible moments as points on a line (see p. 151)—although the Leibnizian principle of continuity seems to have mitigated the atomism of real separate instants Locke suggested.

If one recalls that one prime reason for the new Critical edifice was the grounding of the new Newtonian physics, and that the new science

required an identification of temporal succession with pervasive causal determinism, one can then understand what moved Kant to deny
the applicability of the temporal predicates of the cognitive understanding to the concerns of practical reason. Physics is necessarily
determinist because it explains the present phenomenon in terms of
causal chains in the past leading up to and yielding the content of the
present moment; only in this way is it able to look beyond into the
future and extend that causal chain in prediction.

But moral freedom, responsibility, and purposive thinking in general, cannot be explained without a reference to some kind of open
future. Because this seems to be precluded by mechanistic explanatory schemes, because the possibility of morality presupposes the
possibility of free responsible decision, and because Kant's essential
understanding of time was apparently derived from the sequential
and linear "train of ideas" in which the past 'produces' the future, he
felt obligated to sever the whole domain of practical reason from that
of sequential time. In doing this, however, he effectively propounded
a question which he never really came close to resolving—just how
practical reason can, in fact, find application or entrance into the
chain of sequential time. But this is to ask whether time is *only* a
subjective form, whether there is any sense in which time is 'really
real', whether there is any meaning in the notion of a kind of time that
encompasses both our theoretical understanding and the activity of
moral practical reason which Kant associated with the ultimate reality
of the human self.

It would seem that Kant's working concept of time was what Descartes had already called 'the number of motion that is only thinking,' with what Newton had referred to as "relative, apparent and
common time [which] is some sensible . . . measure of duration by
means of motion." This is the kind of time which we use to understand
the world of nature as it appears to us. For Kant to have discussed, in
the context of comprehending the phenomenal world, Descartes'
"duration" or Newton's "absolute time" would have been to have
gone beyond his own established bounds of all possible human experience. As such they might be thought, in Kantian terminology, as
'noumenal time'; although the nature of this 'noumenal time', or
indeed its existence, is beyond our cognitive competence and thereby

beyond the bounds of meaningful discussion, there is little doubt that Kant presumed some such kind of 'backdrop' for the working quantifiable time we do know and use.

When he turned from the First *Critique* to the Second, the *Critique of Practical Reason,* which is devoted to the concerns of freedom and practical moral reason that provide us a 'glimpse' of ultimate reality, Kant repeatedly referred to a kind of 'noumenal' duration. He grounded the postulate of immortality, for example, on "the practically necessary condition of a *duration* adequate to the perfect fulfillment of the moral law." [21] He posed the problem of comprehending God's "magnitude of existence, duration, which is not in time even though this is the only means by which we can think of the magnitude of existence." [22]

The duration, or enduring of things, does not appear in sensuous experiences, is somehow presumed but beyond our measure or our scientific understanding. Kant's concern with the primacy of time was with the measurable time of theoretical reason. He did not argue that there was no other kind of durational description; he merely insisted that if there is and what it may be is beyond the legitimate area of valid human cognitive experience. The criticism, then, that Kant's notion of time may have been unduly restrictive is to be tied to the companion criticism that his concept of experience was also.

Kant saw himself as propounding what he termed the Copernican Revolution in philosophy—the recognition that we cannot talk meaningfully about the world-in-itself, but only about the ways in which we are able to perceive it and understand our perceptions of it. But this is to imply a second and consequent revolution of thought: if not the world, but our thinking about the world is primary, then time, not space, becomes the fundamental 'dimension'. Though one can, with some justification, argue that Kant never really followed through the full implications of his temporalist revolution, he certainly indicated the path to which it pointed. For his entire theory of knowledge was concerned to work out the primacy of time-structuring as the ground of certainty, validity, meaningfulness, and anticipation in human experience. Whatever the ambiguities and equivocations concerning the nature of time, or the ways it functions, he did set out in forceful terms the thesis that time is the prime area in which man's cognitive problems are to be found and mastered. In this work he may be said to have thought through the full meaning of the primacy of the cogito with which modern philosophy began. Seeking to elucidate its attempt to understand the world in which it finds itself, he was the first to see the need for a fuller understanding of the temporal form of the understanding mind itself. This is to say that he was the first to see that although the mind uses spatial concepts, the mind's activity is one that is essentially and pervasively temporal.

But why didn't he take some of his temporalist arguments across some of the problems they raised? It may well be that, concerned with the use of time rather than with time itself, Kant did not think through the meanings of the concept but worked with the only concept of time he really knew, the dual notion of time he received from Descartes and Newton; one was apparently ascribed to noumenal reality and thereby beyond our cognitive competence; the other became the measurable time which seemed to follow the Lockean "train of ideas." Apparently following out Locke's insight that the formation of temporal concepts arises out of the internal functioning of the mind, Kant internalized the whole concept of measurable time which science uses to describe external things. Apparently accepting the priority of Leibniz's monadological orientation, Kant developed the view that, from the outlook of the experiencer, time predicates are prior to space predicates, that the content of experience always starts from the 'when' of the experience itself. The consequence was to regard time as the prime ground for the perceiving of nature with the necessity of temporalizing the conceptual system we use to understand nature.

Kant's exclusion of quantifiable time predicates from his own equal concern with the foundations of moral reason would indicate his awareness of the limited range of competence which a quantifiable concept of time enjoys. It would also indicate Kant's concern for the integrity of the self which cannot be reduced to or explained by a series of discrete or separable moments along a line. Sequestering the freedom of moral reason from a causal determinism associated with quantifiable time seemed to solve the problem of saving the possibility of morality without endangering the foundations of the new science. For this grounding seemed to be the foremost philosophic problem of the movement initiated by Descartes. A product of his age, Kant subjected its strains and concerns to a systematic examination and an attempted unification. In so doing, he brought an age to an end while initiating a new one. For Kant's work—in what it did and did not do, in the questions it raised and the criticisms it invited—has served as the fulcrum of the continued attempt to understand the nature of time in human experience to our own day.

Prime sets of questions raised by Kant's work prepare the stage for the discussions that have followed:

· Is time ultimately 'real' in any sense, or does Kant's description of time as a subjective form preclude any notion of ultimate becoming or reality? If time is somehow 'really real', is it essentially quantifiable, and how does it relate to the reality of space?
· Is sequential time composed of a line of separable moments authentically experiential? If so, in what kind of language can it be best expressed? Is the experience of sequentiality authentically voiced in terms of before-and-after? Or is there a more fitting mode of description? Is there any question about the equation between the time of things and of our own thinking minds?
· How do we form our temporal concepts in terms of our practical needs for using them? Do our temporal concepts merely apply to the time of the world or to ourselves as well? Is there a specifically human kind of time and, if so, what could it signify?
· How authentically do we understand our own understanding of our time? What does our understanding of time, say, about our understanding of our own experience? Is there a kind of time that joins together the time of theoretical reason and of practical activity, both of which seem to function together in each individual's unity-of-experience? If time is central to our experience, and our concept of experience is broadened from theoretical reason to the whole of our existence, how is the unity of our temporal experience structured?

Questions such as these take us in different directions. But they take us into the prime schools of discussion that mark the broadly conceived contemporary scene. For the schools of contemporary philosophy have been largely shaped by their responses to the Kantian theses, or their criticisms, as they have sought in one way or another to comprehend the import and context of experiential time. Starting with an attempt to understand man's new understanding of nature, they lead to diverse attempts to understand man's understanding of man.

8.

LOCKE:

The Idea of Duration

John Locke (1632–1704) first published his *Essay Concerning the Human Understanding* in 1690 upon returning to England from exile in Holland. The selection comprises most of the thirty-one numbered sections of "Idea of Duration and Its Simple Modes," the fourteenth chapter in Book II (as edited by A. C. Fraser). This, the first of the two chapters concerned with time, explored the way in which the experience of time arises; the following chapter dealt with the relation of time and space.

1. There is another sort of distance, or length, the idea whereof we get not from the permanent parts of space, but from the fleeting and perpetually perishing parts of succession. This we call *duration;* the simple modes whereof are any different lengths of it whereof we have distinct ideas, as *hours, days, years,* &c., *time* and *eternity.*

2. The answer of a great man [Augustine], to one who asked what time was: *Si non rogas intelligo,* (which amounts to this; The more I set myself to think of it, the less I understand it,) might perhaps persuade one that time, which reveals all other things, is itself not to be discovered. Duration, time, and eternity, are, not without reason, thought to have something very abstruse in their nature. But however remote these may seem from our comprehension, yet if we trace them right to their originals, I doubt not but one of those sources of all our knowledge, viz. sensation and reflection, will be able to furnish us with these ideas, as clear and distinct as many others which are thought much less obscure; and we shall find that the idea of eternity itself is derived from the same common original with the rest of our ideas.

3. To understand *time* and *eternity* aright, we ought with attention to consider what idea it is we have of *duration,* and how we came by it. It is evident to any one who will but observe what passes in his own mind, that there is a train of ideas which constantly succeed one

another in his understanding, as long as he is awake. Reflection on these appearances of several ideas one after another in our minds, is that which furnishes us with the idea of *succession:* and the distance between any parts of that succession, or between the appearance of any two ideas in our minds, is that we call *duration.* For whilst we are thinking, or whilst we receive successively several ideas in our minds, we know that we do exist; and so we call the existence, or the continuation of the existence of ourselves, or anything else, commensurate to the succession of any ideas in our minds, the duration of ourselves, or any such other thing co-existent with our thinking.

4. That we have our notion of succession and duration from this original, viz. from reflection on the train of ideas, which we find to appear one after another in our own minds, seems plain to me, in that we have no perception of duration but by considering the train of ideas that take their turns in our understandings. When that succession of ideas ceases, our perception of duration ceases with it; which every one clearly experiments in himself, whilst he sleeps soundly, whether an hour or a day, a month or a year; of which duration of things, while he sleeps or thinks not he has no perception at all, but it is quite lost to him; and the moment wherein he leaves off to think, till the moment he begins to think again, seems to him to have no distance. And so I doubt not it would be to a waking man, if it were possible for him to keep *only one* idea in his mind, without variation and the succession of others. And we see, that one who fixes his thoughts very intently on one thing, so as to take but little notice of the succession of ideas that pass in his mind, whilst he is taken up with that earnest contemplation, lets slip out of his account a good part of that duration, and thinks that time shorter than it is. But if sleep commonly unites the distant parts of duration, it is because during that time we have no succession of ideas in our minds. For if a man, during his sleep, dreams, and variety of ideas make themselves perceptible in his mind one after another, he hath then, during such dreaming, a sense of duration, and of the length of it. By which it is to me very clear, that men derive their ideas of duration from their reflections on the train of the ideas they observe to succeed one another in their own understandings; without which observation they can have no notion of duration, whatever may happen in the world.

5. Indeed a man having, from reflecting on the succession and

number of his own thoughts, got the notion or idea of duration, he can apply that notion to things which exist while he does not think; as he that has got the idea of extension from bodies by his sight or touch, can apply it to distances, where no body is seen or felt. And therefore, though a man has no perception of the length of duration which passed whilst he slept or thought not; yet, having observed the revolution of days and nights, and found the length of their duration to be in appearance regular and constant, he can, upon the supposition that that revolution has proceeded after the same manner whilst he was asleep or thought not, as it used to do at other times, he can, I say, imagine and make allowance for the length of duration whilst he slept. But if Adam and Eve, (when they were alone in the world,) instead of their ordinary night's sleep, had passed the whole twenty-four hours in one continued sleep, the duration of that twenty-four hours had been irrecoverably lost to them, and been for ever left out of their account of time.

6. Thus by reflecting on the appearing of various ideas one after another in our understandings, we get the notion of succession; which, if any one should think we did rather get from our observation of motion by our senses, he will perhaps be of my mind when he considers, that even motion produces in his mind an idea of succession no otherwise than as it produces there a continued train of distinguishable ideas. For a man looking upon a body really moving, perceives yet no motion at all unless that motion produces a constant train of successive ideas: v.g. a man becalmed at sea, out of sight of land, in a fair day, may look on the sun, or sea, or ship, a whole hour together, and perceive no motion at all in either; though it be certain that two, and perhaps all of them, have moved during that time a great way. But as soon as he perceives either of them to have changed distance with some other body, as soon as this motion produces any new idea in him, then he perceives that there has been motion. But wherever a man is, with all things at rest about him, without perceiving any motion at all,—if during this hour of quiet he has been thinking, he will perceive the various ideas of his own thoughts in his own mind, appearing one after another, and thereby observe and find succession where he could observe no motion.

7. And this, I think, is the reason why motions very slow, though

they are constant, are not perceived by us; because in their remove from one sensible part towards another, their change of distance is so slow, that it causes no new ideas in us, but a good while one after another. And so not causing a constant train of new ideas to follow one another immediately in our minds, we have no perception of motion; which consisting in a constant succession, we cannot perceive that succession without a constant succession of varying ideas arising from it.

8. On the contrary, things that move so swift as not to affect the senses distinctly with several distinguishable distances of their motion, and so cause not any train of ideas in the mind, are not also perceived. For anything that moves round about in a circle, in less times than our ideas are wont to succeed one another in our minds, is not perceived to move; but seems to be a perfect entire circle of that matter or colour, and not a part of a circle in motion.

9. Hence I leave it to others to judge, whether it be not probable that our ideas do, whilst we are awake, succeed one another in our minds at certain distances; not much unlike the images in the inside of a lantern, turned round by the heat of a candle. This appearance of theirs in train, though perhaps it may be sometimes faster and sometimes slower, yet, I guess, varies not very much in a waking man: there seem to be certain bounds to the quickness and slowness of the succession of those ideas one to another in our minds, beyond which they can neither delay nor hasten.

10. The reason I have for this odd conjecture is, from observing that, in the impression made upon any of our senses, we can but to a certain degree perceive any succession; which, if exceeding quick, the sense of succession is lost, even in cases where it is evident that there is a real succession. Let a cannon-bullet pass through a room, and in its way take with it any limb, or fleshy parts of a man, it is as clear as any demonstration can be, that it must strike successively the two sides of the room: it is also evident, that it must touch one part of the flesh first, and another after, and so in succession: and yet, I believe, nobody who ever felt the pain of such a shot, or heard the blow against the two distant walls, could perceive any succession either in the pain or sound of so swift a stroke. Such a part of duration as this, wherein we perceive no succession, is that which we call an *instant*, and is that

which takes up the time of only one idea in our minds, without the succession of another; wherein, therefore, we perceive no succession at all.

11. This also happens where the motion is so slow as not to supply a constant train of fresh ideas to the senses, as fast as the mind is capable of receiving new ones into it; and so other ideas of our own thoughts, having room to come into our minds between those offered to our senses by the moving body, there the sense of motion is lost; and the body, though it really moves, yet, not changing perceivable distance with some other bodies as fast as the ideas of our own minds do naturally follow one another in train, the thing seems to stand still; as is evident in the hands of clocks, and shadows of sun-dials, and other constant but slow motions, where, though, after certain intervals, we perceive, by the change of distance, that it hath moved, yet the motion itself we perceive not.

12. So that to me it seems, that the constant and regular succession of *ideas* in a waking man, is, as it were, the measure and standard of all other successions. Whereof, if any one either exceeds the pace of our ideas, as where two sounds or pains, &c., take up in their succession the duration of but one idea; or else where any motion or succession is so slow, as that it keeps not pace with the ideas in our minds, or the quickness in which they take their turns, as when any one or more ideas in their ordinary course come into our mind, between those which are offered to the sight by the different perceptible distances of a body in motion, or between sounds or smells following one another,—there also the sense of a constant continued succession is lost, and we perceive it not, but with certain gaps of rest between.

13. If it be so, that the ideas of our minds, whilst we have any there, do constantly change and shift in a continual succession, it would be impossible, may any one say, for a man to think long of any one thing. By which, if it be meant that a man may have one self-same single idea a long time alone in his mind, without any variation at all, I think, in matter of fact, it is not possible. For which (not knowing how the ideas of our minds are framed, of what materials they are made, whence they have their light, and how they come to make their appearances) I can give no other reason but experience: and I would have any one try, whether he can keep one unvaried single idea in his mind, without any other, for any considerable time together.

14. For trial, let him take any figure, any degree of light or white-ness, or what other he pleases, and he will, I suppose, find it difficult to keep all other ideas out of his mind; but that some, either of another kind, or various considerations of that idea, (each of which consider-ations is a new idea,) will constantly succeed one another in his thoughts, let him be as wary as he can.

16. Whether these several ideas in a man's mind be made by certain motions, I will not here dispute; but this I am sure, that they include no idea of motion in their appearance; and if a man had not the idea of motion otherwise, I think he would have none at all, which is enough to my present purpose; and sufficiently shows that the notice we take of the ideas of our own minds, appearing there one after another, is that which gives us the idea of succession and duration, without which we should have no such ideas at all. It is not then *motion*, but the constant train of *ideas* in our minds whilst we are waking, that furnishes us with the idea of duration; whereof motion no otherwise gives us any perception than as it causes in our minds a constant succession of ideas, as I have before showed: and we have as clear an idea of succession and duration, by the train of other ideas succeeding one another in our minds, without the idea of any motion, as by the train of ideas caused by the uninterrupted sensible change of distance between two bodies, which we have from motion; and there-fore we should as well have the idea of duration were there no sense of motion at all.

17. Having thus got the idea of duration, the next thing natural for the mind to do, is to get some *measure* of this common duration, whereby it might judge of its different lengths, and consider the distinct order wherein several things exist; without which a great part of our knowledge would be confused, and a great part of history be rendered very useless. This consideration of duration, as set out by certain periods, and marked by certain measures or epochs, is that, I think, which most properly we call *time*.

18. In the measuring of extension, there is nothing more required but the application of the standard or measure we make use of to the thing of whose extension we would be informed. But in the measuring of duration this cannot be done, because no two different parts of succession can be put together to measure one another. And nothing

being a measure of duration but duration, as nothing is of extension but extension, we cannot keep by us any standing, unvarying measure of duration, which consists in a constant fleeting succession, as we can of certain lengths of extension, as inches, feet, yards, &c., marked out in permanent parcels of matter. Nothing then could serve well for a convenient measure of time, but what has divided the whole length of its duration into apparently equal portions, by constantly repeated periods. What portions of duration are not distinguished, or considered as distinguished and measured, by such periods, come not so properly under the notion of time; as appears by such phrases as these, viz. 'Before all time,' and 'When time shall be no more.'

19. The diurnal and annual revolutions of the sun, as having been, from the beginning of nature, constant, regular, and universally observable by all mankind, and supposed equal to one another, have been with reason made use of for the measure of duration. But the distinction of days and years having depended on the motion of the sun, it has brought this mistake with it, that it has been thought that motion and duration were the measure one of another. For men, in the measuring of the length of time, having been accustomed to the ideas of minutes, hours, days, months, years, &c., which they found themselves upon any mention of time or duration presently to think on, all which portions of time were measured out by the motion of those heavenly bodies, they were apt to confound time and motion, or at least to think that they had a necessary connexion one with another. Whereas any constant periodical appearance, or alteration of ideas, in seemingly equidistant spaces of duration, if constant and universally observable, would have as well distinguished the intervals of time, as those that have been made use of. For, supposing the sun, which some have taken to be a fire, had been lighted up at the same distance of time that it now every day comes about to the same meridian, and then gone out again about twelve hours after, and that in the space of an annual revolution it had sensibly increased in brightness and heat, and so decreased again,—would not such regular appearances serve to measure out the distances of duration to all that could observe it, as well without as with motion? For if the appearances were constant, universally observable, in equidistant periods, they would serve mankind for measure of time as well were the motion away.

20. For the freezing of water, or the blowing of a plant, returning at

equidistant periods in all parts of the earth, would as well serve men to reckon their years by, as the motions of the sun: and in effect we see, that some people in America counted their years by the coming of certain birds amongst them at their certain seasons, and leaving them at others. For a fit of an ague; the sense of hunger or thirst; a smell or a taste; or any other idea returning constantly at equidistant periods, and making itself universally be taken notice of, would not fail to measure out the course of succession, and distinguish the distances of time. . . .

21. But perhaps it will be said,—without a regular motion, such as of the sun, or some other, how could it ever be known that such periods were equal? To which I answer,—the equality of any other returning appearances might be known by the same way that that of days was known, or presumed to be so at first; which was only by judging of them by the train of ideas which had passed in men's minds in the intervals; [by which train of ideas discovering inequality in the natural days, but none in the artificial days, the artificial days, or $\nu\nu\chi\theta\eta\mu\varepsilon\rho\alpha$, were guessed] to be equal, which was sufficient to make them serve for a measure; though exacter search has since discovered inequality in the diurnal revolutions of the sun, and we know not whether the annual also be not unequal. These yet, by their presumed and apparent equality, serve as well to reckon time by (though not to measure the parts of duration exactly) as if they could be proved to be exactly equal. We must, therefore, carefully distinguish betwixt duration itself, and the measures we make use of to judge of its length. Duration, in itself, is to be considered as going on in one constant, equal, uniform course: but none of the measures of it which we make use of can be *known* to do so, nor can we be assured that their assigned parts or periods are equal in duration one to another; for two successive lengths of duration, however measured, can never be demonstrated to be equal. The motion of the sun, which the world used so long and so confidently for an exact measure of duration, has, as I said, been found in its several parts unequal. And though men have, of late, made use of a pendulum, as a more steady and regular motion than that of the sun, or, (to speak more truly,) of the earth;—yet if any one should be asked how he certainly knows that the two successive swings of a pendulum are equal, it would be very hard to satisfy him that they are infallibly so; since we cannot be sure

that the cause of that motion, which is unknown to us, shall always operate equally; and we are sure that the medium in which the pendulum moves is not constantly the same: either of which varying, may alter the equality of such periods, and thereby destroy the certainty and exactness of the measure by motion, as well as any other periods of other appearances; the notion of duration still remaining clear, though our measures of it cannot (any of them) be demonstrated to be exact. Since then no two portions of succession can be brought together, it is impossible ever certainly to know their equality. All that we can do for a measure of time is, to take such as have continual successive appearances at seemingly equidistant periods; of which seeming equality we have no other measure, but such as the train of our own ideas have lodged in our memories, with the concurrence of other *probable* reasons, to persuade us of their equality.

22. One thing seems strange to me,—that whilst all men manifestly measured time by the motion of the great and visible bodies of the world, time yet should be defined to be the 'measure of motion': whereas it is obvious to every one who reflects ever so little on it, that to measure motion, space is as necessary to be considered as time; and those who look a little farther will find also the bulk of the thing moved necessary to be taken into the computation, by any one who will estimate or measure motion so as to judge right of it. Nor indeed does motion any otherwise conduce to the measuring of duration, than as it constantly brings about the return of certain sensible ideas, in seeming equidistant periods. For if the motion of the sun were as unequal as of a ship driven by unsteady winds, sometimes very slow, and at others irregularly very swift; or if, being constantly equally swift, it yet was not circular, and produced not the same appearances,—it would not at all help us to measure time, any more than the seeming unequal motion of a comet does.

23. Minutes, hours, days, and years are, then, no more necessary to time or duration, than inches, feet, yards, and miles, marked out in any matter, are to extension. For, though we in this part of the universe, by the constant use of them, as of periods set out by the revolutions of the sun, or as known parts of such periods, have fixed the ideas of such lengths of duration in our minds, which we apply to all parts of time whose lengths we would consider; yet there may be other parts of the universe, where they no more use these measures of

ours, than in Japan they do our inches, feet, or miles; but yet
something analogous to them there must be. For without some regular
periodical returns, we could not measure ourselves, or signify to
others, the length of any duration; though at the same time the world
were as full of motion as it is now, but no part of it disposed into
regular and apparently equidistant revolutions. But the different
measures that may be made use of for the account of time, do not at all
alter the notion of duration, which is the thing to be measured; no
more than the different standards of a foot and a cubit alter the notion
of extension to those who make use of those different measures.

27. By the same means, therefore, and from the same original that
we come to have the idea of time, we have also that idea which we call
Eternity; viz. having got the idea of succession and duration, by
reflecting on the train of our own ideas, caused in us either by the
natural appearances of those ideas coming constantly of themselves
into our waking thoughts, or else caused by external objects successively affecting our senses; and having from the revolutions of the sun
got the ideas of certain lengths of duration,—we can in our thoughts
add such lengths of duration to one another, as often as we please, and
apply them, so added, to durations past or to come. And this we can
continue to do on, without bounds or limits, and proceed *in infinitum,*
and apply thus the length of the annual motion of the sun to duration,
supposed before the sun's or any other motion had its being; which is
no more difficult or absurd, than to apply the notion I have of the
moving of a shadow one hour to-day upon the sun-dial to the duration
of something last night, v.g. the burning of a candle, which is now
absolutely separate from all actual motion; and it is as impossible for
the duration of that flame for an hour last night to co-exist with any
motion that now is, or for ever shall be, as for any part of duration,
that was before the beginning of the world, to co-exist with the motion
of the sun now. But yet this hinders not but that, having the *idea* of the
length of the motion of the shadow on a dial between the marks of two
hours, I can as distinctly measure in my thoughts the duration of that
candle-light last night, as I can the duration of anything that does now
exist: and it is no more than to think, that, had the sun shone then on
the dial, and moved after the same rate it doth now, the shadow on the
dial would have passed from one hour-line to another whilst that
flame of the candle lasted.

28. The notion of an hour, day, or year, being only the idea I have of the length of certain periodical regular motions, neither of which motions do ever all at once exist, but only in the ideas I have of them in my memory derived from my senses or reflection; I can with the same ease, and for the same reason, apply it in my thoughts to duration antecedent to all manner of motion, as well as to anything that is but a minute or a day antecedent to the motion that at this very moment the sun is in. All things past are equally and perfectly at rest; and to this way of consideration of them are all one, whether they were before the beginning of the world, or but yesterday: the measuring of any duration by some motion depending not at all on the *real* co-existence of that thing to that motion, or any other periods of revolution, but the having a clear *idea* of the length of some periodical known motion, or other interval of duration, in my mind, and applying that to the duration of the thing I would measure.

29. Hence we see that some men imagine the duration of the world, from its first existence to this present year 1689, to have been 5639 years, or equal to 5639 annual revolutions of the sun, and others a great deal more; as the Egyptians of old, who in the time of Alexander counted 23,000 years from the reign of the sun; and the Chinese now, who account the world 3,269,000 years old, or more; which longer duration of the world, according to their computation, though I should not believe to be true, yet I can equally imagine it with them, and as truly understand, and say one is longer than the other, as I understand, that Methusalem's life was longer than Enoch's. And if the common reckoning of 5639 should be true, (as it may be as well as any other assigned,) it hinders not at all my imagining what others mean, when they make the world one thousand years older, since every one may with the same facility imagine (I do not say believe) the world to be 50,000 years old, as 5639; and may as well conceive the duration of 50,000 years as 5639. Whereby it appears that, to the measuring the duration of anything by time, it is not requisite that that thing should be co-existent to the motion we measure by, or any other periodical revolution; but it suffices to this purpose, that we have the idea of the length of *any* regular periodical appearances, which we can in our minds apply to duration, with which the motion or appearance never co-existed.

31. And thus I think it is plain, that from those two fountains of all knowledge before mentioned, viz. reflection and sensation, we got the ideas of duration, and the measures of it.

For, First, by observing what passes in our minds, how our ideas there in train constantly some vanish and others begin to appear, we come by the idea of *succession.*

Secondly, by observing a distance in the parts of this succession, we get the idea of *duration.*

Thirdly, by sensation observing certain appearances, at certain regular and seeming equidistant periods, we get the ideas of certain *lengths* or *measures of duration,* as minutes, hours, days, years, &c.

Fourthly, by being able to repeat those measures of time, or ideas of stated length of duration, in our minds, as often as we will, we can come to imagine *duration, where nothing does really endure or exist;* and thus we imagine to-morrow, next year, or seven years hence.

Fifthly, by being able to repeat ideas of any length of time, as of a minute, a year, or an age, as often as we will in our own thoughts, and adding them one to another, without ever coming to the end of such addition, any nearer than we can to the end of number, to which we can always add; we come by the idea of *eternity,* as the future eternal duration of our souls, as well as the eternity of that infinite Being which must necessarily have always existed.

Sixthly, by considering any part of infinite duration, as set out by periodical measures, we come by the idea of what we call *time* in general.

9.

Leibniz:

TIME AS RELATIONAL

Gottfried Wilhelm Leibniz (1646–1716) wrote only two books, *The Theodicy,* published in 1710, and *New Essays on the Human Understanding,* which was not published until 1765. His main philosophic work was largely contained in a voluminous number of small essays and letters. Although his discussions of time appear through these, it was most forcefully voiced in an exchange of letters with Samuel Clarke, who defended the Newtonian view against Leibniz's attack; unfortunately for the present purpose, most of that discussion was conducted in terms of the concept of space with the implications of the particular extended argument then being drawn for time as well; the complete exchange was first published in 1717 under Clarke's direction. The selection consists of excerpts from several of Leibniz's papers, as indicated.*

a. On True Method in Philosophy and Theology [c. 1686]

. . . It is an indubitable fact, and one recognized also by Aristotle, that everything in nature is derived from size, figure, and motion. The theory of size and figure has been developed in a pre-eminent way; the innermost nature of motion is not yet patent due to the neglect of First Philosophy from which its laws are derived. For it is the task of Metaphysics to treat of continuous temporal modifications in the universe, since motion is only one kind of modification. In so far as the nature of motion is not understood, important philosophers having attributed the essence of matter only to extension, there has resulted a notion of bodies, previously unheard of, which fails to do justice to either the phenomena of nature or the mysteries of faith. For it can be

* Reprinted by permission of Charles Scribner's Sons from *Leibniz Selections,* edited by Philip P. Wiener. Copyright 1951 Charles Scribner's Sons.

demonstrated that extension without the addition of other qualities is not capable of either action or its passive reception . . .

b. Remarks on M. Arnauld's Letter Concerning My Proposition: That the Individual Concept of Each Person Contains Once for All Everything That will Ever Happen to Him. [1686]

Let a certain straight line A B C represent a certain time, and let a certain individual, say myself, endure or exist during this period. Then let us consider the me which exists during the time A B and the me which exists during the time B C. Now, since we suppose that it is the same individual substance which persists in me during the time A B while I am in Paris and during the time B C while I am in Germany, there must be some reason why we can truly say that I persist or to say that it is the same I who was in Paris and is now in Germany; if there were not a reason, then it would be quite right to say it was not I but another person. To be sure, introspection convinces me *a posteriori* of this identity, but there must also be some *a priori* reason. It is not possible to find any other reason than the fact that my attributes of the preceding time and state as well as the attributes of the succeeding time and state are all predicates of the same subject *(insunt eidem subjecto)*. Now, what is it to say that the predicate is in the subject if not that the concept of the predicate is in some manner involved in the concept of the subject? Since from the very time that I began to exist it could be truly said of me that this or that would happen to me, we must grant that these predicates were principles involved in the subject or in the complete concept of me which constitutes the so-called ego and is the basis of the interconnection of all my different states. . . .

My idea of a true proposition is such that every predicate, necessary or contingent, past, present, or future, is included in the idea of the subject. . . . This is a very important proposition that deserves to be well established, for it follows that every soul is as a world apart, independent of everything else but God; that it is not only immortal and impenetrable but retains in its substance traces of everything that happens to it. It also determines what the relations of communication among substances shall be, and in particular, the union of the soul and body. The latter is not explained by the ordinary hypothesis of the physical influence of one on the other, for each present state of a substance occurs in it spontaneously, and is nothing but a consequence of its preceding state. Nor does the hypothesis of occasional causes explain how it happens, as Descartes and his followers imagine. . . . My hypothesis of concomitant harmony appears to me to demonstrate how it happens. That is to say, every substance expresses the whole sequence of the universe in accordance with its own viewpoint or relationship to the rest, so that all are in perfect correspondence with one another.

c. Metaphysical Foundations of Mathematics [1715]

Given the existence of a multiplicity of concrete circumstances which are not mutually exclusive, we designate them as *contemporaneous* or *co-existing*. Hence, we regard the events of past years as not co-existing with those of this year because they are qualified by incompatible circumstances.

When one of two non-contemporaneous elements contains the ground for the other, the former is regarded as the *antecedent*, and the latter as the *consequent*. My earlier state of existence contains the ground for the existence of the later. And since, because of the connection of all things, the earlier state in me contains also the earlier state of the other thing, it also contains the ground of the later state of the other thing, and is thereby prior to it. All existing elements may be thus ordered either by the relation of *contemporaneity* (co-existence) or by that of being *before or after in time* (succession).

Time is the order of non-contemporaneous things. It is thus the *universal* order of change in which we ignore the specific kind of changes that have occurred.

Duration is the quantity of time. If the quantity of time is continuously and uniformly diminished, the time passes into an *instant* which has zero magnitude.

Space is the order of co-existing things, or the order of existence for all things which are contemporaneous. In each of both orders—in that of time as that of space—we can speak of a *propinquity* or *remoteness* of the elements *according to whether fewer or more connecting links are required to discern their mutual order.* Two points, then, are nearer to one another when the points between them and the structure arising out of them with the utmost definiteness, present something relatively simpler. Such a structure which unites the points between the two points is the simplest, i.e., the shortest and also the most uniform, *path* from one to the other; in this case, therefore, the straight line is the shortest one between two neighboring points.

Extension is the quantity of space. It is false to confound extension, as is commonly done, with extended things, and to view it as substance. If the quantity of space is continuously and uniformly diminished, then it becomes a *point* which has zero magnitude.

Position is a determination of togetherness. It includes, therefore, not only quantity, but also quality.

Quantity or magnitude is that determination of things which *can be known in things only through their immediate contemporaneous togetherness (or through their simultaneous observation)....*

d. Leibniz's Third Paper (Answer to Clarke's Second Reply)

3. These gentlemen maintain, therefore, that *space* is a *real absolute being.* But this involves them in great difficulties; for such a *being* must needs be *eternal* and *infinite.* Hence some have believed it to be *God himself,* or, one of his attributes, his *immensity.* But since space consists of *parts,* it is not a thing which can belong to God.

4. As for my own opinion, I have said more than once, that I hold *space* to be something *merely relative,* as *time* is; that I hold it to be an *order of co-existences,* as *time* is an *order of successions.* For *space* denotes, in terms of possibility, *an order* of things which exist at the same time, considered as existing *together;* without inquiring into their particular manner of existing. And when many things are seen *together,* one perceives *that order of things among themselves.*

5. I have many demonstrations, to confute the fancy of those who take *space* to be a *substance,* or at least an absolute *being.* But I shall only use, at the present, one demonstration, which the author here [i.e., Clarke] gives me occasion to insist upon. I say then, that if *space* were an absolute *being,* there would happen something, for which it would be impossible there should be a *sufficient reason.*

Which is against my Axiom. And I can prove it thus. *Space* is something absolutely *uniform;* and, without the things placed in it, *one point* of space does not absolutely differ in any respect whatsoever from *another point* of space. Now from hence it follows (supposing space to be something in itself, besides the *order of bodies among themselves,*) that it is impossible there should be a *reason*, why God, preserving the same situations of bodies among themselves, should have placed them in space in *one certain particular manner*, and not *otherwise;* why everything was not placed the *quite contrary way,* for instance, by changing *east* into *west.* But if space is nothing else but that *order* or *relation;* and is nothing at all without bodies but the possibility of placing them; then those two states, the *one* such as it now is, the *other* supposed to be the quite contrary way, would not at all differ from one another. *Their difference* therefore is only to be found in our *chimerical* supposition of the *reality* of space in itself. But in truth the *one* would exactly be the same thing as the *other*, they being absolutely *indiscernible;* and consequently there is no room to inquire after a reason for the preference of the one to the other.

6. The case is the same with respect to *time.* Supposing any one should ask, why God did not create everything a *year sooner;* and the same person should infer from thence, that God has done something, concerning which it is *not possible* there should be a *reason*, why he did it *so*, and not *otherwise:* the answer is, that his inference would be right, if *time* were any thing distinct from things existing in time. For it would be *impossible* there should be any *reason*, why things should be applied to such *particular instants*, rather than to *others*, their succession continuing the same. But then the same argument proves, that *instants*, considered without the things, are *nothing at all;* and that they consist only in the successive *order* of things: which order remaining the same, *one* of the two states, *viz.* that of a supposed anticipation, would not at all differ, nor could be discerned from, the *other* which now is.

e. Leibniz's Fourth Paper (Answer to Clarke's Third Reply)

6. To suppose *two* things *indiscernible*, is to suppose the *same thing* under *two names.* And therefore to suppose that the universe could have had at first *another* position of *time* and *place*, than that which it actually had; and yet that all the parts of the universe should have had the same situation among themselves, as that which they actually had; such a supposition, I say, is an *impossible* fiction.

16. If *space* and *time* were anything absolute, that is, if they were anything else, besides certain *orders* of things; then indeed my assertion would be a *contradiction*. But since it is not so, the hypothesis [*that space and time are anything absolute*] is contradictory, that is, 'tis an *impossible* fiction.

f. Leibniz's Fifth Paper (Answer to Clarke's Fourth Reply)

27. The *parts* of *time* or *place*, considered *in themselves*, are *ideal* things; and therefore they perfectly resemble one another, like two *abstract units*. But it is not so with two *concrete ones*, or with two *real times*, or two *spaces filled up*, that is, truly *actual*.

28. I don't say that *two* points of space are *one and the same* point, nor that *two* instants of time are *one and the same* instant, as the author seems to charge me with saying. But a man may fancy, for want of knowledge, that there are two different instants, where there is but one: in like manner as I observed in the 17th paragraph of the foregoing answer, that frequently in geometry we suppose *two*, in order to show up the error of an adversary, when there is really but *one*. If any man should suppose that a straight line cuts another in *two* points; it will be found after all, that those *two* pretended points must coincide, and make but *one* point. This happens also when a straight line becomes a tangent to a curve instead of cutting it.

29. I have demonstrated that *space* is nothing else but an *order* of the existence of things, observed as existing together; and therefore the fiction of a material finite universe, moving forward in an infinite empty space, cannot be admitted. It is altogether unreasonable and *impracticable*. For, besides that there is *no real space* out of the material universe; such an action would be without any design in it; it would be working without doing anything, *agendo nihil agere*. There would happen *no change*, which could be observed by any person whatsoever. These are imaginations of *philosophers who have incomplete notions*, who make space an absolute reality. Mere mathematicians, who are only taken up with the conceits of imagination, are apt to forge such notions; but they are destroyed by superior reasons.

31. I don't grant that *every finite is movable*. According to the hypothesis of my adversaries themselves, a *part* of *space*, though *finite*, is not movable. What is movable must be capable of changing its

situation with respect to *something else,* and to be in a new state *discernible* from the first: otherwise the change is but a fiction. A *movable finite,* must therefore be part of *another* finite, in order that any change may happen which can be *observed.*

33. Since *space* in itself is an *ideal* thing, like *time; space out of the world* must needs be imaginary, as the *schoolmen* themselves have acknowledged. The case is the same with empty space *within* the world; which I take also to be imaginary, for the reasons before alleged.

44. If infinite *space* is God's *immensity,* infinite *time* will be God's *eternity;* and therefore we must say, that what is in space, is in God's immensity, and consequently in his essence; and that what is in time, is also in the essence of God. *Strange* expressions; which plainly show, that the author [i.e., Clarke] makes a wrong use of terms.

45. I shall give another instance of this. God's immensity makes him actually present in all spaces. But now if God is *in* space, how can it be said that space is *in* God, or that it is a property of God? We have often heard, that a property is in its subject; but we never heard, that a subject is in its property. In like manner, God exists *in* all time. How then can time be *in* God; and how can it be a property of God? These are perpetual abuses of words.

46. It appears that the author confounds immensity, or the *extension of things,* with the *space* according to which that extension is taken. Infinite space is not the immensity of God; finite space is not the extension of bodies: as time is not their duration. Things keep their extension, but they do not always keep their space. Everything has its own extension, its own duration; but it has not its own time, and does not keep its own space.

47. I will here show *how* men come to form to themselves the notion of *space.* They consider that many things exist at once, and they observe in them a certain *order* of co-existence, according to which the relation of one thing to another is more or less simple. This order is their *situation* or distance. When it happens that one of those co-existent things changes its *relation* to a multitude of others, which do not change their relation among themselves; and that another thing, newly come, acquires the same relation to the others, as the former had; we then say it is come into the *place* of the former; and this change, we call a *motion* in that body, wherein is the immediate cause

of the change. And though many, or even all the co-existent things, should change according to certain known rules of direction and swiftness; yet one may always determine the relation of situation, which every co-existent acquires with respect to every other co-existent; and even that relation, which any other co-existent would have to this, or which this would have to any other, if it had not changed, or if it had changed any otherwise. And supposing, or feigning, that among those co-existents there is a sufficient number of them which have undergone no change; then we may say, that those which have such a *relation* to those fixed existents, as others had to them before, have now the same *place* which those others had. And that which comprehends *all those places,* is called *space.* Which shows, that in order to have an idea of *place,* and consequently of space, it is sufficient to consider these *relations,* and the rules of their changes, without needing to facing any absolute reality *outside* the things whose situation we consider, and, to give a kind of definition: *place* is that which we say is the same to *A,* and to *B,* when the *relation* of the co-existence of *B,* with *C, E, F, G,* etc., supposing there has been no cause of change in *C, E, F, G,* etc. It might be said also, without entering into any further particularity, that *place* is that, which is the same in different moments to different existent things, when their *relations* of *co-existence* with certain other existents, which are supposed to continue fixed from one of those moments to the other, agree entirely together. And *fixed existents* are those, in which there has been no cause of any change of the *order* of their co-existence with others; or (which is the same thing), in which there has been no *motion.* Lastly, *space* is that which results from *places taken together....*

49. It cannot be said, that (a certain) *duration* is eternal; but that *things,* which continue always, are eternal (by gaining always new duration). Whatever exists of time and of duration (being successive) perishes continually: and how can a thing exist eternally, which (to speak exactly) does never exist at all? For, how can a thing exist, whereof no part does ever exist? Nothing of time ever does exist, but instants; and an instant is not even itself a part of time. Whoever considers these observations, will easily apprehend that time can only be an ideal thing. And the analogy between time and space will easily make it appear that the one is as merely ideal as the other. (However,

if by saying that the duration of a thing is eternal, is merely understood that it last eternally, I have no objection.)

54. ... As for the objection that *space* and *time* are *quantities*, or rather things *endowed with quantity;* and that *situation* and *order* are not so: I answer, that *order* also has its quantity; there is in it, that which goes before, and that which follows; there is distance or interval. *Relative* things have their *quantity*, as well as *absolute* ones. For instance, *ratios* or *proportions* in mathematics, have their *quantity*, and are *measured* by *logarithms;* and yet they are *relations*. And therefore though *time* and *space* consist in *relations*, yet they have their *quantity*.

10.

Kant:

THE PRIMACY OF TIME

In 1770, Immanuel Kant (1724–1804) delivered his *Dissertation on the Form and Principles of the Sensible and Intelligible World,* usually referred to as the "Inaugural Dissertation." * This short work, preceding the *Critique of Pure Reason* by eleven years, was the first statement of the new Critical outlook. Its two central sections, from which the selection is taken, provide a clearly readable statement of Kant's thesis that space and time are— neither things nor relations of things—but the two forms of human sensibility, and thereby the dual condition for all cognitive experience. Here Kant first set out his new thesis that, from the vantage point of the experien*cer* (as distinct from the experien*ced* world), time is more fundamental than space.

Section II.
On the Distinction of Sensibles and Intelligibles in General [§ 3–12.]

§ 10. No intuition of things intellectual, but only a symbolic knowledge of them, is given to man. Intellection is possible to us only through universal concepts in the abstract, not through a singular concept in the concrete. For all our intuition is bound to a certain formal principle under which alone anything can be apprehended by the mind immediately, that is, as *singular,* and not as merely conceived discursively through general concepts. But this formal principle of our intuition (space and time) is the condition under which

* As translated from the Latin original by John Handyside in *Kant's Inaugural Dissertation and Early Writings on Space* (La Salle, Ill.: The Open Court Publishing Co., 1929). By permission.

anything can be an object of our senses; and being thus the condition of sensitive knowledge, it is not a means of *intellectual* intuition. Further, all the matter of our knowledge if given by the senses alone, whereas a noumenon, as such, is not to be conceived through representations derived from sensations. Consequently, an intelligible concept is, as such, destitute of all that is given by human intuition. Thus, in our minds intuition is always passive, and so is possible only so far as something is able to affect our senses. But the divine intuition, which is the ground of its objects, not consequent on them, is, owing to its independence, archetypal, and so is completely intellectual.

§ 11. But although phenomena are, properly, appearances, not ideas, of things, and express no internal and absolute quality of the objects, the knowledge of them is none the less quite genuine knowledge. For, in the first place, so far as they are sensual concepts or apprehensions, they bear witness, as being caused, to the presence of an object—which is opposed to idealism. On the other hand, to take judgments about what is known by sense, the truth of a judgment consists in the agreement of its predicate with the given subject. But the concept of the subject, so far as it is a phenomenon, can be given only by its relation to the sensitive faculty of knowledge; and it is by the same faculty that sensitively observable predicates are also given. Hence it is clear that the representations of subject and of predicate arise according to common laws, and so allow of a perfectly true knowledge.

§ 12. All things which, as objects, are referred to our senses, are phenomena; whatever does not affect the senses but contains merely the special form of sensibility, belongs to pure intuition (i.e., to intuition empty of sensations, but not for that reason intellectual). The phenomena of outer sense are reviewed and expounded in physics, those of inner sense in empirical psychology. But pure intuition (in man) is not a universal or logical concept *under* which, but a singular concept *in* which, all sensibles are apprehended. That is to say, it contains the concepts of space and time. Since these in no way determine the quality of sensibles, they are objects of science only as determining things according to quantity. Pure mathematics considers space in geometry, time in pure mechanics. To these there is added a certain concept which, though itself indeed intellectual, yet demands for its actualization in the concrete the auxiliary notions of time and space (in the successive addition and simultaneous jux-taposition of a plurality), namely, the concept of number, treated of by arithmetic. Thus pure mathematics, in that it deals with the form of all our sensitive knowledge, is the organon of all knowledge which is at once intuitive and distinct; and since its objects are not merely formal principles of all intuition, but themselves original intuitions, it yields us quite genuine knowledge, and at the same time furnishes a model of the highest certainty for knowledge in other fields. There is thus a science of things sensual, although, since they are phenomena, the use of the understanding in reference to them is not real but only logical. From this it is clear in what sense those thinkers who derive their inspiration from the Eleatic School are to be understood as having denied scientific knowledge of phenomena.

Section III
On the Principles of the Form of the
Sensible World [§ 13–15]

§ 13. A principle of the form of the universe is one which contains the ground of a universal connection, whereby all substances and their states belong to one and the same whole, that is, to a world. A principle of the form of the sensible world is one which contains the ground of a universal connection of all things so far as they are phenomena. A form of the intelligible world implies an objective principle, that is, some cause in virtue of which there is a connection between things existing in themselves. But the world, regarded as phenomenon, that is, in relation to the sensibility of the human mind, acknowledges no principle of its form save a subjective one, that is to say, a law of the mind, on account of which all things which can (through their quality) be objects of the senses must necessarily be presented as belonging to the same whole. Thus whatever principle of the form of the sensible world may finally be acknowledged, it yet includes within its range only such existences as are deemed possible objects of sense, and so extends neither to immaterial substances (which, as such, are by definition at once excluded altogether from the outer senses) nor to the cause of the world (which cannot be an object of the senses, since through it the mind itself exists and is endowed with the power of sense). I shall now show that there are two such formal principles of the phenomenal universe which are absolutely primary and universal, and which are, as it were, the schemata and conditions of all human knowledge that is sensitive. I refer to time and space.

§ 14. On time.

1. *The idea of time does not originate in the senses, but is presupposed by them.* For the impressions of sense can be represented either as simultaneous or as successive only through the idea of time; succession does not beget the concept of time, but presupposes it. Thus the notion of time (regarded as acquired through experience) is very badly defined in terms of the series of actual things existing *after* one another. For what the word *after* may signify, I know only by means of an antecedently formed concept of time. Things are one after another when they exist at *different times,* just as things are simultaneous when they exist at the *same time.*

2. *The idea of time is singular, not general.* For no time is apprehended except as part of one and the same boundless time. If we think of two years we cannot represent them save by a determinate dating with regard to one another, and if they do not follow one another immediately, save as joined to one another by some intermediate time. But which of different times is earlier, which later, can by no means be defined by any marks conceivable by the intellect, unless we are to incur a vicious circle. The mind does not distinguish earlier and later except by a *singular* intuition. Further, we conceive all actual things as located *in* time, not as contained *under* its general notion as under a common mark.

3. *The idea of time is therefore an intuition.* And since the idea is conceived prior to all sensation, as a condition of relations exhibited in sensibles, the intuition is not sensual but *pure.*

4. *Time is a continuous quantum.* It is the principle of the laws of continuity in the changes of the universe. For a quantum is continuous which is not composed of simples. But since through time there are apprehended relations alone, without any given things related to one another, it follows that in time as a quantum we have a complex of such a kind that when the complexity is thought away nothing remains. But any complex of which nothing at all remains if its composition is wholly removed, does not consist of simple parts; therefore, etc. Thus any part of time is a time, and the simples which

exist in time, namely, moments, are not parts of time, but limits between which there is a stretch of time. For, given two moments, a time is not given except in so far as in these moments an actual succession takes place. Therefore, besides a given moment it is necessary that there be given a time in whose later part there is another moment.

The metaphysical law of continuity is this: all changes are continuous, or flow; i.e., opposite states do not succeed one another except through an intermediate series of different states. For since two opposite states are in different moments of time, and between two moments of time there is always some intervening time, and in the infinite series of its moments the substance is not in either of the given states, nor yet in no state, it will be in different states, and so *in infinitum.*

The celebrated Kaestner, in proceeding to examine this Leibnizian law, challenges its defenders to prove that the continuous motion of a point along all the sides of a triangle is impossible, a proposition which assuredly requires to be proved if the law of continuity is granted. Here, then, is the desired demonstration. Let the letters *abc* denote the three angular points of a rectilinear triangle. If a moving point traverses with a continuous motion the lines *ab, bc, ca,* i.e., the whole perimeter of the figure, it is necessary that it move through the point *b* in the direction *ab,* and through the same point *b* in the direction *bc.* But since these motions are diverse, they cannot take place simultaneously. Therefore the moment of the presence of the moving point in the vertex *b,* so far as it moves in the direction *ab,* is different from the moment of the presence of the moving point in the same vertex *b,* so far as it moves in the direction *bc.* But between two moments there is a time; therefore the moving point is present during some time in the same point, i.e., is at rest, and thus does not proceed with a continuous motion, which contradicts the hypothesis. The same demonstration holds of motion along any straight lines including an assignable angle. Therefore a body in continuous motion changes its direction only according to a line no part of which is straight, i.e., in a curve, as Leibniz declared.

5. *Time is not something objective and real.* It is neither substance nor accident nor relation, but is a subjective condition, necessary owing to the nature of the human mind, of the co-ordinating of all

sensibles according to a fixed law; and it is a *pure intuition*. For we co-ordinate alike substances and accidents, whether according to simultaneity or according to succession, only through the concept of time; and thus the notion of time, as a formal principle, is prior to the concepts of simultaneity and succession. As for relations of whatever sort, in so far as they come within the scope of the senses (namely, as to their simultaneity or succession), they involve nothing further [than is involved in the co-ordination of substances and accidents] except this, the determining of positions in time, as either in the same point of it, or in different points.

Those who assert the objective reality of time, conceive it in one or other of two ways. Among English philosophers especially, it is regarded as a continuous real flux, and yet as apart from any existing thing—a most egregious fiction. Leibniz and his School declare it to be a real characteristic, abstracted from the succession of internal states. This latter view at once shows itself erroneous by involving a vicious circle in the definition of time, and also by entirely neglecting simultaneity,[1] a most important consequence of time. It thus upsets the whole use of sane reason, in as much as, instead of requiring the laws of motion to be defined in terms of time, it would have time itself defined in respect of its own nature by reference to the observation of moving things or of some series of internal changes—a procedure by which, clearly, all the certainty of our rules is lost. But as for the fact that we cannot estimate *quantity* of time save in the concrete, namely,

---

1. What is simultaneous is not made simultaneous simply by not being successive. For when succession is taken away there is indeed removed a certain conjunction of things within the temporal series, but there does not on that account at once arise another real relation, such as is the conjunction of all in the same moment. For simultaneous things are joined in the same moment of time just as successive things in different moments. Thus though time possesses only one dimension, yet the ubiquity of time (to use Newton's manner of speaking), owing to which all things conceivable by sense are *at some time*, adds to the quantum of actuals a second dimension, so far as they hang, as it were, from the same point of time. For if you represent time by a straight line produced to infinity, and simultaneous things at any point of time by lines drawn perpendicular to it, the plane thus generated will represent the phenomenal world, both as to its substance and as to its accidents.

either by motion or by the series of [our] thoughts, this is because the concept of time rests only on an internal law of the mind. For since the concept is not a connate intuition, the action of the mind, in the co-ordinating of its sensa, is called forth only by the help of the senses. So far is it from being possible that anyone should ever deduce and explain the concept of time by the help of reason, that the very principle of contradiction presupposes it, involving it as a condition. For A and not-A are not incompatible unless they are judged of the same thing *together* (i.e., in the same time); but when they are judged of a thing successively (i.e., at different times), they may both belong to it. Hence the possibility of changes is thinkable only in time; time is not thinkable through changes, but *vice versa*.

6. But though time, posited in itself and absolutely, is an imaginary being, yet so far as it is related to an immutable law of sensibles as such, it is a quite genuine concept, and a condition of intuitive representation, extending *in infinitum* through all possible objects of the senses. For since simultaneous things, as such, cannot be presented to the senses except with the help of time, and changes are thinkable only through time, it is clear that this concept contains the universal form of phenomena, and accordingly that all events observable in the world, all motions and all internal changes, necessarily agree with any axioms, such as we have partly expounded, which can be determined in regard to time, since only under these conditions can objects of the senses be, and be co-ordinated. It is therefore absurd to wish to incite reason against the first postulates of pure time (e.g., continuity, etc.), since these follow from laws than which nothing more primary or original can be found. Reason itself cannot, in the use of the principle of contradiction, dispense with the support of this concept of time, so primitive and original is it.

7. Thus time is an absolutely primary, formal principle of the sensible world. For all things that are in any way sensibles can be apprehended only as at the same time or in successive times, and so as included and definitely related to each other within the course of the one single time. Thus through this concept, primary in the domain of sense, there necessarily arises a formal whole which is not a part of any other, i.e., the phenomenal world.

§ 15. On space.

A. *The concept of space is not abstracted from outer sensations.* . . .

B. *The concept of space is a singular representation* including all spaces *in* itself, not an abstract common notion containing them *under* itself. . . .

C. *The concept of space is thus a pure intuition,* since it is a singular concept. . . .

D. *Space is not something objective and real,* neither substance, nor accident, nor relation, *but subjective and ideal;* and, as it were, a schema, issuing by a constant law from the nature of the mind, for the coordinating of all outer sense whatsoever. . . .

E. Although the concept of space, viewed as an objective and real being or affection, is imaginary, nevertheless relatively to all sensibles, it is not merely altogether true, but the foundation of all truth in outer sensibility. . . . Space, therefore, is an absolutely first formal principle of the sensible world, not only for the reason that the objects composing the universe cannot be phenomena save through the concept of space, but especially for this reason, that, by its essence, it must necessarily be single, embracing absolutely all outer sensibles, and so constitutes a principle of totality, i.e., of a whole which cannot be part of another whole.

Corollary. There, then, are the two principles of sensitive knowledge, not general concepts (as is the case in matters of intellect), but singular intuitions which yet are pure. In these intuitions it is not true that the parts, and in especial the simple parts, contain, as the laws of reason prescribe, the ground of the possibility of the composite; but instead, after the pattern of sensitive intuition, the infinite contains the ground of every thinkable part, and finally of the simple, or rather of the limit. For only if infinite space and time be given, can any definite space or time be marked out by limitation of it; neither a point nor a moment can be thought by itself; they are conceived only as limits in an already given space or time. Thus all primary properties of these concepts are beyond the jurisdiction of reason, and so cannot in any way be intellectually explained. But none the less they are the presuppositions upon which the intellect rests when, with the greatest possible certainty, and in accordance with logical laws, it draws consequences from the primary data of intuition.

Of these concepts, the one [i.e., space] properly concerns the intui-

tion of an *object,* the other [i.e., time] a *state,* namely, that of representation. Thus space is applied as an image to the concept of time
itself, representing it by a line, and its limits (moments) by points. But
time approaches more nearly to a universal, rational concept, in that it
embraces absolutely everything within its survey, namely, space itself,
and in addition the accidents which are not comprehended in
space-relations, such as the thoughts of the soul. Further, though time
does not indeed prescribe laws to reason, it yet establishes the chief
conditions by the help of which the mind can order its notions according to the laws of reason. Thus I cannot decide whether a thing is
impossible, except by predicating A and not-A of the same subject *at
the same time.* Above all, when the intellect is applied to experience,
though the relation of cause and effect, even in outer objects, involves
space relations, none the less in all cases, outer and inner alike, the
mind can be informed what is earlier and what later (i.e., which is
cause and which effect) only by means of the time relation. Indeed
even the quality of space cannot be made intelligible, unless we relate
it to some measure as unit and express it by a number, which is itself
only an aggregate distinctly apprehended by numeration, i.e., by the
process of adding one to one successively in a given time.

Finally, the question naturally arises whether these concepts are
connate or acquired. The latter alternative, it is true, seems already
refuted by our demonstrations, but the former is not to be rashly
admitted, since, in appealing to a first cause, it opens a path for that
lazy philosophy which declares all further research to be vain. Both
concepts are without doubt acquired, as abstracted, not indeed from
the sensing of objects (for sensation gives the matter, not the form, of
human apprehension), but from the action of the mind in co-ordinating its sensa according to unchanging laws—each being, as it were,
an immutable type, and therefore to be known intuitively. For,
though sensations excite this act of the mind, they do not determine
the intuition. Nothing is here connate save the law of the mind,
according to which it combines in a fixed manner the sensa produced
in it by the presence of the object.

IV. On the Principle of the Form of
the Intelligible World [§ 16–22]

Scholium.

If it were legitimate to overstep a little the limits of apodeictic certainty befitting metaphysics, it might be worth while to investigate certain questions concerning not merely those laws but also those causes of sensitive intuition which can only be known through the intellect. The human mind, we might then say, is affected by outer things, and the world lies open to its view *in infinitum,* only in so far as the mind, along with all other things, is upheld by the infinite power of a single cause. For this reason it senses external things only through the presence of the one upholding common cause; and space, which is the universal and necessary condition, sensitively apprehended, of the co-presence of all things, can therefore be entitled *omnipraesentia phaenomenon.* (For it is not because the cause of the universe is in the same place with each and everything that it is present to them all; on the contrary, places, i.e., possible relations of substances, exist, because it, the cause, is inwardly present to all things). Again, the possibility of all changes and successions—the principle of which, so far as it is sensitively known, resides in the concept of time—presupposes the persistence of a subject whose opposite states succeed one another. But that whose states change does not persist unless maintained by another; and thus the concept of time as single, infinite, and immutable,[2] in which are and persist all things, is the *aeternitas phaenomenon* of the general cause. But it seems wiser to hug the shore of the knowledge granted us by the mediocrity of our intellect, than thus, after the manner of Malebranche, whose opinion is little different from that just expounded, namely, that we see all things in God, to push out into the open sea of mystical enquiries.

2. The moments of time do not present themselves as successive to one another; for, in that case another time would have to be presupposed for the succession of the moments; but in sensitive intuition actual things seem to descend, as it were, through a continuous series of moments.

11.

Kant:

THE TEMPORALIZATION OF CONCEPTS

In the *Critique of Pure Reason*, first published in 1781, Immanuel Kant was concerned to show how our cognitive experience is structured in a priori terms. The section entitled "Transcendental Aesthetic" amplified the "Inaugural Dissertation" thesis that all sense experience, with which experience begins, is 'received' in the forms of space and time. The section entitled "Transcendental Analytic" sought to show that all cognitive claims are inherently limited to what can appear in these forms of space and time. First, he was concerned to demonstrate that the internal structuring of our conceptual thinking is necessarily done by certain constitutive logical categories and that these particular categories are drawn from, and are applicable to, the cognitive understanding of an essentially temporal experience. This essential transition—from these logically defined categories of the human understanding to the temporalized applicable principles of human knowledge—was provided in a short and difficult chapter entitled "The Schematism of the Pure Concepts of Understanding"; the following selection is drawn from its last part.* The essential argument here is that *all* the categories of our cognitive thinking, which are justifiably applicable to experienced objects, arise from the four ways in which we structure time experience.

That we may not be further delayed by a dry and tedious analysis of the conditions demanded by transcendental schemata of the pure concepts of understanding in general, we shall now expound them according to the order of the categories and in connection with them.

The pure image of all magnitudes *(quantorum)* for outer sense is

---

* Kant, *Critique of Pure Reason*, trans. N. K. Smith (London: Macmillan & Co., Ltd., New York: St. Martin's Press, Inc.) A 142 = B181–A 147 = B187, pp. 183–187.

space; that of all objects of the senses in general is time. But the pure *schema* of magnitude *(quantitatis)*, as a concept of the understanding, is *number*, a representation which comprises the successive addition of homogeneous units. Number is therefore simply the unity of the synthesis of the manifold of a homogeneous intuition in general, a unity due to my generating time itself in the apprehension of the intuition.

Reality, in the pure concept of understanding, is that which corresponds to a sensation in general; it is that, therefore, the concept of which in itself points to being (in time). Negation is that the concept of which represents not-being (in time). The opposition of these two thus rests upon the distinction of one and the same time as filled and as empty. Since time is merely the form of intuition, and so of objects as appearances, that in the objects which corresponds to sensation is not the transcendental matter of all objects as things in themselves (thinghood,[1] reality). Now every sensation has a degree of magnitude whereby, in respect of its representation of an object otherwise remaining the same, it can fill out one and the same time, that is, occupy inner sense more or less completely, down to its cessation in nothingness ($= o = negatio$). There therefore exists a relation and connection between reality and negation, or rather a transition from the one to the other, which makes every reality representable as a quantum. The schema of a reality, as the quantity of something in so far as it fills time, is just this continuous and uniform production of that reality in time as we successively descend from a sensation which has a certain degree to its vanishing point, or progressively ascend from its negation to some magnitude of it.

The schema of substance os permanence of the real in time, that is, the representation of the real as a substrate of empirical determination of time in general, and so as abiding while all else changes. (The existence of what is transitory[2] passes away in time but not time itself. To time, itself non-transitory[3] and abiding, there corresponds in the [field of] appearance what is non-transitory in its existence, that is,

1. [*Sachheit*]
2. [*des Wandelbaren*]
3. [*unwandelbar*]

substance. Only in [relation to] substance can the succession and coexistence of appearances be determined in time.)

The schema of cause,[4] and of the causality[5] of a thing in general, is the real upon which, whenever posited, something else always follows. It consists, therefore, in the succession of the manifold, in so far as that succession is subject to a rule.

The schema of community or reciprocity, the reciprocal causality of substances in respect of their accidents, is the co-existence, according to a universal rule, of the determinations of the one substance with those of the other.

The schema of possibility is the agreement of the synthesis of different representations with the conditions of time in general. Opposites, for instance, cannot exist in the same thing at the same time, but only the one after the other. The schema is therefore the determination of the representation of a thing at some time or other.

The schema of actuality is existence in some determinate time.

The schema of necessity is existence of an object at all times.

We thus find that the schema of each category contains and makes capable of representation only a determination of time. The schema of magnitude is the generation (synthesis) of time itself in the successive apprehension of an object. The schema of quality is the synthesis of sensation or perception with the representation of time; it is the filling of time. The schema of relation is the connecting of perceptions with one another at all times according to a rule of time-determination. Finally the schema of modality and of its categories is time itself as the correlate of the determination whether and how an object belongs to time. The schemata are thus nothing but *a priori* determinations of time in accordance with rules. These rules relate in the order of the categories to the *time-series,* the *time-content,* the *time-order,* and lastly to the *scope of time*[6] in respect of all possible objects.

It is evident, therefore, that what the schematism of understanding effects by means of the transcendental synthesis of imagination is simply the unity of all the manifold of intuition in inner sense, and so indirectly the unity of apperception which as a function corresponds

4. [*Ursache*]
5. [*Kausalität*]
6. [*Zeitinbegriff*]

to the receptivity of inner sense. The schemata of the pure concepts of understanding are thus the true and sole conditions under which these concepts obtain relation to objects and so possess *significance*. In the end, therefore, the categories have no other possible employment than the empirical. As the grounds of an *a priori* necessary unity that has its source in the necessary combination of all consciousness in one original apperception, they serve only to subordinate appearances to universal rules of synthesis, and thus to fit them for thorough going connection in one experience.

All our knowledge falls within the bounds of possible experience, and just in this universal relation to possible experience consists that transcendental truth which precedes all empirical truth and makes it possible.

But it is also evident that although the schemata of sensibility first realise the categories, they at the same time restrict them, that is, limit them to conditions which lie outside the understanding, and are due to sensibility.... The categories, therefore, without schemata, are merely functions of the understanding for concepts; and represent no object. This [objective] meaning they acquire from sensibility, which realises the understanding in the very process of restricting it.

IV.

Time and Reality

Whatever the import of time in Kant's work, he was initially read as having insisted on the complete subjectivity of time and of space. Fichte's attempt to develop the Critical philosophy into a system of moral idealism, but served to intensify the feeling that Kantian time was merely an aspect of human subjectivity. Working out from the received Kantian legacy, the initial response—which might have been anticipated—was to reassert the independent objectivity of what Kant had termed the two forms of pure intuition. Time, as space, was to be seen as *not* dependent upon human apprehension, as being, in some way, a mark or aspect of the ultimately real. Our temporal experience, it was to be argued, is not open to the charge of subjective illusion, just because it is an experience of reality itself.

This rejoinder took at least four diverse forms, each with something of an irony of its own. First, Hegel turned from the introspective method, which had dominated the movement from Descartes to Kant, and initiated something of a revival of Aristotelian realism: time was tied to motion or change in the world of nature and developed into the essential principle of a dynamic world order—which, somehow, was to be explained as the expression of a timeless logic incarnate within it. Lotze, seeking to de-logicize reality, saw time as the mark of real becoming: in transforming his inherited idealism into process philosophy, he effectively reasserted the Aristotelian notion of time as the measure of motion. Fearing that the timeless had again preempted the field, Bergson revived Descartes' dualism of time and duration, which had served Newtonian science so well. Insisting that duration was fundamental to the real, Bergson argued that science seeks understanding in terms of a derivative spatialized time instead of facing durational reality in direct intuitive experience. Alexander insisted on the inseparability of time and space in the structure of the real: his new realism transformed Kant's two subjective forms into one space-time as *the* fundamental form of reality itself, and the Kantian categories of the human understanding into the objective structuring of existent things.

We can readily understand the current and conflicting urges to deprecate Hegel—the convoluted prose offends the epicure of literary style; the intricacy of argument, encyclopaedic breadth of perspective and dismissal of facile explanation conspire to repel those whose devotion to what they call clarity only too often disguises an inability to distinguish the veneer from the substance. And, the unremitting passion for casting the variegated richness of experience into the mold of a monolithic system disenchants those who had hoped against a retreat from his demonstrated empathy for the dynamic and the variegated in lived experience.

Yet few thinkers have exhibited greater courage in seeking to transcend their contemporary shibboleths by working out new ways of thinking. And few have more forcefully insisted that authentic philosophic thinking must focus its reasoning on the problems of men. Hegel forged a dialectic method of historical thinking and taught that we do not transcend the past by ignoring it or leaving it behind; rather, we take it with us and build it into the new synthesis which creates the ongoing future. Just so, if we are to go beyond Hegel, rather than hopelessly trying to detour around him, we can do no better than to incorporate his often profound insights into our own thinking while we yet seek to organize their unity into a newer frame.

However dubious a place Hegel may have finally accorded to time on a cosmological level, one cannot read his *Phenomenology of Spirit*—not an excerptable section but the whole—without experiencing his deep sensitivity to the pervasive temporality of human experience. Time, Hegel insisted, is itself the concept or 'Notion' *(Bergriff)* of the whole of reality "in the form of existence." [1] Identifying phenomenological time with history, Hegel was "very radical," as Kojève has pointed out, in identifying time with the self-consciousness of mankind's development.[2] Time is the form in which the existent and real world develops as a unity and finds its own rational meaning.[3] But, although the temporal form of existence pervades Hegel's discussions of the many facets of human experience, it is at least 'curious' [4] that his only clear and explicit discussion of the concept of time as such takes place in his *Philosophy of Nature*.

Hegel's prime datum was the pervasiveness of change in every aspect of the experiential world. And change was, in many ways, seen as Heraclitus and Plato had described it, as qualitative change between contrary states. Hegel cast this ancient concept into the form of development and, as such, it became, in the development of self-consciousness itself, the key to meaning in history.

If change and development are crucial, then the principle of the negative assumes prime importance. For as a thing develops, it changes from its initial stage to a subsequent stage; and, each stage is, in turn, negated, as the thing progressively moves on from what it was into what it was not. The acorn develops into the oak, into what it was not when it was an acorn. Change requires time, and time is, then, "the negative element in the sensuous world," [5] as it is continually negating the present. History, as the development of Spirit or Mind in time, then depends on the continuity of rational negation, as historical development proceeds from one stage to the next. But time and change require space to do their work, and history, then, transpires in a space-bound world. Because of mutual 'need', time and space imply each other and a Cartesian kind of spatial description of an essentially dynamic order is meaningless until brought into the unity of the here-and-now.[6]

But this is to say that Hegel's primary conception of time was in terms of the present now—which takes up into itself the no-longer existent past and, in turn, yields to the presently not-yet existent future. Time, then, is the now-series in which each now negates the previous now, and it is thus because of the continuity of negation that change is able to occur. As such, the Absolute—as the eternal and non-temporal presence of the reason of the world—is present in any temporal moment and yet transcends the time and the change which gives it a continuity of expression in the sequential order of rational change in space. Time, then, as the form of existence of the inherent rationality of the world, represents this abiding rationality in each momentary now and the change that is its content.

In his essay on time, Hegel reverted to the old argument about the perception of time itself. All that we can directly perceive, he argued, is not time as such—which is abstract and ideal—but Becoming. Time is, then, to be understood as the form of Becoming, as the form of rational change. Time is understood as essentially continuous, as

Becoming in terms of *Chronos:* the dynamic and creative 'unrest' of the world in a continuity of movement is the time which negates each stage of actual being so that the next one may emerge.

But time, then, as the enabling form of change, is but a manifestation of the 'notion', the concept or idea of the whole, the rational principle of the dynamic unity of the world as such. Truth, then, is the whole and is eternal. With approval, Hegel quotes Aristotle: "Movement can neither come into being, nor cease to be; nor can time come into being, or cease to be." [7] As unending, without temporal genesis or finality, truth is expressed in the time-order of processes of rationally explicable change. Reality, then, on the level of ultimate rationality is distinct from time as somehow supervening; on the level of the existent world which we experience, it is identical with time.

Time, then, is not a container in which things happen. Time is not merely the form of human apprehension. Time is the form of the happening of all that is. It is the process of finite things in ordered change that makes time. Time may, then, perhaps be understood, not so much as the measure of motion but as the principle of motion which is inseparable from it. And the supervening Absolute is to be characterized as absolute timelessness—not as duration in time, but as non-temporal eternal presence. For it is not *in* process and what is not *in* process is not in time or temporal form. It does not become; it just is.

We experience Becoming in terms of its past, which is shown in present memory-recall, and in its future which is marked by present hope or fear. These are acts of subjective imagination concerning real Becoming which is only perceived in the actual present. Thus, only the present truly *is,* is actual; it exists as the present negation of the no-longer actual past and the not-yet-actual future. The experiential present is a moving present—continually determined by the time-principle of negation; it is, in each momentary manifestation, but the expression of the true abiding Present which is not moving because it is eternal reason.

Time is then the "purely formal soul of Nature." It is understood by us because of the motion of matter which is real in space and time and is perceivable Becoming. Matter relates space and time in the unity of that continuity of processes which we understand as a continuity of

Becoming. "Time," then, "is just the notion definitely existent, and presented to consciousness in the form of empty intuition." [8]

Thus working from Kant, Hegel saw time as encompassing the world of phenomenal appearance, the only world in which rational reality itself is continually manifested. But his working from Kant does not appear to be in a forward direction. For Kant had sought to bring reasoning into time so that its dynamic content may be understood. Kant had seen that this necessitated the development of a new transcendental logic because a formal conceptual logic would not do. But, although proclaiming the need for a dialectic logic of development, it would seem that Hegel's move was to a 'pre-Copernican' realism. Like the ancient Greeks, he has seen time and change as manifesting a logically prior atemporal notion of the real, and thus amenable to a reading in terms of a formal conceptual logic. Hegel's move was, at least in this regard, back from Kant to the pre-Critical models of the ancients—to whom he apparently looked for his model of reality and for his understanding of time.

It seems somewhat significant that, like Aristotle, he placed his essay on time in his book on the nature of the physical world. Indeed, his concept of time is roughly similar to Aristotle's—although its ontological rooting seems to be in an active rather than a passive, and immanent rather than transcendent, 'prime mover'.

Just as Aristotle had argued that the now is not time itself but the way we mark the times of changing motions, so Hegel has understood time in terms of the now, the changing present actual in our own experience, the present actuality of the Absolute. But motion, as we have seen, requires not only time but space, and the concept of time that emerges is not only faithful to this empirical requirement but is essentially spatialized. For Hegel's concept of time is little more than the chronology of present moments or now-points conceived on a line; and the now is not defined in terms of the breadth or the how of the movement it points to; it is designed to show us the 'here' of that moment of time that is 'now', the place on the chronological line.

Time, the principle of motion and thereby of the experiential world, is still but the principle of the manifestation of the continuing presence of transcendent reason, of the Absolute itself. Time is, then, in essence still, as in the traditionally received Platonism, the mark of a secondary order of being amenable to analysis by non-temporal formal concepts. The dynamic quality of the world which Hegel had sought to understand has then been reduced to a logical formalism; the sequence of nows which marks the dynamicity of the experiential world is formalized as the "negation of negation." The dynamic quality of time evaporates into the formalized non-temporality of its conceptual logic. The dynamicity, the life, and the essential concreteness and contingency of history disappear into logical abstractions. We are, by a circuitous route, led back to Aristotle's fundamental critique of traditional Platonism: How are we to understand the temporal in terms of the timeless, change in terms of the changeless, dynamicity in terms of the static?

Hegel had set out to take time, to take change, to take history as the center of experiential consciousness. Indeed, Kojève has seen the "aim of Hegel's philosophy" to be an attempt "to give an account of the fact of History." [9] But, as Berdyaev pointed out, "famished for history, nourished on history, Hegel's philosophy without understanding that it did so, yet advocated fasting." [10] Seeking to understand the nuances of dynamic existence, Hegel, in the end, rooted the temporal in the non-temporal, discovered the dynamic in the static, derived contingency from necessity, and grounded time in eternity. Advocating concreteness, he found the existential concrete in the abstractions of logical formulas. Abounding with insights and pregnant suggestion, his work, taken in his own terms as a rationally comprehensive system, yet evaporates these into his apparently consuming yearning for a timeless vision. Subjecting time to logic, we see here, as Gilson once observed, that "logic has eaten up the whole of reality." [11]

Yet, when all is said and done, the thrust of Hegel's influence has largely been to turn attention away from the timeless eternity he postulated as the ground of being to that experiential becoming which he sought to comprehend. His preoccupation with change, with development, with history in man and in nature, has led us to see all these processes in temporal form. However Hegel may have wished to explain this dynamic world of meaningful change, he may have succeeded in spite of himself. For, following in his wake, philosophers who honor him have generally turned their attention away from allegedly non-temporal transcendent grounds to the experience of dynamicity and change itself, and to the form of time which envelops it.

Considering the extraordinary influence which Rudolf Lotze has had on the prime thinkers who shaped the contemporary scene, it is remarkable, not only that he is unread today, but more so, that he is virtually unknown. His open, undogmatic, and critical mind left him without disciples; yet "few philosophers have been so pillaged" and left by the wayside.[12] The teacher of Royce, the subject of Santayana's first book, regarded by James as the prime mind of his time, built on by Bergson, by Alexander and Dewey in diverse ways, Lotze's work was translated into English by T. H. Green and Bernard Bosanquet and as such served as basic text in the philosophic educations of Moore, Whitehead, Russell, and Broad.

Lotze took idealism forward after Hegel. But he did so by rejecting the Hegelian thesis that Reality is exhausted in one supreme notion or formal system. "Reality is richer than Thought," he insisted, "nor can Thought make Reality after it." [13] Reaching back to Leibniz—for his metaphysical pluralism, respect for science, disavowal of mechanism, concern for the integration of moral and esthetic values with the existence of the natural world, and for his approach to questions of space and time—Lotze sought "to reconstruct an idealistic philosophy on a realistic basis." [14]

In many ways, Lotze was the progenitor of what has since come to be referred to as process philosophy. He emphasized the dynamic in nature and sought to work out the mechanical interconnections in the physical world without a materialism that ignores our experiences of values. He argued that organic life is not reducible to mechanistic explanation, that subjectivity is as truly a part of reality as are physical objects. He built into his metaphysics a doctrine of relations conceived as imbedded in that overarching comprehensive system of process or becoming which he regarded as the real.

The burden of his very long essay on time is to explode any notion of time as an absolute 'thing' or 'container' of things which are alleged to happen 'in' it. Time (as space), he urged, is the human mode of understanding the real relations of the continuity of becoming that *is* reality. Defending an amended form of the subjectivity of time (as of space), he remarks that time is "not something found and picked up all ready on its path by our cognitive energy." The time relations to be found in the flow of ideas in consciousness are the "conditions that compel the soul . . . to educe from itself the nature of time." [15] The temporal relations of past, present, future—in terms of which we consciously experience events—are, he argued, constructions of the human imagination. But, in contrast to the Kantian limitation of time as the form of phenomenal order, Lotze urged that these time relations are but the human way of understanding the real processes of dynamic becoming of things-in-themselves. Temporal organization is *our* way of understanding, not timeless noumena, but the dynamic processes of the real. Time is then real in human experience—as the translation of the real into the terms of our finite understanding; it is, to use the Leibnizian analogue, the way in which our finite understanding is able to 'reflect' reality.

At the outset of the essay, he makes his prime question clear: has time "any application to the Real?" He protests the long tradition—going back to Aristotle—which maintains that we have a direct perception of time. He emphatically argues that the common habit of representing time in spatial metaphors robs time of the unique status it has in human experience. Doing so introduces confusion and obscurity; it leads to a concept of 'empty time' as some sort of 'container' of events, a notion which is repeatedly attacked as inherently self-contradictory and meaningless. Early in the essay (sec. 140), he sets out the central line of attack—that empty time cannot have causal efficacy, that it cannot of itself effect any modification of things allegedly 'in' it, that change comes instead from "the nature and inner connexion of things" as they are themselves structured.

Is time, as Kant urged, merely a form of human apprehension? Lotze argues (see sec. 146) for the uniqueness of the concept of time—unique because it takes time to understand time. Unlike space, time "is not merely a product of the soul's activity, but at the same time the condition of the activity by which Time itself as a product is said to have been obtained." Foundational to human experience, whatever the time-status of things-in-process may be, these processes must be "translatable into forms of Time" in order for us to know and understand them. What is real, he argues (sec. 148), is what happens "in determinate forms," and these forms, for human experience, are the modes of time. The "production of Time must be a production *sui generis*" (sec. 149) and is foundational.

The primary notion of time is the present and it is out of the present that we function; the past and the future, which do not actually exist, are mental projections and the notion of time-as-a-whole is a subjective one: we think what we call past as the necessary condition for the present, and we think the present as a necessary condition for what we call future. Utilizing the spatial analogy of an infinite straight line, we then create the imaginary construct of an infinite time along which we plot the events which, as a matter of fact, provide our creation of the time-scale in the first place. Finite intelligence (sec. 152) develops the idea of time as the relations of conditions of things to produce the present experience—in terms of a past which we say produced it, and in terms of the future whose potentials we see before us.

But this is to say that in reality the order of events is primary and the time-scale derivative. Time cannot be primary just because we have to measure the time of any ordered series in terms of another ordered series; the process series, then, appears to be basic. The reality of events is primary: time cannot cause them; the time-order in terms of which we think merely reflects what is going on in the world (see sec. 153).

It has been argued that time is primary because, as Locke and Kant had urged, we are first aware of the 'train of ideas' in our minds. But this succession, Lotze urges, cannot be primary—just because we must be able to transcend separable moments in order to be aware of the sequence. We must be able to see the earlier and later in one "indivisible act of comparison" (sec. 154). Were it not possible for us to transcend the moments we relate, we would necessarily see the separate instants as separately sovereign. We are only able to unify these moments into the perception of events because the presentation of the events that we understand to be in time is truly apprehended in its process of development.

We cannot, therefore, really separate the time of experience and its content. Although the temporal relations we read into events—past, present, future—are subjective, they arise from our reading of things themselves which are essentially relational and also embodiments of an "efficient process" (sec. 156). We take the operational forms of our mental comparisons, abstract them from the concrete process of concern, and then find ourselves postulating both abstract laws of nature, which we expect the world of things to follow, and the "supposition of an empty Time," which we expect it to fill. But what is primary is neither abstraction; what is primary is the process-world itself. As for time, it is "not Time that is the condition of the operation of things but the operation that produces Time . . . [which is] the so-called 'vision' of the Time in the comparing consciousness."

Time, then, is empirically real. It *is* the process of understanding the process of becoming that constitutes the world. We can, Lotze would then argue, go beyond Kant: we know that anything that appears in human experience is such that it is amenable to "translation" into our temporal forms. We are thus not imprisoned in an anchorless phenomenal world; we are in cognitive contract with reality itself.

Yet when all is said and done, it is not clear just what we have learned about the nature of the time in terms of which we read reality, or how authentic our time experience really is. We perceive in terms of time a process that is somehow timeless, and we are to understand its nature by seeking to master the imaginative idea of a timeless moment (see sec. 153)! We perceive in our present—and all else is claimed to be subjective projection. But the nature of the 'present' is never defined. It apparently transcends momentary sequences so that we may take them into awareness and discriminate their 'before' and their 'after'. We understand the content of experience in terms of sequence but, in an unexplicated manner, we rise above the 'spread' of the sequence so as to take it all in.

Only the 'present' is held to be real. Surely a past moment was real when it was present and a future present will be real when it is present. We are reminded of Augustine's lament, "If, therefore, the present (if it is to be time at all) only comes into existence because it is in transition toward the past, how can we say that even the present *is?*" [16] The identification of reality as the present suggests the most momentary 'knife-edge' notion of actuality—except that it must somehow be spread out so that the sequences within it can be noted.

We mark *our* experiences of things in terms of the moving vantage of past, present, future; we mark out the time sequences of experiential content in terms of non-changing descriptions of before-and-after. But we are not given any indication of how the subjective time experience is correlated with the sequence order accorded to things. Lotze's whole notion of process, of becoming, seems imbedded in the dynamic flow of before-and-after; in the end, our time is to measure these sequences. Aristotle also marked time in this way and regarded time as the way we measure the real motion in the world. It is not clear just how much farther Lotze has really taken us.

Lotze's prime argument is against the notion of an ultimately objective time as the 'container' of events and his thesis is that time as such lacks causal efficacy. But to argue a thesis of efficient causation is already to argue from the presumption of the temporal and to employ the time predicates of 'before' and 'after'. On the same ground, one should question his separation of the process of becoming from the before-and-after sequences which are intrinsic to it. It seems impossible to separate such sequence of events in process from the events themselves and still have anything but the barest abstraction left. These predicates of sequential time he attributes to the events themselves and argues that they do not belong to time. The time he repudiates, then, would seem to be denuded of any definition or description and thus appears to be an empty abstraction. This is, indeed, a strange line of reasoning for one who eschewed abstractions and urged us to take only concrete processes as the real.

Time is our way of understanding reality which, now conceived in dynamic terms is still, somehow, essentially timeless. In the end, then, Lotze has carried forward the long tradition of subsuming time under some notion of a somehow timeless eternal, if now dynamic, order. But, in the course of this progression, he set the ground upon which Bergson and Alexander were able to proclaim their new temporal realisms.

Like a subsurface stream, the concern with time has flowed through the history of thought overarched by the presumably solid ground of some kind of timelessness. In a real sense, Kant, taking up some Leibnizian cues, broke the first opening in this covering ground by insisting that time is the fundamental form of our experience and by seeing the consequence—that for the human understanding, our conceptual structure must be brought into concert with the form of time. But his fidelity to the old tradition showed as he pointed this to a noumenal—unchangeable and thereby timeless—transcendent order.

Lotze, following out some Hegelian leads, finally cleared the ground for the next move—by his insistence on the unique *sui generis* nature of time, by his determined attack on the old tradition of reducing time to spatial description, by his peculiarly hedged insistence that the reality of the world (though somehow timeless) is really and pervasively dynamic, and by his repudiation of any attempt to subordinate reality to the logical formulas created out of human thought.

Finally, on the ground thus cleared, Henri Bergson (with whom we move into the twentieth century), opened it wide and brought this subterranean stream of time into the full light of day. Turning from the object-oriented metaphysic of the theoretical understanding, he called philosophy back to that introspective method whereby thinking first unveils its own internal nature before focusing its attention on external objects. Turning from a preoccupation with the problems of man-made physics to the principle of life itself, he developed a temporalism out of what he took to be the 'immediately given in human consciousness'.[17] In doing so he quite explicitly argued that physics cannot provide an insight into the nature of time accordant with the human experience which is permeated by it.

However his understanding of time may be criticized, and whatever problems we may find with some of his reasoning, his essential insight into the primordial character of time remains a crucial landmark for us today. Accorded every honor which France reserves for its intellectual leaders, and accorded international recognition by a Nobel Prize in 1928—the first ever accorded to a philosopher—the historic accomplishment of his pioneering work is virtually unnoticed today. Yet, he stands as a vital impetus for both pragmatism and phenomenology, the two prime if divergent contemporary schools deeply committed to the primordiality of experiential time. If they have, indeed, gone beyond him in many ways, they are still indebted to him for what he gave them with which to begin.

Bergson's philosophic enterprise was molded by a preoccupation with the question of time which he saw as *the* essential question of philosophic inquiry. When we take time seriously, he argued, we are able to resolve many old philosophic disputes; we thus need a critical reassessment of the philosophical tradition which has declined to take time as seriously as its preeminent role in human experience would indicate. Calling philosophy to a new empiricism in which the essential temporal marks of experience would be honored, he argued that most philosophic puzzles have come out of a tradition which insisted on veiling the essential temporality of that human experience from which they spring.

Our metaphysical tradition comes primarily from the ancient Greeks. Concerned to understand the dynamics of the physical world, they could not take time as seriously as its central place in human experience suggests—just because their questions were about the experienc*ed* world and not about the human experienc*er*. They continually thought in terms of the non-temporal—just because they conceived the dynamics of nature in terms of spatial arrangement; using time as the mark of ordered change, or the measure of change in space, time was spatialized at the outset. Using time for the numbering of motion, they connected time to a number system which—and here Bergson departed from Plato and from Kant—is inherently non-temporal; they, then, sought to number motion in terms of now-points which, as Aristotle recognized, are not time, but abstract and arbitrary boundaries interposed by thought to segment the continuity of time. Time was, understandably, understood in terms of what is not time, in terms of abstract boundaries of arbitrarily marked intervals—not in terms of the lived, concrete *intervals* of duration itself. Reading the world of temporal change in terms of an unchanging order of rational concepts, time was taken, in Plato's immortal phrase, as but the "moving likeness of eternity;" despite all criticisms of Platonism, this Platonic thesis, albeit in Aristotelian form, has dominated all subsequent thinking about the nature of time.

But every aspect of our being is marked by duration, sequence, and change. We experience time, not a timeless eternity. Yet philosophy has refused to come to terms with the time of human experience; it has persisted in reading the world that is experienced in terms of what is not experienced. Lacking all empirical authenticity, it is only surprising that more puzzles, problems, and disputes have not been created. If we can clarify our understanding of time, Bergson argued, in an authentic way, we can then start to resolve those puzzlements which arose out of the refusal to come to terms with the primordiality of experiential time.

Effectively marking the distinction between the useful and the true, his work is a plea for an end to confounding them. He readily conceded the pragmatic value of the practice whereby the scientific understanding seeks to understand motion in space by numbering abstract and timeless now-points. Useful as this may be, it does not give us insight into the essential continuity of actually unsegmented duration in which all changing occurs; it does not give us insight into our own internal experiencing of lived time.

We must, he reasoned, make a crucial distinction between time as conceptualized and time as actually lived. The duration of reality is an indivisible continuity of 'flow' which our intellect, for practical reasons of manipulation and control, separates or 'freezes' into definable and rationally manageable pieces. It does this by translating time into space—the dynamic into the static—and seeking to understand time by taking the metaphor of interchangeable points on a line as if it were literal truth. This spatialization permits the development of nontemporal concepts and a formalized logic to control their use; theoretical understanding forgets that this conceptual structure is but a set of tools by which to cope with the flux, novelty, and essential continuity revealed in primary experience. Abstracted from experience, these conceptual tools refer to nothing directly in experience and certainly do not face the living experience, out of which they are created, in its own terms.

It is to life that Bergson turned for revelations of reality. And life is more than logic. A logical system of stable concepts is a useful abstraction from the concreteness of lived experience; it substitutes symbols, essentially derived from static spatialization, for the concreteness of lived duration which is intuitively apprehended within our own consciousness and within our experience in and of the world.

His first book, published in 1889, was translated into English under the title, *Time and Free Will*. As such it set out to demonstrate the thesis that an authentic conception of time unlocks the key to many traditional philosophic problems and melts others away. Its essential argument was that reality be recognized as duration, as an indivisible pervasive continuity of time; that so understood, philosophic conundrums such as Zeno's paradoxes of the 'impossibility' of motion and the disputes concerning the reality of experiential free will are both readily resolved. The original French title, however, is itself significant; translatable into English as "Essay on what is immediately given in consciousness," it points us to Bergson's own original development of a phenomenological stance.

His most famous book, and the one which he regarded as his most important, *Creative Evolution* (1907), constituted his reexamination of the philosophic tradition; it portrayed the reality of duration as inextricably tied to a continuity of development and the emergence of genuine novelty in the continuity of pervasive becoming. The result is a repudiation of attempts to explain change in terms of mechanistic systems whether they be formulated as an order of prior causes or eternal entitites. A genuine science will accept the primacy of temporal duration in ourselves and in the world we seek to understand. It will welcome novelty and change as marks of an evolutionary reality which is not to be dissolved in the abstractions of an unexperiential timeless order of conceptual relations. It will, therefore, not succumb to a new scholasticism which uses symbols to hide from change and transforms the utility of a number system into an attempt to guarantee the primacy of a static permanence.

The flux of time is the reality itself, and the things which we study are the things which flow. It is true that of this flowing reality we are limited to taking instantaneous views. But, just because of this, scientific knowledge must appeal to another knowledge to complete it. While the ancient conception of scientific knowledge ended in making time a degradation, and change the diminution of a form given from all eternity—on the contrary, by following the new conception to the end, we should come to see in time a progressive growth of the absolute, and in the evolution of things a continual invention of forms ever new.

... it is within the evolutionary movement that we place ourselves, in order to follow it to its present results, instead of recomposing these results artificially with fragments of themselves. Such seems to us to be the true function of philosophy. So understood, philosophy is not only the turning of the mind homeward, the coincidence of human consciousness with the living principle whence it emanates, a contact with creative effort: it is the study of becoming in general, it is true evolutionism and consequently the true continuation of science—provided that we understand by this word a set of truths either experienced or demonstrated, and not a certain new scholasticism that has grown up during the latter half of the nineteenth century around the physics of Galileo, as the old scholasticism grew up around Aristotle.[18]

In *Matter and Memory* (1908), he contended that memory participates *in* present decision, that our entire experience of acting and deciding presupposes the indivisible continuity of durational flow; rejecting the intellectualism of the Cartesian tradition, he yet made it plain that he was carrying forward a neo-Cartesian dualism as the structure of his metaphysics.

The most developed statement of his conception of time was worked out in *Duration and Simultaneity* (1922), which is a sympathetic but critical examination of certain aspects of the new relativity theory of modern physics. Defending its essential thrust against some of its popularizers, he reasoned that within the areas of its competence it is extremely useful but, by the nature of the case, it cannot come to authentic terms with the duration of becoming which is "real time, perceived and lived" (p. 222).

The essential contrast which he drew is that of the duration of experience and of measured time. The first is experienced in the actual "unfold*ing*" of a present experience; in listen*ing* to a melody, we do not measure its beginning and end but enjoy the *interval* between them as an indivisible unity. The second is the measuring of beginning and end as "unfold*ed;*" we do this by use of an independent measure such as a set of spatial positions on a clock, one of which was simultaneous with the first note and another with the last. As we listen, we may say that the melody is slow or fast, long or short—as *a quality* of what we are hearing. After hearing it, we say, by comparing the extent of its duration with the movements of spatial indicators, simultaneous with it, that it took, say, 'five minutes'; but to do this we necessarily look outside of the melody after it is over, note its terminal now-points as marks on a line which we simultaneously consider, and ignore the interval between them, the melody itself.

Lived experience, Bergson reasoned, is the experienc*ing* of the duration of the melody itself; scientific understanding is the external measure of its terminals in terms of a parallel spatial indicator—which then provides a symbolic abstraction from the playing of the melody in terms of spatial translation. To treat this measuring as *the* real thing is to confuse measure and measured, and to forget that the measurer, himself, is living and "unfold*ing*" *while* he is using visual perception of movement in static space to measure the continuity of the experienced duration.

Echoing Augustine, he rejects the thesis that we would not be able to know whether or not all things were speeded up or slowed down together; its problem really arises from this long persistence in confusing the measure and the measuring. The measurement of terminal points would not show the difference if the measuring instrument were speeded or slowed together with the melody. But the experience of the melody would, indeed, change with its duration, for its duration *is* its reality and a change of its duration would be a change *in* it. To take the spatialized quantitative statement of its extension for its duration is to confuse spatial extension with temporal duration. "Real duration is *experienced;* we *learn* that time unfolds and, moreover, we are unable to measure it without converting it into space and without assuming all we know of it to be unfolded." (p. 234).

"Time is succession" (p. 236), but the quantitative measuring of time is in terms of space—which Leibniz had defined for us as 'simultaneous co-existence'. To measure duration, then, is to translate indivisibility into instants, conceived as points on a line which we take in at one glance. Real duration is directly intuitional; the translation into static space is useful for some purposes but it is not the original experience—just as the translation of a poem does not claim to be the poet's original creation. This duration of the original experience we directly intuit but cannot conceptualize; it is part of us; it is integral to our experience and to the aspects of the world which appears in that experience. It is the reality within which we dwell.

Time, as the continuity of duration, Bergson insisted, is reality itself. But how do we structure our experience of this duration? Bergson found our source of knowledge of our experiential world in the living present—but how broad is this moving present and how is it formed? Is it only the union of the past melting into the present 'moment', or does it take in the present anticipations of the future as well? How do we construe the duration of the world as distinguished from our experience of it? How do we, indeed, develop from indivisible duration the conceptual tools of scientific understanding by means of which we seek to manipulate and control it? And, is the human intellect really so contratemporal as he suggested? Questions such as these arise from Bergson's thesis of the primordiality of experiential time; in different ways which were to become the fulcrum of discussion as realists, pragmatists, and phenomenologists, following him in both similar and contrasting ways, sought to come to terms with the time of human experience, while respecting the science which tells us about the things which appear within that experience.

On an intuitive level, Bergson has used a reality theory of perception—we do directly (intuitively) perceive the durational reality of the world. He has perhaps taken Descartes' distinction between real duration and mathematical time further and in a direction radically different from any that a Cartesian might have imagined. He has attacked the long tradition from Plato to Kant, which has tied time and number together; in effect, he has repudiated the Kantian thesis that arithmetic is the 'science' of time as geometry is the 'science' of space, and insisted that they are both inherently spatial in character.

He has carried the Cartesian tradition forward by retaining its priority of dualistic categories. For Bergson has generally reasoned by urging dualistic distinctions—the dualism of experienced and symbolized time has been largely discussed in terms of the dualism of time and space. He carried this through by tying auditory examples to the experience of time and citing visual examples as inherently spatial. But this may be an overstatement of the case: Don't we *see* things in succession? Don't we *hear* sounds as near or far? Can we experience time or space, temporality or spatiality, without the other? Isn't the experience of each an equally authentic constituent of our experience? Questions such as these seem to have impelled Samuel Alexander to take up Bergson's temporal realism and to recast it in a different perspective.

Alexander worked out a detailed and systematic metaphysic which sought to deal directly with time as ultimately real while avoiding some of Bergson's difficulties and also facing some of the technical philosophical questions of his day. His prime work was accorded the honor of presentation as Gifford Lectures. Working with a realist epistemology and in strong concert with the theory of "emergent evolution" being developed by Lloyd Morgan, Alexander presented a philosophy of temporal realism. His method of presentation was not one of deductive argumentation; in common with much of Bergson, Whitehead, and many other contemporary thinkers, his views were offered in the form of reasoned coherent description.

In many ways, he can be seen as carrying forward Bergson's emphasis on the ultimacy of time—but he did so in a different philosophic context and the differences between them are crucial. Like Bergson, Alexander claimed to start from the facts of ordinary experience; but ordinary experience, he pointed out, presents the inseparability of space and time. We can think of one only with reference to the other. If the new realism was correct in claiming that experience must be understood as reflecting the nature of the real, then space is linked with time as primordial reality.

Bergson had protested the reading of time in terms of space; Alexander replied that unless we read each in terms of the other, we have only irreducible mystery. Time and space are not only inextricably connected in our experience and in our scientific measurings of change; they are ontologically inseparable. Together, they constitute the elemental 'stuff' of the universe—the matter *and* mind of the universe, that out of which matter and mind, as we conceive them, have evolved. We experience time as the duration of succession together with space as co-existence or simultaneity; we could not experience one without the other; we thus experience only 'space-time'.

If we are to understand time—as succession, duration, irreversibility, and transitivity—Alexander contends that it is necessary to recognize the inextricable internal relationship of space and time as presenting the two distinguishable faces of the one underlying reality. But this impels us, Alexander argued, echoing Bergson, to reject the notion that time is merely the fourth dimension of space. Time is the *only* dimension of space; every instant of time encompasses the whole of space, just as every point of space endures through infinite time. Thus, in language that would have appalled Bergson, Alexander referred to the pure events or 'units' of space-time as 'point-instants'. But he agreed with Bergson, in accord with their fundamentally evolutionary approaches, that while space is the principle of stability in the world, time is the generator of change, novelty, and development. Both thus accorded to time a causal character. Both rejected the disvaluation of time as the great destroyer. As Alexander stated it in a way that Bergson would have endorsed, time is not to be pictured primarily as the man with the scyth, the grim reaper; time is much more the planter, as the "abiding principle of impermanence which is the real creator."

Against the later English neo-Hegelians, such as Bradley and McTaggart, who had argued that time includes contradictory predicates and must thereby be unreal, Alexander urged that time does indeed comprise the notions of succession and duration, of decay as well as creation, of difference as well as sameness—but all this does not mean that time is thereby unreal. Time is, with space, the essential fundament of reality. Time and space are not human illusions or neutral 'containers' in which things happen; they are ingredient to *every* existent; every thing, insofar as it is, incorporates the principles of succession and co-existence, of time and space. Things and events are 'made' of space-time and thereby incorporate the categorial structure of space-time in themselves. This is why the universe is not only dynamic but ordered, not only changing but developing.

Working out this general thesis in considerable detail takes up the larger portion of Alexander's two-volume work. In a way which is helpful, one can see Alexander developing something of a 'reverse reading' out of Kant, by eliminating the distinction of phenomena and noumena, and by ontologizing the Kantian description of the nature of human knowledge.

Kant had argued, in effect, that the categories of the human understanding, in terms of which we understand the flow of events of which we are aware, are but facets of the mind's operational procedure and have no cognitive force beyond the realm of possible human experience. Kant had formulated a 'table of categories' (as abstractive logical statements of the temporal principles by which human knowledge operates) but he had never really explained why these, instead of other, categories were to be taken as constitutive of cognition. Nor did Kant really explain *why* he took time and space as absolutely foundational, not as categories but as the comprehensive 'forms' of all experience within which the categories were to be 'applied'.

Alexander's general thesis may be read as something of a reply. Space-time, his work asserts, is not merely the two-sided nature of our 'pure intuition'; we receive all data of experience in the dual form of space-and-time *just because* space-time is the elemental 'stuff' out of which the universe, and what is in it, is composed. The cognitive categories, in terms of which we understand the content of experience—which is essentially spatio-temporal—are not merely subjective; they are the intrinsic structural modes of the being of space-time itself and of everything 'in' it. The categories are universally applicable because they are the structure of the one universal reality; they thus apply to every real entity. They are a priori in the sense that they necessarily apply, thereby, to any possible existent just because any possible existent must incorporate the structure of space-time which is its essential substance. The categories, then, are features of reality and only *thereby* are features of our knowledge—just because knowledge is knowledge only when it is knowledge of reality. The categories are the essential procedures, not merely of our cognitive thinking, but of true becoming itself. They are the principles of order that unify all development within the world and of the world as one evolving entity. They apply to *all* events, mental as well as physical. They are the characters of the true being that is space-time and only thereby are they the signs of truth in the cognitive judgments we make about those aspects of reality with which we are concerned.

Time, Kant had urged, is *the* form of human apprehension, the elemental form of the apprehending mind. Alexander's argument, if true, would go far to explaining why Kant felt constrained to regard, without need of explanation, space and time as the only two forms of receptivity, why he made time prior, and why he was never really able to deal satisfactorily with space in his own terms. For, the consequence of Alexander's ontologizing of the categories is to say: Space and time are not merely the two forms through which the human mind receives data for understanding; they are the two discernible ways in which our minds participate in, and reflect, the nature of the data itself; they are the two ways in which the mind receives information and becomes aware of external reality which is so constituted. Alexander summarized this well:

. . . Time as a whole and in its parts bears to Space as a whole and in its corresponding parts, a relation analogous to the relation of mind to its equivalent bodily or nervous basis, or to put the matter shortly, that Time is the mind of Space, and Space is the body of Time . . .

Rather than hold that Time is a form of mind, we must say that mind is a form of Time.[19]

Alexander was aware of the problems that his view raises, particularly with reference to any conception of mental activity. In the opening sections of *Space, Time, and Deity,* he quickly turned to the admitted difficulties. Having sketched out the conception of space-time on the physical level, he faced a number of questions concerning our experience of it. Explicitly he assumed a realist epistemology of direct apprehension of reality itself. Awareness of an event, he argued, must occur in the same time as the event itself; we experience our own minds as 'in time', but this time must be the same as the physical time of the events of which we are aware. Similarly, he reasoned, our mental space belongs to the same space as physical space.

In the course of these analyses, Alexander subjected an impressive number of topics pertaining to our experience, as spatial and temporal, to a careful series of considerations. These include the role of memory, the notion of 'specious present', the status of the past and the future, the developed concept of mental space-time and its relation to physical space-time. "The main result," of this extended discussion, he pointed out in explicit contrast to Bergson, "has been to show that Time is really laid out in Space and is intrinsically spatial." [20] His writing is clear and needs no summary or restatement; his argument is too detailed for extensive comment here. Some salient points which arise from the main thrust of his argument, however, must be noted.

One essential ground thesis which Alexander advanced is the simultaneity of perceiving and perceived. But, if we take this in any but a metaphoric sense, it cannot be literally true. In his own terms, time is essentially 'ingredient' to being—but this means that *every* process essentially takes time, even if it be so miniscule as to be beyond the bounds of our capacity to measure it. Indeed, our eyes depend for perception on the speed of light as our ears depend upon the speed of sound. If we are dealing with objects and events in our immediate proximity, these time-lapses may be so minute as to justify the practical dismissal of their significance. But, if the astronomer is justified in depending upon the speed of light for his perception of distant stars, so must we recognize, in principle, the fact that light (as sound) *takes time* to travel. If we are to follow Alexander in resolving time into instants, then it is literally true that no external event takes place in the same instant in which it is perceived. In the most literal sense, we are always seeing the past of the perceived object, not its present; we are perceiving what is literally its past *in* our present. Time-lapse is, then, necessarily involved in the physical aspect of perception itself, and there can be no one-to-one correspondence of the instant of the perceiving and the perceived.

It is also difficult to see how we can describe our own perceiving in terms of miniscule instants. If we then expand them and introduce the notion of 'spread' into the instant, as Alexander has done in discussing the 'specious present', it is not at all clear just what an 'instant' of time is supposed to mean or to what it is supposed to refer. Yet it is, finally, in terms of 'point-instants' that he examined our experience of space-time and plotted the nature of space-time as well.

Time-lapse and duration of process are in our thinking, in physical perception, and in nature as well. Both Bergson and Alexander have been impressed by this pervasiveness of time characteristics. Both have been concerned to escape from the notion of a disembodied mind somehow contemplating the physical world, a notion that has plagued philosophy at least since Descartes. Both have recognized that the mind's relation to body involves a spatial as well as a temporal function in mental processes. Both were concerned to recognize and account for the mind's essential involvement with its environment. Both tried to start from our actual experience—but that experience was differently conceived. Bergson started from the process of experienc*ing* itself, Alexander apparently from the pervasive characteristics of the world as it is experienc*ed*.

Bergson would have charged that Alexander took the notion of the instant from the spatial analogy of an indivisible point on a chronological line. Although Alexander tried to literalize the analogy, he found that the present of human experience cannot be reduced to a point devoid of past and future. His argument, based on analogy, proceeds by analogy. Yet we are never really clear as to just what time is to be taken to be. Alexander repeatedly offers us analogies in a chain of careful metaphoric reasoning but, seemingly forgetting that an analogy is an illustration and never an argument, he neither defended the analogies themselves nor gave us any reason for taking them as literally true.

Alexander pursued his spatial analogy of temporal instants, succeeded, despite his continuing assertion of their co-equality, in reducing time to space, placing time *in* space and temporalizing spatial existence. But this is not to explain just what time (or space) may be, why time is its own dimension and cannot be reduced to spatial measurement, or why space-time functions as a unity in terms of their constituting categories. For, if they do function in terms of the same categories, if they are inseparable in reality, what is the real difference between them? Indeed, there is presumably no real difference and that is why our historic linguistic confusion is honored while the new truth is pointed out in referring to them conjointly as 'space-time'. But, then, much of the discussion presumes that the experience of time is somehow different from the experience of space—even if, in fact, we cannot experience them except together. Space-time, then, cannot be a unity although it can be a system of unification.

Our perception of time, he has argued, is only a duration of a few seconds; but, then, why or how does this fragmentary experience give us an insight into the totality of reality, its nature, and the direction of its development? Space-time, in the end, appears to be little more than a concept of Pure Motion which is a sequential occupation of spatial points. But Aristotle had effectively done this a long time ago. What Alexander seems to have done is to take up again an Aristotelian cosmology, introduce the notion of progress, and point the ultimate source of the motion of the world, not to an ever-present 'prime mover' as its continuing efficient cause, but to the development of deity in an evolving universe. Just what this does to questions of cosmogenesis remains obscure. Presumably he would have opted for the Greek thesis of 'no beginning'—but with the notion of the development of meaningful significance rather than that of unending redundancy.

In any event, in his own way, Alexander has, despite all of the fundamental divergences from Bergson, joined him in the effective repudiation of a causal metaphysic which seeks to explain the present in terms of the past. For, if we take Alexander's metaphysic of evolution in full force, he is pointing us—for explanation of the present to the future—to the possibilities of the future which the present is presenting to us.

These post-Kantian attempts, to go beyond the priority of time as subjective experiential form by seeking to establish the relationship of time and ultimate reality, really urge us to retrace our steps. Despite important and often brilliant insights along the way, we are left, in each case, with new puzzles:

- Hegel, recognizing the pervasiveness of temporal change in experience of the existential, sought to ground it in a transcendent Reason that was itself conceived as non-temporal.
- Lotze, recognizing the uniqueness of time in human experience, sought to explain it as a finite perspective onto a somehow non-temporal but yet dynamic world order.
- Bergson, seeking to treat durational experience in its authenticity, opened to us the pervasiveness of temporal process in ourselves *and* in the world in which we find ourselves—but he never really explained the concomitant pervasiveness of spatiality and left space, at best, as an inexplicably created stage on which time was to do its work.
- Alexander, reasserting the ultimate reduction of time to spatial description, could not really explain the durational spread of the human experience of the temporal which he tried to account for in terms of a synthetic union of time and space—just because, in the end, he ignored the distinctly temporal aspect of time-lapse inherent in any process of perception or judgment regardless of the spatiality of its object of focus.

Bergson and Alexander, particularly, have taken us far—just because each tried in his own way to deal with the temporality of experience as something not to be explained away. Perhaps they did not succeed in taking us farther just because they tried to cover too much ground too soon! Each of these four, both separately and taken together, points to the possibility of going beyond them; but it seems that the only way in which we might succeed in doing so is to retrace our steps and look to a new beginning. Perhaps by taking greater care of the nature of time as it is revealed in our experience, we can better determine its nature, its potentials for relevance to our areas of concern, and the reasons its demonstrates for new and hopefully more comprehensive and insightful perspectives.

By turning back to a more careful preparation for investigation of this question of time which has dogged our philosophic tradition for some three thousand years, we might be better prepared for a more meaningful inquiry into the nature of time and its significance for us. Better prepared, we might prosper better on a more carefully conceived journey.

Three prime questions now emerge as central to the new discussions of our own day; hopefully they will conspire to clear the ground for a new and more fruitful departure:

- How is our experience of time revealed in the conceptual structures we use to express our temporal involvements?
- How do our practical experiences relate to, or illuminate, the intentional significance which time has for us?
- How do we actually develop temporal concepts and structure the time in terms of which we seek to understand ourselves and the world of which we find ourselves a part?

Demonstrating, perhaps, that sequential chronology is not necessarily the first face of time, and reinforcing Whitehead's observation that durations overlap, some of these new questions were being raised while older ones were still being discussed. Different thinkers in much the same period, operating in different traditions, and often reading and citing each other, took these questions as definitional to the continuing discussion, although they vehemently disagreed as to which of these questions was basic and in which directions they should be taken. However these differences may be, these three questions, taken together, shape the diversity and underlying unity of the attempt to come to terms with the nature and meaning of time in this, our, century.

12.

Hegel:

TIME AND BECOMING

The concept of time plays a crucial role throughout the metaphysical system of G. W. F. Hegel (1770–1831). Its systematic
statement, however, is to be found in the second part of the
tripartite *Encyclopaedia of the Philosophical Sciences,* entitled
The Philosophy of Nature.* The selection is taken from the new
translation by A. V. Miller (Section One, "Mechanics," Part A) in
which Hegel had discussed, in addition to the notion of time,
those of space, place, and motion.

(2) Time

§ 257

Negativity, as point, relates itself to space, in which it develops its
determinations as line and plane; but in the sphere of self-externality,
negativity is equally *for itself* and so are its determinations; but, at the
same time, these are posited in the sphere of self-externality, and
negativity, in so doing, appears as indifferent to the inert side-by-
sideness of space. Negativity, thus posited for itself, is Time.

Zusatz. Space is the immediate existence of Quantity in which
everything subsists, even the limit having the form of subsistence;
this is the defect of space. Space is this contradiction, to be
infected with negation, but in such wise that this negation falls
apart into indifferent subsistence. Since space, therefore, is only
this inner negation of itself, the self-sublating of its moments is its
truth. Now time is precisely the existence of this perpetual self-

---

* G. W. F. Hegel, *The Philosophy of Nature,* trans. A. V. Miller. Copyright (C) 1970,
Oxford University Press, and by permission of The Clarendon Press, Oxford.

sublation; in time, therefore, the point has actuality. Difference has stepped out of space; this means that it has ceased to be this indifference, it is for itself in all its unrest, is no longer paralysed. This pure Quantity, as self-existent difference, is what is negative in itself, Time; it is the negation of the negation, the self-relating negation. Negation in space is negation attached to an Other; the negative in space does not therefore yet receive its due. In space the plane is indeed a negation of the negation, but in its truth it is distinct from space. The truth of space is time, and thus space becomes time; the transition to time is not made subjectively by us, but made by space itself. In pictorial thought, space and time are taken to be quite separate: we have space and *also* time; philosophy fights against this 'also.'

§ 258

Time, as the negative unity of self-externality, is similarly [to Space] an out-and-out abstract, ideal being. It is that being which, inasmuch as it *is,* is *not,* and inasmuch as it is *not, is:* it is Becoming directly *intuited;* this means that differences, which admittedly are purely *momentary,* i.e. directly self-sublating, are determined as *external,* i.e. as external to *themselves.*

Remark

Time, like space, is a *pure form* of *sense* or *intuition,* the nonsensuous sensuous; but, as in the case of space, the distinction of objectivity and a subjective consciousness confronting it, does not apply to time. If these determinations were applied to space and time, the former would then be abstract objectivity, the latter abstract subjectivity. Time is the same principle as the I = I of pure self-consciousness, this principle, or the simple Notion, still in its uttermost externality and abstraction—as intuited mere *Becoming,* pure being-within-self as sheer coming-out-of-self.

Time is *continuous,* too, like space, for it is the negativity abstractly relating self to self, and in this abstraction there is as yet no real difference.

Everything, it is said, *comes to be* and *passes away* in time. If abstraction is made from *everything,* namely from what fills time, and also from what fills space, then what we have left over is empty time and empty space: in other words, these abstractions of externality are posited and represented as if they were for themselves. But it is not *in* time that everything comes to be and passes away, rather time itself is the *becoming,* this coming-to-be and passing away, the *actually existent abstraction, Chronos,* from whom everything is born and by whom its offspring is destroyed. The real is certainly distinct from time, but is also essentially identical with it. What is real is limited, and the Other to this negation is *outside* it; therefore the determinateness in it is self-external and is consequently the contradiction of its being; the abstraction of this externality and and unrest of its contradiction is time itself. The finite is perishable and *temporal* because, unlike the Notion, it is not in its own self total negativity; true, this negativity is immanent in it as its universal essence, but the finite is not adequate to this essence: it is *one-sided,* and consequently it is related to negativity as to the power that dominates it. The Notion, however, in its freely self-existent identity as I = I, is i and for itself absolute negativity and freedom. Time, therefore, has no power over the Notion, nor is the Notion in time or temporal; on the contrary, *it* is the power over time, which is this negativity only *qua* externality. Only the natural, therefore, is subject to time in so far as it is finite; the Ture, on the other hand, the Idea, Spirit, is *eternal*. But the notion of eternity must not be grasped negatively as abstraction from time, as existing, as it were, outside of time; nor in a sense which makes eternity come *after* time, for this would turn eternity into futurity, one of the moments of time.

Zusatz. Time is not, as it were, a receptacle in which everything is placed as in a flowing stream, which sweeps it away and engulfs it. Time is only this abstraction of destruction. It is because things are finite that they are in time; it is not because they are in time that they perish; on the contrary, things themselves are the temporal, and to be so is their objective determination. It is therefore the process of actual things themselves which makes time; and though time is called omnipotent, it is also completely impotent. The Now has a tremendous right; it *is* nothing as the individual Now, for as I pronounce it, this proudly exclusive Now dissolves, flows away and falls into dust. *Duration* is the universal of all these Nows, it is the sublatedness of this process of things which do not endure. And though things endure, time still passes away and does not rest; in this way, time appears to be independent of, and distinct from, things. But if we say that time passes away even though things endure, this merely means that although some things endure, change is nevertheless apparent in others, as for example in the course of the sun; and so, after all, things are in time. A final shallow attempt to attribute rest and duration of things is made by representing change as gradual. If everything stood still, even our imagination, then we should endure, there would be no time. But all finite things are temporal, because sooner or later they are subject to change; their duration is thus only relative.

Absolute timelessness is distinct from duration; the former is eternity, from which natural time is absent. But in its Notion, time itself is eternal; for time as such—not any particular time, nor Now—is its Notion, and this, like every Notion generally, is eternal, and therefore also absolute Presence. Eternity will not come to be, nor was it, but it *is*. The difference therefore between eternity and duration is that the latter is only a relative sublating of time, whereas eternity is infinite, i.e. not relative, duration but duration reflected into self. What is not in time is that in which there is no process. The worst and the best are not in time, they endure. The former because it is an abstract universality: such are space, time itself, the sun, the Elements, stones, mountains, inorganic Nature generally, and even works of man, e.g. the pyramids; their duration is no virtue. What endures is esteemed higher than what soon perishes, but all bloom, all beautiful vitality, dies early. But what is best also endures, not only the lifeless, inorganic universal, but the other universal, that which is concrete in itself, the Genus, the Law, the Idea, Spirit. For we must distinguish between what is the whole process and what is only a moment of the process. The universal, as law, also has a process within it and lives only as a process; but it is not a *part* of the process, is not in process, but contains its two sides, and is itself processless. On its phenomenal side, law enters into the time-process, in that the moments of the Notion have a show of self-subsistence; but in their Notion, the excluded differences are reconciled and co-exist in peace again. The Ideal Spirit, transcends time because it is itself the Notion of time; it is eternal, in and for itself, and is not dragged into the time-process because it does not lose itself in one side of the process. In the individual as such, it is otherwise, for on one side it is the genius; the most beautiful life is that in which the universal and its individuality are completely united in a single form. But the individual is then also separated from the universal, and as such it is only one side of the process, and is subject to change; it is in respect of this moment of mortality that it falls into time. Achilles, the flower of Greek life, Alexander the Great, that infinitely powerful individuality, do not survive. Only their deeds, their effects remain, i.e. the world which they brought into being. Mediocrity endures, and in the end rules the world. There is also a mediocrity of thought which beats down the contemporary world, extinguishes spiritual vitality, converts it into mere habit and in this way endures. Its endurance is simply this: that it exists in untruth, does not receive its due, does not honour the Notion, that truth is not manifest in it as a process.

§ 259

The dimensions of time, *present, future,* and *past,* are the *becoming* of externality as such, and the resolution of it into the differences of being as passing over into nothing, and of nothing as passing over into being. The immediate vanishing of these differences into *singularity* is the present as *Now* which, as singularity, is *exclusive* of the other moments, and at the same time completely *continuous* in them, and is only this vanishing of its being into nothing and of nothing into its being.

Remark

The finite present is the *Now* fixed as *being* and distinguished as the concrete unity, and hence as the affirmative, from what is *negative,* from the abstract moments of past and future, but this being is itself only abstract, vanishing into nothing. Furthermore, in Nature where time is a *Now,* being does not reach the *existence* of the difference of these dimensions; they are, of necessity, only in subjective imagination, in *remembrance* and *fear* or *hope.* But the past and future of time as *being* in Nature, are space, for space is negated time; just as sublated space is immediately the point, which developed for itself is time.

There is no *science of time* corresponding to the *science of space,* to *geometry.* The differences of time have not this *indifference* of self-externality which constitutes the immediate determinateness of space, and they are consequently not capable of being expressed, like space, in configurations. The principle of time is only capable of being so expressed when the Understanding has paralysed it and reduced its negativity to the *unit.* This inert One, the uttermost externality of thought, can be used to form external combinations, and these, the numbers of *arithmetic,* can in turn be brought by the Understanding under the categories of equality and inequality, of identity and difference.

One could also conceive the idea of a philosophical mathematics knowing by Notions, what ordinary mathematics deduces from hypotheses according to the method of the Understanding. However, as mathematics is the science of finite determinations of magnitude which are supposed to remain fixed and valid in their finitude and not to pass beyond it, mathematics is essentially a science of the Understanding; and since it is able to be this in a perfect manner, it is better that it should maintain this superiority over other sciences of the kind, and not allow itself to become adulterated either by mixing itself with the Notion, which is of a quite different nature, or by empirical applications. There is nothing in that to hinder the Notion from establishing a more definite consciousness alike of the leading principles of the Understanding and also of order and its necessity in arithmetical operations (see § 102) as well as in the theorems of geometry.

Furthermore, it would be a superfluous and thankless task to try to express *thoughts* in such a refractory and inadequate medium as spatial figures and numbers, and to do violence to these for this purpose. The simple elementary figures and numbers, on account of their simplicity, can be used for *symbols* without fear of misunderstanding; but even so, these symbols are too heterogeneous and cumbersome to express thought. The first efforts of pure thought had recourse to such aids, and the Pythagorean system of numbers is the famous example of this. But with richer notions these means become completely inadequate, because their *external* juxtaposition and their contingent combination do not accord at all with the nature of the Notion, and it is altogether ambiguous which of the many possible relationships in complex numbers and figures should be stuck to. Besides, the fluid character of the Notion is dissipated in such an external medium, in which each determination is indifferent to and outside the others. This ambiguity could be removed only by an *explanation;* but then the essential expression of the thought is this explanation, so that the representation by symbols becomes a worthless superfluity.

Other mathematical determinations such as the *infinite and its relationships,* the *infinitesimal, factors, powers,* etc., have their true notions in philosophy itself; it is inept to employ these determinations in philosophy, borrowing them from mathematics where they are employed in a notionless, often meaningless way; rather must they await their justification and meaning from philosophy. It is only indolence which, to spare itself the labour of thought and notional determination, takes refuge in formulae which are not even an immediate expression of thought, and in their ready-made schemata.

The truly philosophical science of mathematics as *theory of magnitude,* would be the science of *measures;* but this already presupposes the real particularity of things, which is found only in concrete Nature. On account of the *external* nature of magnitude, this would certainly also be the most difficult of all sciences.

Zusatz. The dimensions of time complete the determinate content of intuition in that they posit for intuition the Notion of time, which is *becoming,* in its totality or reality; this consists in positing each of the abstract moments of the unity which becoming is, as the whole, but under opposite determinations. Each of these two determinations is thus itself a unity of being and nothing; but they are also distinguished. This difference can only be that of coming-to-be and passing away. In the one case, in the Past (in Hades), being is the foundation, the starting point; the Past has been actual as history of the world, as natural events, but posited under the category of non-being which is added to it. In the other case, the position is reversed; in the Future, non-being is the first determination, while being is later, though of course not in time. The middle term is the indifferent unity of both, so that neither the one nor the other is the determinant. The Present *is,* only because the Past is not; conversely, the being of the Now is determined as not-being, and the non-being of its being is the Future; the Present is this negative unity. The non-being of the being which is replaced by the Now, is the Past; the being of the non-being which is contained in the Present, is the Future. In the positive meaning of time, it can be said that only the Present *is,* that Before and After are not. But the concrete Present is the result of the Past and is pregnant with the Future. The True Present, therefore, is eternity.

The name of mathematics could also be used for the philosophical treatment of space and time. But if it were desired to treat the forms of space and the unit philosophically, they would lose their peculiar significance and pattern; a philosophy of them would become a matter of logic, or would even assume the character of another concrete philosophical science, according as a more concrete significance was imparted to the notions. Mathematics deals with these objects only *qua quantitative,* and among them it does not—as we noted—include time itself but only the unit variously combined and linked. No doubt in the theory of motion time *is* an object considered, but applied mathematics is, on the whole, not an immanent science, simply because it is the application of pure mathematics to a given material and to its empirically derived determinations.

(3) Place and Motion

§ 261

Remark

Zusatz ... Just as Time is the purely formal soul of Nature, and Space, according to Newton, is the sensorium of God, so Motion is the Notion of the veritable soul of the world. We are accustomed to regard it as a predicate or a state; but Motion is, in fact, the Self, the Subject as Subject, the abiding of vanishing. The fact that Motion appears as a predicate is the immediate necessity of its self-extinction. Rectilinear motion is not Motion in and for itself, but Motion in subjection to an Other in which it has become a predicate or a sublated moment. The restoration of the duration of the Point, as opposed to its motion, is the restoration of Place as unmoved. But this restored Place is not immediate Place, but Place which has returned out of alteration, and is the result and ground of Motion. As forming a dimension, i.e. as opposed to the other moments, it is the centre. This return of the line is the circle; it is the Now, Before and After which have closed together in a unity in which these dimensions are indifferent, so that Before is equally After, and vice versa. It is in circular motion that the necessary paralysis of these dimensions is first posited in space. Circular motion is the spatial or subsistent unity of the dimensions of time. The point proceeds towards a place which is its future, and leaves one which is the past; but what it has left behind is at the same time what it has still to reach: it has been already at the place which it is reaching. Its goal is the point which is its past; and this is the truth of time, that the goal is not the future but the past. The motion which relates itself to the centre is not the future but the past. The motion which relates itself to the centre is itself the *plane,* motion as the synthetic whole in which exist its moments, the extinction of the motion in the centre, the motion itself and its relation to its extinction, namely the radii of the circle. But this plane itself moves and becomes the other of itself, a complete space; or the reversion-into-self, the immobile centre, becomes a universal point in which the whole is peacefully absorbed. In other words, it is motion in its essence, motion which has sublated the distinctions of Now, Before and After, its dimensions or its Notion. In the circle, there are in a unity; the circle is the restored Notion of duration, Motion extinguished within itself. There is posited *Mass,* the persistent, the self-consolidated, which exhibits motion as its possibility.

Now this is how we conceive the matter: since there is motion, something moves; but this something which persists is matter. Space and Time are filled with Matter. Space does not conform to its Notion; it is therefore the Notion of Space itself which gives itself existence in Matter. Matter has often been made the starting-point, and Space and Time have then been regarded as forms of it. What is right in this standpoint is that Matter is what is real in Space and Time. But these, being abstract, must present themselves here as the First, and then it must appear that Matter is their truth. Just as there is no Motion without Matter, so too, there is no Matter without Motion. Motion is the process, the transition of Time into Space and of Space into Time: Matter, on the other hand, is the relation of Space and Time as a peaceful identity. Matter is the first reality, existent being-for-self; it is not merely the abstract being, but the positive existence of Space, which, however, excludes other spaces. The point *should* also be exclusive: but because it is only an abstract negation, it does not yet exclude. Matter is exclusive relation-to-self, and is thus the first real limit in Space. What is called the filling of Space and Time, the palpable and tangible, what offers resistance and what, in its being-for-other, is also for itself, all this is attained simply in the unity of Space and Time.

13.

Lotze:

TIME AND PROCESS

Perhaps no other currently unread thinker has had as much influence on contemporary thought, in all of its warring schools, as Rudolf Hermann Lotze (1817–1881). The selection is taken from his chapter "Of Time" (as translated by T. H. Green) in his *Metaphysic* (translated under the editorship of Bernard Bosanquet, who is presumably responsible for the section titles).* This chapter, from the section on "Cosmology," follows after a general defense of the Kantian thesis of the subjectivity of space; here he attacks the notion of empty time (in part by a consideration of the contrary argument presented by Kant in the First Antinomy—which is largely excised), and critically examines the Kantian thesis of the subjectivity of time. In so doing, he emerges as the first thinker to protest the usual description of time in the language of spatial description and to argue for the uniqueness of time itself.

The Psychologist may if he pleases make the gradual development of our ideas of Time the object of his enquiry, though, beyond some obvious considerations which lead to nothing, there is no hope of his arriving at any important result. The Metaphysician has to assume that this development has been so far completed that the Time in which, as a matter of fact, we all live is conceived as one comprehensive form in which all that takes place between things as well as our own actions are comprehended. The only question which he has to ask is how far Time, thus conceived, has any application to the Real or admits of being predicated of it with any significance.

---

*[Rudolf] Hermann Lotze, *Metaphysic*, trans. and ed. Bernard Bosanquet, Second Edition, Oxford, 1887.

138. [Spatial representation of Time.] In regard to the conception I must in the first place protest against the habit, which since the time of Kant has been prevalent with us, of speaking of a direct perception of Time, co-ordinate with that of space and with it forming a connected pair of primary forms of our presentative faculty. On the contrary we have no primary and proper perception of it at all. The character of direct perception attaching to our idea of Time is only obtained by images which are borrowed from Space and which, as soon as we follow them out, prove incapable of exhibiting the characteristics necessary to the thought of Time. We speak of Time as a line, but however large the abstraction which we believe ourselves able to make from the properties of a line in space in order to the subsumption of Time under the more general conception of the line, it must certainly be admitted that the conception of a line involves that of a reality belonging equally to all its elements. Time however does not correspond to this requirement. Thought of as a line, it would only possess one real point, namely, the present. From it would issue two endless but imaginary arms, each having a peculiar distinction from each other and from simple nullity, viz. Past and Future. The distinction between these would not be adequately expressed by the opposition of directions in space. Nor can we stop here. Even though we leave out of sight the relation in which empty Time stands to the occurrences which fall within it, still even in itself it cannot be thought of as at rest. The single real point which the Present constitutes is in a state of change and is ceaselessly passing over to the imaginary points of the Past while its place is taken by the realisation of the next point in the Future.

Hence arises the familiar representation of Time as a stream. All however that in this representation can be mentally pictured originates in recollections of space and leads only to contradictions. We cannot speak of a stream without thinking of a bed of the stream: and in fact, whenever we speak of the stream of Time, there always hovers before us the image of a plain which the stream traverses, but which admits of no further definition. In one point of it we plant ourselves and call it the Present. On one side we represent to ourselves the Future as emerging out of the distance and flowing away into the Past, or conversely—to make the ambiguity of this imagery more manifest—we think of the stream as issuing from the Past and running on into an endless Future. In neither case does the image correspond to the thought. For this never-ending stream is and remains of equal reality throughout, whether as it already flows on the side where we place the future or as it is still flowing on that which stands for the past; and the same reality belongs to it at the moment of its crossing the Present. Nor is it this alone that disturbs us in the use of the image. Even the movement of the stream cannot be presented to the mind's eye except as having a definite celerity, which would compel us to suppose a second Time, in which the former (imaged as a stream) might traverse longer or shorter distances of that unintelligible background.

139. [The conception of empty Time.] Suppose then that we try to dispense with this inappropriate imagery, and consider what empty time must be supposed to be, when it is merely thought of, without the help of images presented to the mind's eye. Nothing is gained by substituting the more abstract conception of a series for the un-available image of a line. It would only be the order of the single moments of Time in relation to each other that this conception would determine. It is, no doubt, involved in the conception of Time that there is a fixed order of its constituents and that the moment m has its place between $m+1$ and $m-1$: also that its advance is uniform and that the interval between two of its members is the sum of the intervals between all the intervening members. Thus we might say that if Time is to be compared with a line at all, it could only be with a straight line. Time itself could not be spoken of as running a circular course. There may be a recurrence of events in it, but this would not be a recurrence if the points of Time, at which what is intrinsically the same event occurs, were not themselves different. So far the conception of a series serves to explain what Time is, but it does so no further. Time does not consist merely in such an order as has been described. That is an order in virtue of which the moment m would have its place eternally between $m+1$ and $m-1$. The characteristic of Time is that this order is traversed and that the vanishing m is constantly replaced by $m+1$, never by $m-1$. Our thoughts thus turn to that motion of our con-sciousness in which it ranges backwards and forwards at pleasure over a series which is in itself at rest. If Time were itself a real existence, it would correspond to this motion, with the qualification of being a

process directed only one way, in which the reality of every stage would be the offspring of the vanished or vanishing reality of the preceding one and itself in turn the cause of its own cessation and of the commencing reality of the next stage. We might fairly acquiesce in an impossibility of learning what the moments properly are at which these occurrences take place and what are the means by which existence is transferred from one to the other. In the first place it would be maintained that Time is something *sui generis,* not to be defined by conceptions proper to other realities: and secondly we know that the demand for explanation must have its limit and may not insist on making a simplest possible occurrence intelligible by constructions which would presuppose one more complex. But without wanting to know how Time is made, it would still be the fact that we were bringing it under the conception of a process and we should have to ask whether to such a conception of it any complete and consistent sense could be given.

We cannot think of a process as occurring in which nothing proceeds, in which the continuation would be indistinguishable from the beginning, the result produced from the condition producing it. This however would be the case with empty Time. Every moment in it would be exactly like every other. While one passed away, another would take its place, without differing from it in anything but its position in the series. This position however it would not itself indicate by a special nature, incompatible with its occupying another. It would only be the consciousness of an observer, who counted the whole series, that would have occasion to distinguish it by the number of places counted before it was reached from other moments with which it might be compared. But if so, there would not in Time itself be any stream, bringing the new into the place of the old. Nor can appeal be made to the view previously stated, according to which even the unchanged duration of a certain state is to be regarded as the product of a process of self-maintenance in constant exercise and thus as a permanent event, though there would be no outward change to make this visible. If this view were applied to Time, it would only help us to the idea of a Time for ever stationary, not flowing at all. A distinction of earlier and later moments in it would only be possible on the basis of the presentation to thought of a second Time, in which we should be compelled to measure the extent in a definite direction of the first Time, the Time supposed to be at rest.

140. [The connexion of 'Time' with events *in* it.] Such is the obscurity which attaches to the notion of a stream of empty Time, when taken by itself. The same obscurity meets us when we enquire into the relation of Time to the things and events which are said to exist and take place *in* it. Here too the convenient preposition only disguises the unintelligibleness of the relation which it has the appearance of enabling us to picture to the mind. There would be no meaning in the statement that things exist in Time if they did not incur some modification by so existing which they would not incur if they were not in Time. What is this? To say that the stream of Time carries them along with it would be a faulty image. Not only would it be impossible to understand how empty Time could exercise such a force as to compel what is not empty but real to a motion not its own. The result too would be something impossible to state. For even supposing the real to be thus carried along by the stream of Time, it would be in just the same condition as before, and thus our expression would contradict what we meant it to convey. For it is not a mere change in the place of something which throughout retains its reality, but an annihilation of one reality and an origination of another, that we mean to signify by the power at once destructive and creative of the stream of Time. But, so understood, this power would involve a greater riddle still. Its work of destruction would be unintelligible in itself, nor would it be possible to conceive the relation between it and that vital power of things to which must be ascribed the greater or less resistance which they offer to their annihilation. Empty Time would be the last thing that could afford an explanation of the selection which we should have to suppose it to exercise in calling events, with all their variety, into existence in a definite order of succession.

But if, aware of this impossibility, we transfer the motive causes of this variety of events to that to which they really belong, viz. to the nature and inner connexion of things, what are we then to make of the independent efflux of empty Time, with which the development of things would have to coincide without any internal necessity of doing so? There would be nothing on this supposition to exclude the adventurous thought that the course of events run counter to time and brings the cause into reality after the effect. In short, whichever way we look at the matter, we see the impossibility of this first familiar view, according to which an empty Time has an existence of its own, either as something permanent or in the way of continual flux, including the sum of events within its bounds, as a power prior to all reality and governed by laws of its own. But the certainty with which we reject this view does not help us to the affirmation of any other.

141. [Kant's view of Time as subjective.] Doubts have indeed been constantly entertained in regard to the reality which is commonly ascribed to Time and many attempts have been made, in the interests of a philosophy of religion, to establish the real existence of a Timeless Being as against changeable phenomena. A more metaphysical basis was first given to this exceptional view by the labours of Kant. He was led by the contradictions, which the supposition of the reality of Time seemed to introduce even into a purely speculative theory of the world, to regard it equally with space as a merely subjective form of our apprehension. This is not the line which I have myself taken. It seemed to me a safer course to show that Time in itself, as we understand it and as we cannot cease to understand it without a complete transformation of the common view, excludes every attribute which would have to be supposed to belong to it if it had an independent existence prior to other existence. On the other hand I cannot find in the assumption of its merely phenomenal reality a summary solution of difficulties, which only *seem* to arise out of the application of Time to the Real but in truth are inseparable from the intrinsic nature of the Real. . . .

142. [Kant's proof that the world has a beginning in Time.]

143. [The endlessness of Time not self-contradictory.]

144. [The past need not be finite because each event is finished.] . . . I would dispose of the difficulty which may be suggested by Kant's expression, that '*up to* any moment of the present an infinite series of Time must have elapsed.' It seems to me improper to represent the Present as the end of this series. It is not the stream of Time of which

the direction can be described by saying that it flows out of the Past, through the Present, into the Future. It is only that which fills Time—the concrete course of the world—that conditions what is contained in the later by what is contained in the earlier. Empty Time itself, if there were such a thing, would take the opposite direction. The Future would pass unceasingly into the Present and this into the Past. In presenting it to ourselves we should have no occasion to seek the source of this stream in the past.

This correction, however, only alters the form of the above objection, which might be repeated thus:—If the Past is held to be infinite, then there must be considered to have elapsed an infinite repetition of that mysterious process, by which every moment of the empty Future becomes the Present, and again pushes the Present before it as a Past. The true ground, however, of the misunderstanding is as follows. Future and Past alike are *not;* but the manner of their not-being is not the same. It is true that in regard to empty Time, though we would fain make this distinction, we cannot show that it obtains, for one point of the elapsed void is exactly like every point of the void that has still to come. But if we think of that course of the world which fills Time, then the Future presents itself to us as that which, for us at any rate, is shapeless, dubious, still to be made, while the Past alone is definitely formed and ready-made. Only the Past—which indeed *is* not, but still has known what Being is—we take as given, and as in a certain way belonging to reality. For every moment of what has been the series of conditions is finished—the conditions which must have been thought or must have been active in order to make it the definite object which it is. This character of what has been, since it belongs to every moment of the past, is shared by the whole past of the world's history, and is transferred by us to empty Time. Thus, as a matter of course, when we speak of an endless Past, we take it to be the same thing as saying that this endless Past *'has been'*. But it is quite a different notion that Kant conveys by his expression 'gone by'. This is the term used of a stream, of which it is already known or assumed that it has an end and exhausts itself in its lapse. But there is nothing in the essential character of the Past to justify this assumption. Nothing is finished but the sum of conditions which made each single moment what it has been. . . .

145. [An infinite series may be given.] . . . I find nothing to prove that in the conception of reality, as such, there is anything to hinder us from recognising, beside finite values which we are forced to admit, the reality of the infinite, as soon as the necessary connexion of our thoughts compels us to do so.

Now for those who consider a stream of empty Time, as such, possible, such a necessity lies not merely in the fact that no moment of this time has any better title than another to form the beginning. On the contrary, try as we may, an independent stream of Time cannot be regarded as anything but a process, in which every smallest part has the condition of its reality in a previous one. There thus arises the necessity of an infinite progression—a necessity equally unavoidable if, on the other hand, we look merely to the real process of events and regard this as producing in some way the illusion of there being an empty Time. It is impossible to think of any first state of the world, which contains the first germ of all the motion that takes place in the world in the form of a still motionless existence, and yet more impossible to suppose a transition out of nothing, by means of which all reality, together with the motive impulses contained in it, first came into being.

146. [Time as a mode of our apprehension.] All these remarks, however, have only been made on supposition that a stream of empty Time is in itself possible. Since we found it impossible, we will try how far we are helped by the opposite view, that Time is merely a subjective way of apprehending what is not in Time. A difficulty is here obvious, which had not to be encountered by the analogous view of Space. Ideas, *ex parte nostra,* do not generally admit of that which forms their content being predicated of them. The idea of Red is not itself red, nor that of choler choleric, nor that of a curve curved. These instances make that clear and credible to us which in itself, notwithstanding, is most strange; the nature, namely of every intellectual presentation, not itself to be that which is presented in it. It may indeed be difficult for the imagination, when the expanse of Space spreading before our perception announces itself so convincingly as present outside us, to regard it as a product, only present for us, of an activity working in us which is itself subject to no conditions of Space.

Still, in the conception of an activity there is nothing to make us look for extension in Space on the part of the activity itself as a condition of its activity. On the contrary, had we believed that the impressions of Space in our inner man could themselves have position in Space, we should have been obliged to seek out a new activity of observation which had converted this inner condition into a knowledge of it, and to look to this activity for that strange apprehension of what is in Space which must do its work without being in Space itself.

If, on the other hand, we try to speak in a similar way of a timeless presentation of what is in time, the attempt seems to break down. The thought that Time is only a form or product of our presentative susceptibility, cannot take away from the presentation itself the character of an activity or at least of an event, and an event seems inconceivable without presupposition of a lapse of time, of which the end is distinguishable from the beginning. Thus Time, unlike Space, is not merely a product of the soul's activity, but at the same time the condition of the exercise of the activity by which Time itself as a product is said to have been obtained, and the presentation to consciousness of any change seems impossible without the corresponding real change on the part of the presenting mind. Now it must be borne in mind that in no case could Time be a subjective form of apprehension in such a sense as that the process of events, which we present to ourselves in it, should be itself opposed to the form of apprehension as being of a completely alien nature. Whatever basis in the way of timeless reality we may be disposed to supply to phenomena in Time, it must at any rate be such that its own nature and constitution remain translateable into forms of Time. To this hidden timeless reality, it may be suggested, that activity of thought would itself belong, of which the product in our consciousness would be that course of occurrences and of our ideas which is seemingly in Time. Of it, and by consequence of every activity as such, it must be sought to show, according to the view which takes Time to be merely our form of apprehension, that while not itself running a course in a time already present, it may yet present itself to sense in its products as running such a course. Let us pursue the consideration by which it may be attempted to vindicate this paradoxical notion.

147. [Empty Time not even a condition of Becoming.] No one will maintain that the stream of empty Time brings forth events in the sense of being that which determines their character and the succession of the various series of them. It would be admitted that all this is decided by the actual inner connexion of things. . . .

For the present my concern is to show that for the very process of Becoming in question the mere lapse of Time can afford no means, any possible application of which could be necessary to bringing it about. . . . For a reality, which was to take account of the lapse of Time in order to direct its becoming accordingly, there would be needed the constant summing of the impressions received by it from another real process, by means of which it itself or its own condition had been so changed as to be able to serve as indicator of the length of Time elapsed. The conclusion plainly is that a process of becoming, *B*, which required a lapse of time in order to come about, must have already traversed in itself a succession of different stages, in order to feel in that succession the lengths of the periods according to which it is supposed to direct itself, and which it is supposed to employ for the purpose of effecting the transition from one stage to another.

148. [Time as an abstraction from occurrence.] These considerations do not lead us at once to the end of our task. For the present I may put their result, which I shall not again discuss, as follows. It is quite unallowable to put the system of definite causes and effects, which gives its character to any occurrence, on one side and on the other side to suppose a stream of empty Time, and then to throw the definitely characterised event into the stream in expectation that its fabric of simultaneous conditions will in the fluidity of this stream melt into a succession, in which each of the graduated relations of dependence will find its appropriate point of time and the period of its manifestation. It is only in the actual content of what happens, not in a form present outside it into which it may fall, that the reason can be found for its elements being related to each other in an order of succession, and at the same time for the times at which they succeed each other.

The other view therefore begins to press itself upon us—the view that it is not Time that precedes the process of Becoming and Activity, but this that precedes Time and brings forth from itself either the real course of Time or the appearance in us of there being such a thing.

The constant contradiction to this reversal of the habitual way of looking at the matter which our imagination would present, we could no more get rid of than we could of the habit of saying that the sun rises and sets. What we might hope to do would be to understand one illusion as well as the other. It is also our habit to speak of general laws, standing outside things and occurrences and regulating their course; yet we have been forced to the conviction that these have no reality except in the various particular cases of their application. Only that which happens and acts in determinate forms is the real. The general law is the product of our comparison of the various cases. After we have discovered it, it appears to us as the first, and the realities, out of the consideration of which it arose, as dependent on its antecedence. In just the same way, after the manifold web of occurrence has in countless instances assumed for us forms of succession in Time, we misunderstand the general character of these forms, which results from our comparison of them—the empty flowing Time—and take it for a condition antecedent, to which the occurrence of events must adjust itself in order to be possible. That we are mistaken in so doing and that the operation of such a condition is unthinkable—this 'reductio ad impossibile,' which I have sought to make out, is, it must be admitted, the only thing which can be opposed to this unavoidable habit of our mental vision.

149. [Time as an infinite whole is subjective.] The positive view, which we found emerging in place of the illusion rejected, is still ambiguous. Is it a real Time that the process of events, in its process, produces or only the appearance of Time in us? In answering this question we cannot simply affirm either of the alternatives. One thing is certainly clear, that the production of Time must be a production *sui generis*. Time does not remain as a realised product behind the process that produces it. As little does it lie before that process as a material out of which the process can constantly complete itself. Past and future are *not*, and the representation of them both as dimensions of Time is in fact but an aritficial projection, which takes place only for our mind's eye, of the unreal upon the plane which we think of as containing the world's real state of existence.

Undoubtedly therefore Time, conceived as an infinite whole with its two opposite extensions, is but a subjective presentation to our mind's eye; or rather it is an attempt by means of images borrowed from space, to render so presentable a thought which we entertain as to the inner dependence of the individual constituents of that which happens. What we call Past, we regard primarily as the condition '*sine qua non*' of the Present, and in the Present we see the necessary condition of the Future. This one-sided relation of dependence, abstracted from the content so related and extended over all cases which it in its nature admits of, leads to the idea of an infinite Time, in which every point of the Past forms the point of transition to Present and Future, but no point of Present or Future forms a point of transition to the Past. That this process must appear infinite scarcely needs to be pointed out. The condition of that which has a definite character can never lie in a complete absence of such character. Every state of facts, accordingly, of which we might think for a moment as the beginning of reality, would immediately appear to us either as a continuation of a previous like state of facts, or as a product of one unlike; and in like manner every state of facts momentarily assumed to be an end would appear as the condition of the continuance of the same state of facts, or in turn as the beginning of a new one. If finally the course of the world were thought of as a history, which really had a beginning and end, still beyond both alike we should present to ourselves the infinite void of a Past and Future, just as two straight lines in space which cut each other at the limit of the real, still demand an empty extension beyond in which they may again diverge.

150. [No mere systematic relation explains 'Present' and 'Past'.] It will be felt, however, that we have not yet reached the end of our doubts. It will be maintained that though the process of Becoming does indeed make no abiding Time, it yet does really bring into being or include the course of Time, by means of which the various parts of the content of what happens, standing to each other in the relation of dependence described above, having been at first only something future, acquire *seriatim* the character of the Present and the Past. If we chose to confine ourselves simply to highly developed thought, and to regard the dimensions of Time merely as expressions for conditionedness or the power of conditioning, then the whole content of the world would again change into a motionless systematic whole, and everything would depend on the position which a consciousness capable of viewing the whole might please to take up facing, so to

speak, some one part of it, *m*. From this point of departure, *m*, the contemplator would reckon everything as belonging to the Past, $m-1$, in which he had recognised the conditions that make the content of *m* what it is, while he would assign to the Future, $m+1$, all the consequences which the necessities of thought compelled him to draw from it: and this assignment of names would change according as *m* or *n* might be made the point of departure for this judgment. This however does not represent the real state of the case. This capacity of tracing out the connexion of occurrences in both directions—forwards and backwards—would only be possible to a consciousness standing outside the completed course of the world. It belongs to us only in relation to the past, so far as the past has become known to us through tradition. Immediate experience is confined to a definite range, and neither does the recollection of the past reproduce for experience its actual duration, nor does the sure foresight of the future, in the few cases where it is possible, take place for experience of the real occurrence of the foreseen event.

What then is the proper meaning of the Reality, which in this connexion of thought we ascribe only to the Present? Or conversely, what constitutes this character of the present, which we suppose to belong successively in unalterable series of the events of which each has its cause in the other, and to be equivalent to reality? I will not attempt to prepare the way for an answer to this question, or to lead up to it as a discovery. I will merely state what seems to me the only possible answer to it. It is not the mere fact that they happen which attaches *this* character to the content of events. On the contrary the import of the statement that they happen is only explained by the expression 'the Present,' in which Language aptly makes us aware of the necessity of a subject, in relation to which alone the thinkable content of the world's course can be distinguished either as merely thinkable and absent on the one hand, or on the other as real and present. To explain this, however, I am obliged to go into detail to an extent for which I must ask indulgence and patience. . . .

151. [Indication of 'Present' to a Subject.] . . . The question indeed as to the foundation of this faculty of distinguishing a represented absent object from one experienced as present is a question upon which any psychological or physiological explanation may be thank-

fully accepted in its place. Here however it would be useless. What we
are now concerned with is merely the fact itself, that we are able to
make this distinction and to represent to ourselves what we *have*
experienced without experiencing it again. This alone renders it possible for ideas of a proper succession to be developed in us, in which
the member n has a different kind of reality from $n + 1$. It would have
been more convenient to arrive at this result otherwise than by this
tedious process of development. I thought the process indispensable,
however, because it leads to some peculiar deductions, which require
further patient consideration.

152. [Subjective Time need not make the Past still exist.] . . . There
would not indeed on our view be that kind of past into which the
conditioning stage of development would be supposed to vanish,
instead of illegitimately continuing in the present alongside of the
consequence conditioned by it—that consequence to which it ought to
have transferred the exclusive possession of the quality of being present. The histories of the past would not continue to live in this
present, petrified in each of their phases, alongside of that which
further proceeded to happen in the course of things. It would not be
the case that s_1 really existed earlier than s_2 and strangely continued
along with it, but rather that it had reality only so far as it was
contained in s_2 and was presented by the latter to itself as earlier. It
will be with Time as with Space. As we saw, there is no such thing as a
Space in which things are supposed to take their places. The case
rather is that in spiritual beings there is formed the idea of an extension, in which they themselves seem to have their lot and in which they
spatially present to themselves their non-spatial relations to each
other. In like manner there is no real Time in which occurrences run
their course, but in the single elements of the Universe which are
capable of a limited knowledge there develops itself the idea of a
Time in which they assign themselves their position in relation to their
more remote or nearer conditions as to what is more or less long past,
and in relation to their more remote or nearer consequences as to a
future that is to be looked for more or less late.

It is not out of wantonness that I have gone so far in delineating this
paradoxical way of looking at things. It is what we must come to if we
wish to put clearly before us the view of the merely subjective validity

of Time in relation to a timeless reality. It is vexatious to listen to the mere asseveration of this antithesis without the question being asked whether, when adopted, it intrinsically admits of being in any way carried out, and whether it would be a sufficient guide to the understanding of that experience from which we all start. The description which has been given will be enough to raise a doubt whether the latter is the case. The reasons for this doubt, however, are not all of equal value. In regard to them again, while passing to the consideration of this contradiction, I must ask to be allowed some detail.

153. [Absence of real succession conceivable by approximation.] In order to find a point of departure in what is familiar, I will first repeat the objection which will always recur. Pointing to the external world the objector will enquire—'Is it not then the case that something is forever happening? Do not things change? Do they not operate on each other? And is all this imaginable without a lapse of time?' Imaginable it certainly is not, and we have never maintained that it is so. But in what relation do the lapse of Time and this happening stand to each other, which might enable us to maintain the correctness of this imagination of ours? That it is only in what is contained in a sufficient cause, G, that there lies a necessity for the consequence, F—that the necessity, if otherwise lacking, could not supervene through lapse of a time, T—this we found obviously true. It was admitted also that, G being given, it would neither be intelligible where the hindrance should come from which should retard its transition into F, nor how the lapse of empty Time could overcome that hindrance. Thus constrained to confess that our habit of thinking the effect as *after* the cause does not point to anything which in the things themselves contributes to the production of the effect, what other conclusion can we draw than this, that succession in Time is something which our mode of apprehension alone introduces into things—introduces in a way absolutely inevitable for us, so that our thought about things remains constantly in contradiction with our habit of presenting them to the mind's eye?

One may attempt to make this thought clear to oneself by gradual approximation. To a definite period of Time it is our habit in common apprehension to ascribe a certain absolute quantity. If we ask ourselves, however, how long a century or an hour properly lasts we at once recollect that the time filled by one series of events we always measure simply according to its relation to another series, with the ends of which those of the first series do or do not coincide. Our ordinary impression of the duration of periods of time is itself the uncertain result of such a comparison, in which we are not clearly conscious of the standard of our measurement. Hence the same period may appear long or short in memory. The multiplicity of the events contained in it gives it greater extent for the imagination. Poverty of events makes it shrink into nothing. It has itself no extensive quantity which is properly its own. Therefore no hindrance meets us in the attempt to suppose as short a time as we will for the collective course of events. However small we think it, still it is not in it but in the dependence of events on each other that the reason lies of the order in which events occur; and the entire history which fills centuries admits of being presented in a similar image, as condensed into an infinitely small space of Time through proportional diminuition of all dimensions.

With this admission however it will be thought necessary to come to a stop. However small, it will be said, still this differential of Time must contain a distinction of before and after, and thus a lapse, though one infinitely small. But we want to know exactly why. Undoubtedly the transition to a moment completely without extension would deprive History of the character of succession in Time; but then our question is just this, whether the real needed this succession on its own part in order to its appearance as successive to us. And in regard to this we must consantly repeat what has been already said; that neither could the order of events be constituted by Time, if it were not determined by the inner connexion of things, nor is it intelligible how Time should begin to bring that which already has a sufficient cause to reality, if that reality is still lacking to it. On the other hand, we believe that we do understand how a presentative faculty such as to derive from its own nature the habit of viewing the world as in time, should find occasion in the inner connexion between the constituents of that world, as conditioning and conditioned by each other, to treat its parts as following each other in a definite order and as assuming lengths—definite in relation to each other but, apart from such relation, quite arbitrary—of this imagined Time. Thus even upon this method, by help of the idea of an infinitely small moment, we should have mastered the thought of a complete timelessness on the part of what fills the world. For in that case we should certainly not go out of our way to think of that extension of time, within which this moment would seem of a vanishing smallness, and so bring on the world the reproach of a short and fleeting existence, as compared with the duration which expansion into infinite Time would have *promised it.*

154. [Even thought cannot consist of a mere succession.] After all, it will be objected, we have not yet touched the proper difficulty. If all that we had to take account of were an external course of the world, then it would indeed cost us little effort to regard all that it contains as timeless, and to hold that it is only in relation to our way of looking at it that it unfolds itself into a succession. But the motion, which we should thus have excluded from the outer world, would so much the more surely have been transferred into our Thought, which, on the given supposition, must itself pass from one of the elements which constitute the world to another, in order to make them successive for its contemplation. For the unfolding, by which what is in itself time-less comes to be in time, cannot take place in us without a real lapse of Time; the appearance of succession cannot take place without a succession of images in consciousness, nor an apparent transition of a into b without the real transition which we should in such a case effect from the image of a to that of b.

But convincing as these assertions are, they are as far from con-taining the whole truth. On the contrary, without the addition of something further, the doctrine which they allege would be fatal to the possibility of that which it is sought to establish. If the idea of the later b in fact merely followed on that of the earlier a, then a change of ideas would indeed take place, but there would still be no idea of this change. There would be a lapse of time, but not an appearance of such change to any one. In order to a comparison in which b shall be known as the later it is necessary in turn that the two presentations of a and b should be objects, throughout simultaneous, of a relating knowledge, which, itself completely indivisible, holds them together in a single indivisible act. If there is a belief on the part of this knowledge that it passes from one of its related points to another, it will not itself form this idea of its transition through the mere fact of the transition taking place. In order that the idea may be possible, the points with which its course severally begins and ends, being separate in time, must again be apprehended in a single picture by the mind as the limits between which that course lies. All ideas of a course, a distance, a transition —all, in short, which contain a comparison of several elements and the relation between them—can as such only be thought of as products of a timelessly comprehending knowledge. They would all be impossible, if the presentative act itself were wholly reducible to that succession in Time which it regards as the peculiarity of the objects presented by it. Nay if we go further and make the provisional admission that we really had the idea of *a* before we had that of *b,* still *a* can only be known as the earlier on being held together with *b* in an indivisible act of comparison. It is at this moment, at which *a* is no longer the earlier nor *b* the later that for knowledge *a* appears as the earlier and *b* as the later. In assigning these determinate places, however, to the two, the soul can only be guided by some sort of qualitative differences in their content—by temporal signs, if we like to say so, corresponding to the local signs in accordance with which the *non*-spatial consciousness expands its impressions into a system of spatial juxtaposition.

Such could not but be the state of the case even *if* there were a lapse of Time in which our ideas successively formed themselves. The real lapse of Time would not, immediately as such, be a sufficient cause to that which combines and knows of the succession in Time which it presents to itself. It would be so only mediately through signs derived by each constituent element of the world from that place in the order of Time into which it had fallen. But such various signs could not be stamped on the various elements by empty time, even though it elapsed, since one of its elements is exactly like every other. They could only be derived from the peculiar manner in which each element is inwoven into the texture of conditions which determine the content of the world. But just for that reason there was no need of a real sequence in Time to annex them to our ideas as characteristic incidental distinctions. Thus it would certainly be possible for a presentative consciousness, without any need of Time, to be led by means of temporal signs, which in their turn need not have their origin in Time, to arrange its several objects in an apparent succession in the way of Time.

155. [But future cannot become present without succession.] I am painfully aware that my reader's patience must be nearly exhausted. Granted, he will say, that in every single case in which a relation or comparison is instituted this timeless faculty of knowing is active: it remains none the less true that numberless repetitions of such action really succeed each other. Yesterday our timeless faculty of knowledge was employed in presenting the succession of a and b, to-day it presents that from c to d. There are thus, it would seem, many instances of Timeless occurrence which really succeed each other in Time. I venture, however, once again to ask, Whence are we to know that this is so? And if it were so, in what way could we know of it? That consciousness, to which the comparison made yesterday appears as earlier than that made to-day, must yet be the consciousness which we have to-day, not that which may have been yesterday and have vanished in the course of Time. That which appears to us as of yesterday cannot so appear to us because it is *not* in our consciousness, but because it is in it; while at the same time it is somehow so qualitatively determined, that our mental vision can assign in its place only in the past branch of apparent Time.

I will allow, however, that this last reply yields no result. The Past indeed, of which we believe ourselves already to have had living experience, one may try to exhibit as a system of things which has never run a course in Time, and which only consciousness, for its own benefit, expands into a preceding history in Time. But how then would the case stand with the Future, which we suppose ourselves still on the way to meet? Let s_3, according to the symbols previously used, stand for this Ego, which s_2 and s_1 never really preceded but always seem to have preceded, what then is s_4 which s_3 in turn will thus seem to have preceded? What could prevent s_3 from being conscious also of s_4, its own future, if the temporal signs which teach us to assign to simple impressions their position in Time, depended only on the systematic position which belongs to their causes in the complex conditions of a timeless universe? It may be that the content of s_4 which follows systematically upon s_3, is not determined by the conditions, which are contained in s_3 and previously in s_2 and s_1, but jointly by others, resting on the states of other beings which do not cross those of S till a later stage of the system. For that reason s, might be obscure to s_3 and this might constitute the temporal character which gives it in the consciousness of s_3 and this might constitute the temporal character which gives it in the consciousness of s_3 the stamp of something future. But if this were the case, the process would have to stop at this point. It would only be for another being, s_4, that what was Future to s_3 could, owing to its later place in the system, be present. On the other hand in a timeless system there would be no possibility of the change by means of which s_3 would be moved out of its place into that of s_4: if to one and the same yet this would be necessay consciousness that is to become Present which was previously Future to it. If one and the same timeless being by its timeless activity of intellectual preservation gives to one constituent of its existence the Past character of a recollection, to another the significance of the Present, to a third unknown element that of the Future, it could never, it it is to be really timeless, change this distribution of characters. The recollection could never have been Present, the Present could never become Past and the Future would have to remain without change the same unknown obscurity. But if there is a change in this distribution of light; if it is the case that the indefinite burden of the Future gradually enters the presence of living experience and passes through it into the other absence of the Past; and finally if it is impossible for the activity of intellectual presentation to alter this order of sequence; then it follows necessarily that not merely this activity, but the content of the reality which it presents to itself, is involved in a succession of determinate direction.

This being so, we must finally decide as follows: Time, as a whole, is without doubt merely a creation of our presentative intellect. It neither is permanent nor does it elapse. It is but the fantastic image which we seek, rather than are able, to project before the mind's eye, when we think of the lapse of time as extended to all the points of relation which it admits of *ad infinitum,* and at the same time make abstraction of the content of these points of relation. But the lapse of events in time we do not eliminate from reality, and we reckon it a perfectly hopeless undertaking to regard even the idea of this lapse as an *a priori* merely subjective form of apprehension, which developes itself within a timeless reality, in the consciousness of spiritual beings.

156. [Empty 'Time' Subjective, but succession inseparable from Reality.] Thus, at the end of a long and troublesome journey, we come back, as it will certainly appear, to complete agreement with the ordinary view. I fear however that remnants of an error still survive

which call for a special attack—remnants of an error with which we are already familiar and which have here needed to be dealt with only in a new form, viz. the disintegration of the real into its content and its reality. We are unavoidably led by our comparison of the manifold facts given to us to the separation of that on the one hand which distinguishes one real object from another—its peculiar content which our thought can fix in abstraction from its existence—and on the other hand of that in which every thing real resembles every other—the reality itself which, as we fancy, has been imparted to it. For this is just what we go on to imagine—that this separation, achieved in our thoughts, represents a metaphysical history; I do not mean a history which has been completed once for all, but one which perpetually completes itself; a real relation, that is to say, of such a kind that that content, apart from its reality, is something to which this reality comes to belong. The prevalence of this error is evidenced by the abundant use which philosophy, not least since the time of Kant, has made of the conception of a 'Position,' which meeting with the thinkable content establishes its reality. In an earlier part of this work we declared ourselves against this mistake. We were convinced that it was simply unmeaning to speak of being as a kind of placing which may simply supervene upon that intelligible content of a thing, without changing anything in that content or essence or entering as a condition into its completeness. As separate from the energy of action and passion, in which we found the real being of the thing to consist, it was impossible even to think of that essence, impossible to think of it as that to which this reality of action and passion comes from without, as if it had been already, in complete rest, the same essence which it is under this motion.

It is the same impossible separation that we have here once again, in consideration of the prevalence of the misunderstanding, carefully pursued to its consequences in the form of the severance of the thing which happens from its happening. It was thus that we were led to the experiment of seeking the essence [1] of what happens—that by which

---

1. [This is still 'the content'—'that which distinguishes one real object from another.' A verbal difficulty is caused by the distinction being here, *per accidens*, between the actual world and an imaginary world, so that but for the context we might take 'essence' to be used in just the opposite sense to that explained a few lines before, and to refer to that which distinguishes what is real from what is unreal.]

the actual history of the world is distinguished from another which might happen but does not—in a complex system of relations of dependence on the part of a timeless content of thought; while the motion in this system, which alone constitutes the process of becoming and happening, was regarded as a mode of setting it forth which might simply be imposed on this essential matter, or on the other hand, might be wanting to it without changing the distinctive character of the essence. We could not help noticing, indeed, the great difference between reality and that system of intelligible contents. In the latter the reason includes its consequence as eternally coexisting with it. In the former the earlier state of things ceases to be in causing the later. Then began the attempts to understand this succession, which imposes itself like an alien fate on the system in its articulation. They were all in vain. When once the lapse of empty time and the timeless content had been detached from each other, nothing could enable the set nature of the latter to resolve itself into a constant flux in the former. It was clear that in this separation we had forgotten something which forced that content—involving as it did, if it moved, the basis of an order of time—to pass in fact into such a state of motion. I will not suppose that crudest attempt to be made at supplying the necessary complement—the reference to a power standing outside the world which laid hold on the eternal content of things, as a store, of material, in order to dispose its elements in Time in such a way as their inner order, to which it looked as a pattern, directed it to do. Let us rather adopt the view and in the content itself lies the impulse after realisation which makes its manifold members issue from each other. Still, even on that view it would be a mistake, as I hold, to think of the measure and kind of that timeless conditionedness, which might obtain between two elements of the world's content, as the antecedent cause which commanded or forbade that operative impulse to elicit the one element from the other. What I am here advancing is only a further application of a thought which I have previously expressed. Every relation, I have said, exists only in the spirit of the person instituting the relation and for him. When we believe that we find it in things themselves, it is in every case more than a mere relation: it is itself already an efficient process instead of being merely preliminary to effects.

On the same principle we say—It is not the case that there is *first* a relation of unchanging conditionedness between the elements of the world, and that afterwards in accordance with this relation the productive operation, even though it may not come from without but may lie in the things themselves, has to direct itself in order to give reality to legitimate consequences and avoid those that are illegitimate. On the contrary first and alone is there this full living operation itself. Then, when we compare its acts, we are able in thought and abstraction to present to ourselves the constant *modus agendi,* self-determined, which in all its manifestations has remained the same. This abstraction made, we can subordinate each single product of the operation, as we look backward, to this mode of procedure as to an ordaining *prius* and regard it as determined by conditions which are in truth only the ordinary habit of this operation itself. This process of comparison and abstraction leads us in one direction to the idea of general laws of nature, which are first valid and to which there then comes a word, which submits itself to them. In another direction it leads to the supposition of an empty Time, in which the series of occurrences succeed each other and which, in the character of an antecedent *conditio sine qua non,* makes all operation possible. But this last way of looking at the matter we have found as untenable as would be the attempt to represent velocities as prior to motions (somewhat as if each motion had to choose an existing velocity), and to interpret the common expression, according to which the motion of a body *assumes* this or that velocity, as signifying an actual fact; whereas in truth the motion is nothing but the velocity as following a definite direction.

In this sense we may find more correctness in the expressions that may be often heard, according to which it is not Time that is the condition of the operation of things, but this operation that produces Time. Only what it brings forth, while it takes its course, is not an actually existing Time as an abiding product, somehow existing or flowing or influencing things, but only the so-called 'vision' of this Time in the comparing consciousness. Of this—the empty total image of that order in which we place events as a series—it is thus true that it is only a subjective form of apprehension; while of the succession belonging to that operation itself, which makes this arrangement of events possible, the reverse is true, namely that it is the most proper nature of the real.

157. [Existence of Past and Future.] I should not be surprised if the view which I thus put forward met with an invincible resistance from the imagination. The unconquerable habit, which will see nothing wonderful in the primary grounds of things but insists on explaining them after the pattern of the latest effects which they alone render possible, must here at last confess to being confronted by a riddle which cannot be thought out. What exactly happens—such is the question which this habit will prompt—when the operation is at work or when the succession takes place, which is said to be characteristic of the operative process? How does it come to pass—what makes it come to pass—that the reality of one state of things ceases, and that of another begins? What process is it that constitutes what we call perishing, or transition into not-being, and in what other different process consists origin or becoming?

That these questions are unanswerable—that they arise out of the wish to supply a *prius* to what is first in the world—this I need not now repeat: but in this connexion they have a much more serious background than elsewhere, for here they are ever anew excited by the obscure pressure of an unintelligibility, which in ordinary thinking we are apt somewhat carelessly to overlook. We lightly repeat the words 'bygones are bygones'; are we quite conscious of their gravity? The teeming Past, has it really ceased to be at all? Is it quite broken off from connexion with the world and in no way preserved for it? The history of the world, is it reduced to the infinitely thin, for ever changing, strip of light which forms the Present, wavering between a darkness of the Past, which is done with and no longer anything at all, and a darkness of the Future, which is also nothing? Even in thus expressing these questions, I am ever again yielding to that imaginative tendency, which seeks to soften the 'monstrum infandum' which they contain. For these two abysses of obscurity, however formless and empty, would still be there. They would always form an environment which in its unknown within would still afford a kind of local habitation for the not-being, into which it might have disappeared or from which it might come forth. But let any one try to dispense with these images and to banish from thought even the two voids, which limit being: he will then feel how impossible it is to get along with the naked antithesis of being and not-being, and how unconquerable is the demand to be able to think even of that which is not as some unaccountable constituent of the real.

Therefore it is that we speak of the Past and of the Future, covering under this spatial image the need of letting nothing slip completely from the larger whole of reality, though it belong not to the more limited reality of the Present. For the same reason even those unanswerable questions as to the origin of Becoming had their meaning. So long as the abyss from which reality draws its continuation, and that other abyss into which it lets the precedent pass away, shut in that which is on each side, so long there may still be a certain law, valid for the whole realm of this heterogeneous system, according to the determinations of which that change takes place, which on the other hand becomes unthinkable to us, if it is a change from nothing to being and from being to nothing. Therefore, though we were obliged to give up the hopeless attempt to regard the course of events in Time merely as an appearance, which forms itself within a system of timeless reality, we yet understand the motives of the efforts which are ever being renewed to include the real process of becoming within the compass of an abiding reality. They will not, however, attain their object, unless the reality, which is greater than our thought, vouchsafes us a Perception, which, by showing us the mode of solution, at the same time persuades us of the solubility of this riddle. I abstain at present from saying more on the subject. The ground afforded by the philosophy of religion, on which efforts of this kind have commonly begun, is also that on which alone it is possible for them to be continued.

14.

Bergson:

TIME AS LIVED DURATION

The first book by Henri Bergson (1859–1941) was published in 1889; it enunciated the preoccupation with the import of experiential time which was to mark his philosophic career. A fully developed statement of his temporalism is to be found in one of his last, and least known, works a sympathetically critical examination of relativity theory, entitled *Duration and Simultaneity.** Originally published in French in 1922, the title juxtaposes 'duration', Descartes' and Bergson's name for 'real time' with 'simultaneity', the temporal term Leibniz had used to designate spatial relationships. The selection comprises the entire third chapter ("Concerning the Nature of Time") in which he spelled out his developed theory of time and his reasons for believing that modern science is essentially unable to come to terms with experiential time.

Succession and consciousness; origin of the idea of a universal time; real duration and measurable time; concerning the immediately perceived simultaneity: simultaneity of flow and of the instant; concerning the simultaneity indicated by clocks; unfolding time; unfolding time and the fourth dimension; how to recognize real time

There is no doubt but that for us time is at first identical with the continuity of our inner life. What is this continuity? That of a flow or passage, but a self-sufficient flow or passage, the flow not implying a thing that flows, and the passing not presupposing states through which we pass; the *thing* and the *state* are only artificially taken

* From Henri Bergson; *Duration and Simultaneity*, trans. Leon Jacobson. Copyright (C) 1965 by The Bobbs-Merrill Company, Inc., reprinted by permission of the publisher.

snapshots of the transition; and this transition, all that is naturally experienced, is duration itself. It is memory, but not personal memory, external to what it retains, distinct from a past whose preservation it assures; it is a memory without change itself, a memory that prolongs the before into the after, keeping them from being mere snapshots appearing and disappearing in a present ceaselessly reborn. A melody to which we listen with our eyes closed, heeding it alone, comes close to coinciding with this time which is the very fluidity of our inner life; but it still has too many qualities, too much definition, and we must first efface the difference among the sounds, then do away with the distinctive features of sound itself, retaining of it only the continuation of what precedes into what follows and the uninterrupted transition, multiplicity without divisibility and succession without separation, in order finally to rediscover basic time. Such is immediately perceived duration, without which we would have no idea of time.

How do we pass from this inner time to the time of things? We perceive the physical world and this perception appears, rightly or wrongly, to be inside and outside us at one and the other same time; in one way, it is a state of consciousness; in another, a surface film of matter in which perceiver and perceived coincide. To each moment of our inner life there thus corresponds a moment of our body and of all environing matter that is "simultaneous" with it; this matter then seems to participate in our conscious duration.[1] Gradually, we extend this duration to the whole physical world, because we see no reason to limit it to the immediate vicinity of our body. The universe seems to us to form a single whole; and, if the part that is around us endures in our manner, the same must hold, we think, for that part by which it, in turn, is surrounded, and so on indefinitely. Thus is born the idea of a duration of the universe, that is to say, of an impersonal consciousness

1. For the development of the views presented here, see *Essai sur les données immédiates de la conscience (Time and Free Will)* (Paris: F. Alcan, 1889), mainly Chaps. II and III; *Matière et mémoire (Matter and Memory)*, Chaps. I and IV; *L'Evolution créatrice (Creative Evolution)*, *passim*. Cf. *Introduction à la métaphysique (Introduction to Metaphysics);* and *La perception du changement (The Perception of Change)* (Oxford: Oxford University Press, 1911). [The last-named title was republished in Paris in 1934, along with several other essays, under the title *La pensée et le mouvant* and was translated as *The Creative Mind.*]

that is the link among all individual consciousnesses, as between these consciousnesses and the rest of nature.[2] Such a consciousness would grasp, in a single, instantaneous perception, multiple events lying at different points in space; simultaneity would be precisely the possibility of two or more events entering within a single, instantaneous perception. What is true and what illusory, in this way of seeing things? What matters at the moment is not allotting it shares of truth or error but seeing clearly where experience ends and theory begins. There is no doubt that our consciousness feels itself enduring, that our perception plays a part in our consciousness, and that something of our body and environing matter enters into our perception.[3] Thus, our duration and a certain felt, lived participation of our physical surroundings in this inner duration are facts of experience. But, in the first place, the nature of this participation is unknown, as we once demonstrated; it may relate to a property that things outside us have, without themselves enduring, of manifesting themselves in our duration in so far as they act upon us, and of thus scanning or staking out the course of our conscious life.[4] Next, in assuming that this environment "endures," there is no strict proof that we may find the same duration again when we change our surroundings; different durations, differently rhythmed, might coexist. We once advanced a theory of that kind with regard to living species. We distinguished durations of higher and lower tension, characteristic of different levels of consciousness, ranging over the animal kingdom. Still, we did not perceive then, nor do we see even today any reason for extending this theory of a multiplicity of durations to the physical universe. We had left open the question of whether or not the universe was divisible into independent worlds; we were sufficiently occupied with our own world and the particular impetus that life manifests there. But if we had to decide the question, we would, in our present state of knowledge, favor the hypothesis of a physical time that is one and universal. This is only a hypothesis, but it is based upon an argument by analogy that we must regard as conclusive as long as we are offered nothing more satisfactory. We believe this scarcely conscious argument

2. Cf. those of our works we have just cited.
3. See *Matière et mémoire (Matter and Memory)*, Chap. I.
4. Cf. *Essai sur les données immédiates de la conscience (Time and Free Will)*, especially pp. 82ff.

reduces to the following: All human consciousnesses are of like nature, perceive in the same way, keep in step, as it were, and live the same duration. But, nothing prevents us from imagining as many human consciousnesses as we please, widely scattered through the whole universe, but brought close enough to one another for any two consecutive ones, taken at random, to overlap the fringes of their fields of outer experience. Each of these two outer experiences participates in the duration of each of the two consciousnesses. And, since the two consciousnesses have the same rhythm of duration, so must the two experiences. But the two experiences have a part in common. Through this connecting link, then, they are reunited in a single experience, unfolding in a single duration which will be, at will, that of either of the two consciousnesses. Since the same argument can be repeated step by step, a single duration will gather up the events of the whole physical world along its way; and we shall then be able to eliminate the human consciousnesses that we had at first laid out at wide intervals like so many relays for the motion of our thought; there will be nothing more than an impersonal time in which all things will pass. In thus formulating humanity's belief, we are perhaps putting more precision into it than is proper. Each of us is generally content with indefinitely enlarging, by a vague effort of imagination, his immediate physical environment, which, being perceived by him, participates in the duration of his consciousness. But as soon as this effort is precisely stated, as soon as we seek to justify it, we catch ourselves doubling and multiplying our consciousness, transporting it to the extreme limits of our outer experience, then, to the edge of the new field of experience that it has thus disclosed, and so on indefinitely—they are really multiple consciousnesses sprung from ours, similar to ours, which we entrust with forging a chain across the immensity of the universe and with attesting, through the identity of their inner durations and the contiguity of their outer experiences, the singleness of an impersonal time. Such is the hypothesis of common sense. We maintain that it could as readily be considered Einstein's and that the theory of relativity was, if anything, meant to bear out the idea of a time common to all things. This idea, hypothetical in any case, even appears to us to take on special rigor and consistency in the theory of relativity, correctly understood. Such is the conclusion that will emerge from our work of analysis. But that is not the important point at the moment. Let us put aside the question of a single time. What we wish to establish is that we cannot speak of a reality that endures without inserting consciousness into it. The metaphysician will have a universal consciousness intervene directly. Common sense will vaguely ponder it. The mathematician, it is true, will not have to occupy himself with it, since he is concerned with the measurement of things, not their nature. But if he were to wonder what he was measuring, if he were to fix his attention upon time itself, he would necessarily picture succession, and therefore a before and after, and consequently a bridge between the two (otherwise, there would be only one of the two, a mere snapshot); but, once again, it is impossible to imagine or conceive a connecting link between the before and after without an element of memory and, consequently, of consciousness.

We may perhaps feel averse to the use of the word "consciousness" if an anthropomorphic sense is attached to it. But to imagine a thing that endures, there is no need to take one's own memory and transport it, even attenuated, into the interior of the thing. However much we may reduce the intensity of our memory, we risk leaving in it some degree of the variety and richness of our inner life; we are then preserving the personal, at all events, human character of memory. It is the opposite course we must follow. We shall have to consider a moment in the unfolding of the universe, that is, a snapshot that exists independently of any consciousness, then we shall try conjointly to summon another moment brought as close as possible to the first, and thus have a minimum of time enter into the world without allowing the faintest glimmer of memory to go with it. We shall see that this is impossible. Without an elementary memory that connects the two moments, there will be only one or the other, consequently a single instant, no before and after, no succession, no time. We can bestow upon this memory just what is needed to make the connection; it will be, if we like, this very connection, a mere continuing of the before into the immediate after with a perpetutally renewed forgetfulness of what is not the immediately prior moment. We shall nonetheless have introduced memory. To tell the truth, it is impossible to distinguish between the duration, however short it may be, that separates two instants and a memory that connects them, because duration is essentially a continuation of what no longer exists into what does exist. This is real time, perceived and lived. This is also any conceived time, because we cannot conceive a time without imagining it as perceived and lived. Duration therefore implies consciousness; and we place consciousness at the heart of things for the very reason that we credit them with a time that endures.

However, the time that endures is not measurable, whether we think of it as within us or imagine it outside of us. Measurement that is not merely conventional implies, in effect, division and superimposition. But we cannot superimpose successive durations to test whether they are equal or unequal; by hypothesis, the one no longer exists when the other appears; the idea of verifiable equality loses all meaning here. Moreover, if real duration becomes divisible, as we shall see, by means of the community that is established between it and the line symbolizing it, it consists in itself of an indivisible and total progress. Listen to a melody with your eyes closed, thinking of it alone, no longer juxtaposing on paper or an imaginary keyboard notes which you thus preserved one for the other, which then agreed to become simultaneous and renounced their fluid continuity in time to congeal in space; you will rediscover, undivided and indivisible, the melody or portion of the melody that you will have replaced within pure duration. Now, our inner duration, considered from the first to the last moment of our conscious life, is something like this melody. Our attention may turn away from it and, consequently, from its indivisibility; but when we try to cut it, it is as if we suddenly passed a blade through a flame—we divide only the space it occupied. When we witness a very rapid motion, like that of a shooting star, we quite clearly distinguish its fiery line divisible at will, from the indivisible mobility that it subtends; it is this mobility that is pure duration. Impersonal and universal time, if it exists, is in vain endlessly prolonged from past to future; it is all of a piece; the parts we single out in it are merely those of a space that delineates its track and becomes its equivalent in our eyes; we are dividing the unfolded, not the unfolding. How do we first pass from the unfolding to the unfolded, from pure duration to measurable time? It is easy to reconstruct the mechanism of this operation.

If I draw my finger across a sheet of paper without looking at it, the motion I perform is, perceived from within, a continuity of consciousness, something of my own flow, in a word, duration. If I now open my eyes, I see that my finger is tracing on the sheet of paper a line that is preserved, where all is juxtaposition and no longer succession; this is the unfolded, which is the record of the result of motion, and which will be its symbol as well. Now, this line is divisible, measurable. In dividing and measuring it, I can then say, if it suits me, that I am dividing and measuring the duration of the motion that is tracing it out.

It is therefore quite true that time is measured through the intermediary of motion. But it is necessary to add that, if this measurement of time by motion is possible, it is, above all, because we are capable of performing motions ourselves and because these motions then have a dual aspect. As muscular sensation, they are a part of the stream of our conscious life, they endure; as visual perception, they describe a trajectory, they claim a space. I say "above all" because we could, in a pinch, conceive of a conscious creature reduced to visual perception who would yet succeed in framing the idea of measurable time. Its life would then have to be spent in the contemplation of an outside motion continuing without end. It would also have to be able to extract from the motion perceived in space and sharing the divisibility of its trajectory, the "pure mobility," the uninterrupted solidarity of the before and after that is given in consciousness as an indivisible fact. We drew this distinction just before when we were speaking of the fiery path traced out by the shooting star. Such a consciousness would have a continuity of life constituted by the uninterrupted sensation of an external, endlessly unfolding mobility. And the uninterruption of unfolding would still remain distinct from the divisible track left in space, which is still of the unfolded. The latter is divisible and measurable because it is space. The other is duration. Without the continual unfolding, there would be only space, and a space that, no longer subtending a duration, would no longer represent time.

Now, nothing prevents us from assuming that each of us is tracing an uninterrupted motion in space from the beginning to the end of his conscious life. We could be walking day and night. We would thus complete a journey coextensive with our conscious life. Our entire history would then unfold in a measurable time.

Are we thinking of such a journey when we speak of an impersonal time? Not entirely, for we live a social and even cosmic life. Quite naturally we substitute any other person's journey for the one we would make, then any uninterrupted motion that would be contemporaneous with it. I call two flows "contemporaneous" when they are equally *one* or *two* for my consciousness, the latter perceiving them together as a single flowing if it sees fit to engage in an undivided act of attention, and, on the other hand, separating them throughout if it prefers to divide its attention between them, even doing both at one and the same time if it decides to divide its attention and yet not cut it in two. I call two instantaneous perceptions "simultaneous" that are apprehended in one and the same mental act, the attention here again being able to make one or two out of them at will. This granted, it is easy to see that it is entirely in our interest to take for the "unfolding of time" a motion independent of that of our own body. In truth, we find it already taken. Society has adopted it for us. It is the earth's rotational motion. But if we accept it, if we understand it as time and not just space, it is because a journey of our own body is always virtual in it, and *could have been* for us the unfolding of time.

It matters little, moreover, what moving body we adopt as our recorder of time. Once we have exteriorized our own duration as motion in space, the rest follows. Thenceforth, time will seem to us like the unwinding of a thread, that is, like the journey of the mobile entrusted with computing it. We shall say that we have measured the time of this unwinding and, consequently, that of the universal unwinding as well.

But all things would not seem to us to be unwinding along with the thread, each actual moment of the universe would not be for us the tip of the thread, if we did not have the concept of simultaneity at our disposal. We shall soon see the role of this concept in Einstein's theory. For the time being, we would like to make clear its psychological origin, about which we have already said something. The theoreticians of relativity never mention any simultaneity but that of two instants. Anterior to that one, however, is another, the idea of which is more natural: the simultaneity of two flows. We stated that it is of the very essence of our attention to be able to be divided without being split up. When we are seated on the bank of a river, the flowing of the water, the gliding of a boat or the flight of a bird, the ceaseless murmur in our life's deeps are for us three separate things or only one, as we choose. We can interiorize the whole, dealing with a single perception that carries along the three flows, mingled, in its course; or we can leave the first two outside and then divide our attention between the inner and the outer; or, better yet, we can do both at one and the same time, our attention uniting and yet differentiating the three flows, thanks to its singular privilege of being one and several. Such is our primary idea of simultaneity. We therefore call two external flows that occupy the same duration "simultaneous" because they both depend upon the duration of a like third, our own; this duration is ours only when our consciousness is concerned with us alone, but it becomes equally theirs when our attention embraces the three flows in a single indivisible act.

Now from the simultaneity of two flows, we would never pass to that of two instants, if we remained within pure duration, for every duration is thick; real time has no instants. But we naturally form the idea of instant, as well as of simultaneous instants, as soon as we acquire the habit of converting time into space. For, if a duration has no instants, a line terminates in points.[5] And, as soon as we make a line correspond to a duration, to portions of this line there must correspond "portions of duration" and to an extremity of the line, an "extremity of duration"; such is the instant—something that does not exist actually, but virtually. The instant is what would terminate a duration if the latter came to a halt. But it does not halt. Real time cannot therefore supply the instant; the latter is born of the mathematical point, that is to say, of space. And yet, without real time, the point would be only a point, not an instant. Instantaneity thus involves two things, a continuity of real time, that is, duration, and a spatialized time, that is, a line which, described by a motion, has thereby become symbolic of time. This spatialized time, which admits of points, ricochets onto real time and there gives rise to the instant. This would not be possible without the tendency—fertile in illusions—which leads us to apply the motion *against* the distance traveled, to make the trajectory coincide with the journey, and then to decompose the motion over the line as we decompose the line itself; if it has suited us to single out points on the line, these points will then become "positions" of the moving body (as if the latter, moving, could ever *coincide* with something at rest, as if it would not thus stop

5. That the concept of the mathematical point is natural is well known to those who have taught geometry to children. Minds most refractory to the first elements imagine immediately and without difficulty lines without thickness and points without size.

moving at once!). Then, having dotted the path of motion with positions, that is, with the extremities of the subdivisions of the line, we
have them correspond to "instants" of the continuity of the motion—mere virtual stops, purely mental views. We once described the
mechanism of this process; we have also shown how the difficulties
raised by philosophers over the question of motion vanish as soon as
we perceive the relation of the instant to spatialized time, and that of
spatialized time to pure duration. Let us confine ourselves here to
remarking that no matter how much this operation appears learned, it
is native to the human mind; we practice it instinctively. Its recipe is
deposited in the language.

Simultaneity of the instant and simultaneity of flow are therefore
distinct but complementary things. Without simultaneity of flow, we
would not consider these three terms interchangeable: continuity of
our inner life, continuity of a voluntary motion which our mind
indefinitely prolongs, and continuity of any motion through space.
Real duration and spatialized time would not then be equivalent, and
consequently time in general would no longer exist for us; there would
be only each one's duration. But, on the other hand, this time can be
computed thanks only to the simultaneity of the instant. We need this
simultaneity of the instant in order (1) to note the simultaneity of a
phenomenon with a clock moment, (2) to point off, all along our own
duration, the simultaneities of these moments with moments of our
duration which are created in the very act of pointing. Of these two
acts, the first is the essential one in the measurement of time. But
without the second, we would have no particular measurement, we
would end up with a figure t representing anything at all, we would
not be thinking of time. It is therefore the simultaneity between two
instants of two motions outside of us that enables us to measure time;
but it is the simultaneity of these moments with moments pricked by
them along our inner duration that makes this measurement one of
time.

We shall have to dwell upon these two points. But let us first open a
parenthesis. We have just distinguished between two "simultaneities
of the instant"; neither of the two is the simultaneity most in question
in the theory of relativity, namely, the simultaneity between readings
given by two separated clocks. Of that we have spoken in our first
chapter; we shall soon be especially occupied with it. But it is clear

that the theory of relativity itself cannot help acknowledging the two simultaneities that we have just described; it confines itself to adding a third, one that depends upon a synchronizing of clocks. Now we shall no doubt show how the readings of two separated clocks C and C', synchronized and showing the same time, are or are not simultaneous according to one's point of view. The theory of relativity is correct in so stating; we shall see upon what condition. But it thereby recognizes that an event E occurring beside clock C is given in simultaneity with a reading on clock C in a quite different sense—in the psychologist's sense of the word simultaneity. And likewise for the simultaneity of event E' with the reading on its "neighboring" clock C'. For if we did not begin by admitting a simultaneity of this kind, one which is absolute and has nothing to do with the synchronizing of clocks, the clocks would serve no purpose. They would be bits of machinery with which we would amuse ourselves by comparing them with one another; they would not be employed in classifying events; in short, they would exist for their own sake and not to serve us. They would lose their *raison d'être* for the theoretician of relativity as for everyone else, for he too calls them in only to designate the time of an event. Now, it is very true that simultaneity thus understood is easily established between moments in two flows only if the flows pass by "at the same place." It is also very true that common sense and science itself until now have, a priori, extended this conception of simultaneity to events separated by any distance. They no doubt imagined, as we said further back, a consciousness coextensive with the universe, capable of embracing the two events in a unique and instantaneous perception. But, more than anything else, they applied a principle inherent in every mathematical representation of things and asserting itself in the theory of relativity as well. We find in it the idea that the distinction between "small" and "large," "not far apart" and "very far apart," has no scientific validity and that if we can speak of simultaneity outside of any synchronizing of clocks, independently of any point of view, when dealing with an event and a clock not much distant from one another, we have this same right when the distance is great between the clock and the event or between the two clocks. No physics, no astronomy, no science is possible if we deny the scientist the right to represent the whole universe schematically on a piece of paper. We therefore implicitly grant the possibility of reducing without distorting. We believe that size is not an absolute, that there are only relations among sizes, and that everything would turn out the same in a universe made smaller at will, if the relations among parts were preserved. But in that case how can we prevent our imagination, and even our understanding, from treating the simultaneity of the readings of two very widely separated clocks like the simultaneity of two clocks slightly separated, that is, situated "at the same place"? A thinking microbe would find an enormous interval between two "neighboring" clocks. And it would not concede the existence of an absolute, intuitively perceived simultaneity between their readings. More Einsteinian than Einstein, it would see simultaneity here only if it had been able to note identical readings on two microbial clocks, synchronized by optical signals, which it had substituted for our two "neighboring" clocks. Our absolute simultaneity would be its relative simultaneity because it would refer our absolute simultaneity to the readings on its two microbial clocks which it would, in its turn, perceive (which it would, moreover, be equally wrong to perceive) "at the same place." But this is of small concern at the moment; we are not criticizing Einstein's conception; we merely wish to show to what we owe the natural extension that has always been made of the idea of simultaneity, after having actually derived it from the ascertainment of two "neighboring" events. This analysis, which has until now hardly been attempted, reveals a fact that the theory of relativity could make use of it. We see that if our understanding passes here so easily from a short to a long distance, from simultaneity between neighboring events to simultaneity between widely-separated events, if it extends to the second case the absolute character of the first, it is because it is accustomed to believing that we can arbitrarily modify the dimensions of all things on condition of retaining their relations. But it is time to close the parenthesis. Let us return to the intuitively perceived simultaneity which we first mentioned and the two propositions we had set forth: (1) it is the simultaneity between two instants of two motions outside us that allows us to measure an interval of time; (2) it is the simultaneity of these moments with moments dotted by them along our inner duration that makes this measurement one of time [pp. 225-227].

The first point is obvious. We saw above how inner duration exteriorizes itself as spatialized time and how the latter, space rather than time, is measurable. It is henceforth through the intermediary of space that we shall measure every interval of time. As we shall have divided it into parts corresponding to equal spaces, equal by definition, we shall have at each division point an extremity of the interval, an instant, and we shall regard the interval itself as the unit of time. We shall then be able to consider any motion, any change, occurring beside this model motion; we shall point off the whole length of its unfolding with "simultaneities of the instant." As many simultaneities as we shall have established, so many units of time shall we record for the duration of the phenomenon. Measuring time consists therefore in counting simultaneities. All other measuring implies the possibility of directly or indirectly laying the unit of measurement over the object measured. All other measuring therefore bears upon the interval between the extremities even though we are, in fact, confined to counting these extremities. But in dealing with time, we can only count extremities; we merely *agree* to say that we have measured the interval in this way. If we now observe that science works exclusively with measurements, we become aware that, with respect to time, science counts instants, takes note of simultaneities, but remains without a grip on what happens in the intervals. It may indefinitely increase the number of extremities, indefinitely narrow the intervals; but always the interval escapes it, shows it only its extremities. If every motion in the universe were suddenly to accelerate in proportion, including the one that serves as the measure of time, something would change for a consciousness not bound up with intracerebral molecular motions; it would not receive the same enrichment between sunup and sundown; it would therefore detect a change; in fact, the hypothesis of a simultaneous acceleration of every motion in the universe makes sense only if we imagine a spectator-consciousness whose completely qualitative duration admits of a more or less without being thereby accessible to measurement.[6] But the change would

6. It is obvious that our hypothesis would lose its meaning if we thought of consciousness as an "epiphenomenon" added to cerebral phenomena of which it would merely be the result or expression. We cannot dwell here upon this theory of consciousness-as-epiphenomenon, which we tend more and more to consider arbitrary. We have discussed it in detail in several of our works, notably in the first three chapters of *Matière et*

exist only for that consciousness able to compare the flow of things with that of the inner life. In the view of science nothing would have changed. Let us go further. The speed of unfolding of this external, mathematical time might become infinite; all the past, present, and future states of the universe might be found experienced at a stroke; in place of the unfolding there might be only the unfolded. The motion representative of time would then have become a line; to each of the divisions of this line there would correspond the same portion of the unfolded universe that corresponded to it before in the unfolding universe; nothing would have changed in the eyes of science. Its formulae and calculations would remain what they were.

It is true that exactly at the moment of our passing from the unfolding to the unfolded, it would have been necessary to endow space with an extra dimension. More than thirty years ago,[7] we pointed out that spatialized time is really a fourth dimension of space. Only this fourth dimension allows us to juxtapose what is given as succession: without it, we would have no room. Whether a universe has three, two, or a single dimension, or even none at all and reduces to a point, we can always convert the indefinite succession of all its events into instantaneous or eternal juxtaposition by the sole act of granting it an additional dimension. If it has none, reducing to a point that changes quality indefinitely, we can imagine the rapidity of succession of the qualities becoming infinite and these *points of quality* being given all at once, provided we bring to this world without dimension a line upon which the points are juxtaposed. If it already had one dimension, if it were linear, two dimensions would be needed to juxtapose the *lines of quality*—each one indefinite—which were the successive

mémoire (Matter and Memory) and in different essays in L'Energie spirituelle (Mind-Energy). Let us confine ourselves to recalling: (1) that this theory in no way stems from facts, (2) that its metaphysical origins are easily made out, (3) that, taken literally, it would be self-contradictory. (Concerning this last point and the oscillation, which the theory implies between two contrary assertions, see L'Energie spirituelle (Mind-Energy) (Paris: F. Alcan, 1919), pp. 203–223. In the present work, we take consciousness as experience give it to us, without theorizing about its nature and origins.

7. *Essai sur les données immédiates de la conscience (Time and Free Will)*, p. 83.

moments of its history. The same observation again if it had two dimensions, if it were a surface universe, an indefinite canvas upon which flat images would indefinitely be drawn, each one covering it completely; the rapidity of succession of these images will again be able to become infinite, and we shall again go over from a universe that unfolds to an unfolded universe, provided that we have been accorded an extra dimension. We shall then have all the endless, piled-up canvasses giving us all the successive images that make up the entire history of the universe; we shall possess them all together; but we shall have had to pass from a flat to a volumed universe. It is easy to understand, therefore, why the sole act of attributing an infinite speed to time, of substituting the unfolded for the unfolding, would require us to endow our solid universe with a fourth dimension. Now, for the very reason that science cannot specify the "speed of unfolding" of time, that it counts simultaneities but necessarily neglects intervals, it deals with a time whose speed of unfolding we may as well assume to be infinite, thereby virtually conferring an additional dimension upon space.

Immanent in our measurement of time, therefore, is the tendency to empty its content into a space of four dimensions in which past, present, and future are juxtaposed or superimposed for all eternity. This tendency simply expresses our inability mathematically to translate time itself, our need to replace it, in order to measure it, by simultaneities which we count. These simultaneities are instantaneities; they do not partake of the nature of real time; they do not endure. They are purely mental views that stake out conscious duration and real motion with virtual stops, using for this purpose the mathematical point that has been carried over from space to time.

But if our science thus attains only to space, it is easy to see why the dimension of space that has come to replace time is still called time. It is because our consciousness is there. It infuses living duration into a time dried up as space. Our mind, interpreting mathematical time, retraces the path it has traveled in obtaining it. From inner duration it had passed to a certain undivided motion which was still closely bound up with it and which had become the model motion, the generator or computer of time; from what there is of pure mobility in this motion, that mobility which is the link between motion and duration, it passed to the trajectory of the motion, which is pure space;

dividing the trajectory into equal parts, it passed from the points of division of this trajectory to the corresponding or "simultaneous" points of division of the trajectory of any other motion. The duration of this last motion was thus measured; we have a definite number of simultaneities; this will be the measure of time; it will henceforth be time itself. But this is time only because we can look back at what we have done. From the simultaneities staking out the continuity of motions, we are always prepared to reascend the motions themselves and, through them, the inner duration that is contemporaneous with them, thus replacing a series of simultaneities of the instant, which we count but which are no longer time, by the simultaneity of flows that leads us back to inner, real duration.

Some will wonder whether it is useful to return to it, and whether science has not, as a matter of fact, corrected a mental unperfection, brushed aside a limitation of our nature, by spreading out "pure duration" in space. These will say: "Time, which is pure duration, is always in the course of flowing; we apprehend only its past and its present, which is already past; the future appears closed to our knowledge, precisely because we believe it open to our action—it is the promise or anticipation of unforeseeable novelty. But the operation by which we convert time into space for the purpose of measuring it informs us implicitly of its content. The measurement of a thing is sometimes the revealer of its nature, and precisely at this point mathematical expression turns out to have a magical property: created by us or risen at our bidding, it does more than we asked of it; for we cannot convert into space the time already elapsed without treating all of time the same way. The act by which we usher the past and present into space spreads out the future there without consulting us. To be sure, this future remains concealed from us by a screen; but now we have it there, all complete, given along with the rest. Indeed, what we called the passing of time was only the steady sliding of the screen and the gradually obtained vision of what lay waiting, globally, in eternity. Let us then take this duration for what it is, for a negation, a barrier to seeing all, steadily pushed back; our acts themselves will no longer seem like a contribution of unforeseeable novelty. They will be part of the universal weave of things, given at one stroke. We do not introduce them into the world; it is the world that introduces them ready-made into us, into our consciousness, as we reach them. Yes, it is we who are passing when we say time passes; it is the motion before our eyes which, moment by moment, actualizes a complete history given virtually." Such is the metaphysic immanent in the spatial representation of time. It is inevitable. Clear or confused, it was always the natural metaphysic of the mind speculating upon becoming. We need not discuss it here, still less replace it by another. We have explained elsewhere why we see in duration the very stuff of our existence and of all things, and why, in our eyes, the universe is a continuity of creation. We thus kept as close as possible to the immediate; we asserted nothing that science could not accept and use; only recently, in an admirable book, a philosopher-mathematician affirmed the need to admit of an "advance of Nature" and linked this conception with ours.[8] For the present, we are confining ourselves to drawing a demarcation line between what is theory, metaphysical construction, and what is purely and simply given in experience; for we wish to keep to experience. Real duration is *experienced;* we *learn* that time unfolds and, moreover, we are unable to measure it without converting it into space and without assuming all we know of it to be unfolded. But, it is impossible mentally to spatialize only a part; the act, once begun, by which we unfold the past and thus abolish real succession involves us in a total unfolding of time; inevitably we are then led to blame human imperfection for our ignorance of a future that is present and to consider duration a pure negation, a "deprivation of eternity." Inevitably we come back to the Platonic theory. But since this conception *must* arise because we have no way of limiting our spatial representation of elapsed time to the past, it is *possible* that the conception is erroneous, and in any case *certain* that it is purely a mental construction. Let us therefore keep to experience.

If time has a positive reality, if the delay of duration at instantaneity represents a certain hesitation or indetermination inherent in a certain part of things which holds all the rest suspended within it; in short, if

8. Alfred North Whitehead, *The Concept of Nature* (Cambridge: Cambridge University Press, 1920). This work (which takes the theory of relativity into account) is certainly one of the most profound ever written on the philosophy of nature. [The relevant passage occurs (on p. 332, see selection No. 19,) of Whitehead's work and reads as follows: "It is an exhibition . . . in full accord with Bergson, though he uses 'time' for the fundamental fact which I call the 'passage of nature.'"]

there is creative evolution, I can very well understand how the portion of time already unfolded may appear as juxtaposition in space and no longer as pure succession; I can also conceive how every part of the universe which is mathematically linked to the present and past—that is, the future unfolding of the inorganic world—may be representable in the same schema (we once demonstrated that in astronomical and physical matters *prevision* is really a *vision*). We believe that a philosophy in which duration is considered real and even active can quite readily admit Minkowski's and Einstein's space-time (in which, it must be added, the fourth dimension called time is no longer, as in our examples above, a dimension completely similar to the others). On the other hand, you will never derive the idea of a temporal flow from Minkowski's schema. Is it not better, in that case, to confine ourselves, until further notice, to that one of the two points of view which sacrifices nothing of experience, and therefore—not to prejudge the question—nothing of appearances? Besides, how can a physicist wholly reject inner experience if he operates with perceptions and, therefore, with the data of consciousness? It is true that a certain doctrine accepts the testimony of the senses, that is, of consciousness, in order to obtain terms among which to establish relations, then retains only the relations and regards the terms as nonexistent. But this is a metaphysic grafted upon science, it is not science. And, to tell the truth, it is by abstraction that we distinguish both terms and relations: a continual flow from which we simultaneously derive both terms and relations and which is, over and above all that, fluidity; this is the only immediate datum of experience.

But we must close this overly long parenthesis. We believe we have achieved our purpose, which was to describe the salient features of a time in which there really is succession. Abolish these features and there is no longer succession, but juxtaposition. You can say that you are still dealing with time—we are free to give words any meaning we like, as long as we begin by defining that meaning—but we shall know that we are no longer dealing with an experienced time; we shall be before a symbolic and conventional time, an auxiliary magnitude introduced with a view to calculating real magnitudes. It is perhaps for not having first analyzed our mental view of the time that flows, our feeling of real duration, that there has been so much trouble in determining the philosophical meaning of Einstein's theories, that is, their relation to reality. Those whom the paradoxical appearance of the theories inconvenienced have declared Einstein's multiple times to be purely mathematical entities. But those who would like to dissolve things into relations, who regard every reality, even ours, as a confusedly perceived mathematics, are apt to declare that Minkowski's and Einstein's space-time is reality itself, that all of Einstein's times are equally real, as much and perhaps more so than the time that flows along with us. We are too hasty in both instances. We have just stated, and we shall soon demonstrate in greater detail, why the theory of relativity cannot express all of reality. But it is impossible for it not to express some. For the time that intervenes in the Michelson-Morley experiment is a real time—real again is the time to which we return with the application of the Lorentz formulae. If we leave real time to end with real time, we have perhaps made use of mathematical artifices in between, but these must have some connection with things. It is therefore a question of allotting shares to the real and to the conventional. Our analyses were simply intended to pave the way for this task.

But we have just uttered the word "reality"; and in what follows, we shall constantly be speaking of what is real and not real. What shall we mean by that? If it were necessary to define reality in general, to say by what sign we recognize it, we could not do so without classifying ourselves within a school; philosophers are not in agreement, and the problem has received as many solutions as there are shades of realism and idealism. We would, besides, have to distinguish between the standpoints of philosophy and science; the former rather regards the concrete, all charged with quality, as the real; the latter extracts or abstracts a certain aspect of things and retains only size or relation among sizes. Very happily, we have only to be occupied, in all that follows, with a single reality, time. This being so, it will be easy for us to follow the rule we have imposed upon ourselves in the present essay, that of advancing nothing that cannot be accepted by any philosopher or scientist—even nothing that is not implied in all philosophy and science.

Everyone will surely agree that time is not conceived without a *before* and an *after*—time is succession. Now we have just shown that where there is not some memory, some consciousness, real or virtual, established or imagined, actually present or ideally introduced, there cannot be a before *and* an after; there is one *or* the other, not both; and both are needed to constitute time. Hence, in what follows, whenever we shall wish to know whether we are dealing with a real or an imaginary time, we shall merely have to ask ourselves whether the object before us can or cannot be perceived, whether we can or cannot become conscious of it. The case is privileged; it is even unique. If it is a question of color, for example, consciousness undoubtedly intervenes at the beginning of the study in order to give the physicist the perception of the thing; but the physicist has the right and the duty to substitute for the datum of consciousness something measurable and numerable with which he will henceforward work while granting it the name of the original perception merely for greater convenience. He can do so because, with this original perception eliminated, something remains, or at the very least, is deemed to remain. But what will be left of time if you take succession out of it? And what is left of succession if you remove even the possibility of perceiving a before and an after? I grant you the right to substitute, say, a line for time, since to measure it is quite in order. But a line can be called time only when the juxtaposition it affords is convertible into succession; otherwise you are arbitrarily and conventionally giving that line the name of time. We must be forewarned of this so as not to lay ourselves open to a serious error. What will happen if you introduce into your reasoning and figuring the hypothesis that the thing you called "time" *cannot*, on pain of contradiction, be perceived by a consciousness, either real or imaginary? Will you not then be working, by definition, with an imaginary, unreal time? Now such is the case with the times with which we shall often be dealing in the theory of relativity. We shall meet with perceived or perceptible ones—those will be considered real. But there are others that the theory prohibits, as it were, from being perceived or becoming perceptible; if they became so, they would change in scale, so that measurement, correct if it bears upon what we do not perceive, would be false as soon as we do perceive. Why not declare these latter unreal, at least as far as their being "temporal" goes? I admit that the physicist still finds it convenient to call them time; we shall soon see why. But if we liken these times to the other, we fall into paradoxes that have certainly hurt the theory of relativity, even if they have helped popularize it. It will therefore be no surprise if, in the present study, we require the property of being perceived or perceptible for everything held up as real. We shall not be deciding the question of whether all reality possesses this salient feature. We are only dealing here with the reality of time.

15.

Alexander:

TIME AND SPACE

Born in Australia, Samuel Alexander (1859–1938) went to England to study at Oxford and remained to teach at Manchester. He regarded first Bergson and then himself as the two philosophers who first sought to take time seriously. In conjunction with the epistemological realism of the period, he worked out the implications of his temporal realism in *Space, Time and Deity.** Delivered as the Gifford Lectures in 1916–1918, it was first published in 1920. The selection is taken from the first four chapters of Book I which was concerned to spell out the fundamental character of what he called "Space-Time."

Chapter I

Physical Space-Time

It is not, I believe, too much to say that all the vital problems of philosophy depend for their solution on the solution of the problem what Space and Time [1] are and more particularly how they are related to each other. We are to treat it empirically, describing Space and Time and analysing them and considering their connection, if any, as we do with other realities. We do not ask whether they are real in their

* The selection is reprinted with the permission of Macmillan, London and Basingstoke.
 1. I use for convenience capital letters for Space and Time when I am speaking of them in general or as wholes. Small letters are used for any portion of them (thus a space means a portion of Space); or in adjectival phrases like 'in space' or 'in time'. The practice is not without its disadvantages, and I am not sure that I have followed it rigorously.

own right or not, but assume their reality, and ask of what sort this reality is. . . .

Space and Time are presented to us as infinite and continuous wholes of parts. I shall call these parts points and instants, availing myself of the conceptual description of them, and meaning by their connectedness or continuity at any rate that between any two points or instants another can be found. To me, subject to what may be said hereafter, this is a way of saying that the points and instants are not isolated. But if any reader jibs, let him substitute lengths and durations; he will find that nothing is said in what follows except what follows equally from the notion of parts.

Other features will declare themselves as we proceed, some obvious, some less so. But they will be found to require for their understanding the understanding of how Space and Time are related to each other. These are often thought, perhaps commonly, to be independent and separate (whether treated as entities as here or as systems of relations). But a little reflective consideration is sufficient to show that they are interdependent, so that there neither is Space without Time nor Time without Space; any more than life exists without a body or a body which can function as a living body exists without life; that Space is in its very nature temporal and Time spatial. The most important requirement for this analysis is to realise vividly the nature of Time as empirically given as a succession within duration. We are, as it were, to think ourselves into Time. I call this taking Time seriously. Our guides of the seventeenth century desert us here. Besides the infinite, two things entranced their intellects. One was Space or extension; the other was Mind. But entranced by mind or thought, they neglected Time. Perhaps it is Mr. Bergson in our day who has been the first philosopher to take Time seriously.

Empirically Time is a continuous duration, but it is also empirically successive. Physical Time is a succession from earlier to later. As Mr. Russell points out,[2] the succession from past through present to future belongs properly to mental or psychical time. But so long as we take care to introduce no illegitimate assumption we may conveniently speak of past, present, and future in physical Time itself, the present

2. [Alexander cites Russell's paper. See selection no. 17, p. 297—C. M. S.]

being a moment of physical Time fixed by relation to an observing mind and forming the boundary or section or cut between earlier and later, which then may be called past and future. In a manner, earlier and later are, as it were, the past and future of physical Time itself. I shall therefore use liberty of phrase in this matter. Now if Time existed in complete independence and of its own right there could be no continuity in it. For the essence of Time in its purely temporal character is that the past or the earlier is over before the later or present. The past instant is no longer present, but is dead and gone. Time's successiveness is that which is characteristic of it as empirically experienced, in distinction, say, from Space, which also is continuous. This is the plain conclusion from taking Time seriously as a succession. If it were nothing more than bare Time it would consist of perishing instants. Instead of a continuous Time, there would be nothing more than an instant, a now, which was perpetually being renewed. But Time would then be for itself and for an observer a mere now, and would contain neither earlier nor later. And thus in virtue of its successiveness it would not only not be continuous but would cease even to be for itself successive. If we could suppose an observer and events occurring in time, that observer could distinguish the two 'nows' by the different qualities of the events occurring in them. But not even he could be aware that the two 'nows' were continuous, not even with the help of memory. For memory cannot tell us that events were connected which have never been together.

Descartes did, in fact, declare the world to be perpetually re-created. For him the idea of a Creator presented no problem or difficulty, and with his imperfect grasp of the real nature of Time the step he took was inevitable and imperative. For us the case is not the same, even if re-creation at each moment by a Creator left no difficulty unsolved. But in any case the universe at the stage of simplicity represented by mere Time and Space has no place for so complex an idea as creation, still less for that of a supreme Creator. Time and Space are on our hypothesis the simplest characters of the world, and the idea of a Creator lies miles in front.

Chapter II

Perspectives and Sections of Physical Space–Time

The physical universe is thus through and through historical, the scene of motion. Since there is no Space without Time, there is no such thing as empty Space or empty Time and there is no resting or immoveable Space. Space and Time may be empty of qualitative events or things, and if we are serious with Time there is no difficulty in the thought of a Space-Time which contained no matter or other qualities but was, in the language of Genesis, without form and void before there was light or sound. But though empty of qualities Space and Time are always full. Space is full of Time and Time is full of Space, and because of this each of them is a complete or perfect continuum. If this might seem a quibble of words, which it is not, let us say that Space-Time is a *plenum.* Its density is absolute or complete. There is no vacuum in Space-Time, for that vacuum would be itself a part of Space-Time. A vacuum is only an interval between bodies, material or other, which is empty of body; but it is full with space-time. Hence the old difficulty that if there were no vacuum motion would be impossible is without foundation, and was disposed of by Leibniz in answering Locke.[3] If it were completely full of material bodies with their material qualities there would be no room for locomotion of those bodies with their qualities. But it is only full with itself. Material bodies can move in this absolute plenum of Space-Time, because their motion means merely that the time-coefficients of their spatial outlines change.

In the next place, there is no immoveable Space. In one sense, indeed, Space is neither immoveable (or at rest) nor in motion. Space as a whole is neither immoveable nor in motion. For that would suppose there was some Space in which it could rest or move and would destroy its infinitude. Even when we speak of Space as a whole we must observe that it is not a completed whole at any moment, for this would omit its temporality. Under a certain condition, to be explained presently, we may indeed contemplate Space as an infinite

3. *Nouveaux Essais,* Preface (Erdmann, p. 199*b*, Latta, p. 385).

whole when we consider only the points it contains. Directly we allow for its Time, we realise that while there may be a complete whole of conceived timeless points there cannot be one of real point-instants or events. For incompleteness at any moment is of the essence of Time. Neither strictly can the universe be said to be in motion as a whole. It *is* motion, that is in so far as it is expressed in its simplest terms.

But it is not Space as a whole which is understood to be immoveable. The immoveable or absolute Space of Newton is the system of places which are immoveable. Now since every point is also, or rather as such, an instant, a resting place is only a place with its time left out. Rest, as we shall see more clearly presently, is only a relative term.

With this conception of the whole Space-Time as an infinite continuum of pure events or point-instants let us ask what the universe is at any moment of its history. The meaning of this obscure phrase will become clearer as we proceed. The emphasis rests upon the word history. Space-Time or the universe in its simplest terms is a growing universe and is through and through historical. If we resolve it into its phases, those phases must express its real life, and must be such as the universe can be reconstructed from in actual reality, they must be phases which of themselves grow each into the next, or pass over into each other. We are to take an instant which occupies a point and take a section of Space-Time through that point-instant in respect of its space or time.

Chapter III

Mental Space and Time

By mental or psychological time I mean the time in which the mind experiences itself as living, the time which it enjoys; by mental space I mean, assuming it to exist, the space in which the mind experiences itself as living or which it enjoys. They are contrasted provisionally with the space and time of the objects of mind which the mind contemplates. I hope to show on the strength of experience that mental space and time possess the same characters and are related in the same intimacy of relation as physical Space and Time; that the time of mental events is spatial and their space temporal precisely as with physical Space and Time, and further that mental time, the time in which the mind lives its life or minds its mind, is a piece of the Time in which physical events occur; and similarly of mental space. In many respects it would have made the task of analysing physical Space and Time in the preceding chapters easier if, following the method of the angels and assuming mind to be an existence alongside of physical existence, I had examined first the simplest elements in mind rather than in physical objects, and with the results of the analysis of the familiar thing mind, had passed to the analysis of the less familiar external world. But I felt myself precluded from this procedure because it would have meant before approaching physical Space and Time that we should need to accept two very disputable propositions, first that the mind is spatial, that is, is enjoyed in space, and second, that this enjoyed space is at any instant occupied not merely by the mind's present but also by its enjoyed past and future. Accordingly I have endeavoured to examine physical Space and Time without encumbrance by these difficulties.

That the mind as the experienced continuum of mental acts (the nature of what underlies this continuum is a subject for later inquiry) is a time-series, and in that sense is in time, or has Time in its very constitution, would be admitted on all hands. By continuity is meant mental or felt continuity, so that, by memory or other means, in a normal mind no event occurs which is disconnected with the rest. There may be intervals of time, as in sleep without dreams, or in narcosis, when the mind apparently ceases to act: "the mind thinks not always." But consciousness, as William James puts it,[4] bridges these gaps, so long as it is normal, and it feels itself one. The elements of this continuum are conscious events or processes. There is no rest in mind but a relative one. We only think there is, because with our practical interests we are concerned with the persistent objects—the trees and men, which we apprehend in what James calls "substantive" conditions of mind. If we overlook the transitions between these objects, their repugnances and likenesses, how much more easy is it to overlook the transitions in our minds, the feelings of 'and,' and 'but,' and 'because,' and 'if' or 'like'—the "transitive" states. We catch them

4. [See selection no. 20.—C. M. S.]

for notice when we happen to be arrested in our thinking, when we leave off, for instance, in a sentence with a 'because,' when the forward and defeated movement of the mind is directly made the centre of our attention.[5] The sense we have in such cases that the flow of our meaning is stopped is accompanied by caught breath or tense forward bending of the head or other bodily gestures, but it is not to be confused with the consciousness of these gestures. They are but the outward bodily discharge of the mental arrest. It is these transitive conditions which betray the real nature of the mind. The substantive states are but persistence in movements which have the same character and correspond to objects of the same quality. In itself the mind is a theatre of movement or transition, motion without end. Like all other things it has the glory of going on.

But not only is mind experienced in time, but the direct deliverance of our consciousness of external events is that the time in which we enjoy our mind is part of the same Time in which those external events occur. It is only when philosophy steps in with its hasty interpretations, that we can say that Time belongs, as Kant believed, to external events because they have a mental or internal side in our experiencing of them. On the contrary, to be aware of the date or duration of physical events is the most glaring instance, derived from direct experience, of how an enjoyed existent and a contemplated existent can be compresent with one another. In this case the compresence is a time-relation which unites both terms within the one Time (I am assuming, let me remind my reader, the hypothesis of direct apprehension of the external object). In memory or expectation we are aware of the past or future event, and I date the past or future event by reference to the act of remembering or expecting which is the present event. An event five years past occurred five years before my present act of mind. We have seen, in fact, that physical Time is only earlier or later, and that the instants in it are only past, present, or future in relation to the mind which apprehends. Now without doubt, when I remember that a friend called at my house an hour ago, I mean that that event occurred an hour before my present condition of myself in

5. There are many happy examples in Humpty Dumpty's poem in *Through the Looking-Glass:* "I'd go and wake them, if——", "We cannot do it, Sir, because——"

the act of remembering that event, and that the mental and the physical event are apprehended within the one Time.

Only in regard of present physical events does doubt arise. We are accustomed to call those physical events present which are contemplated by us in sensory form in the present moment of consciousness. Now it is certain that the physical events which I contemplate precede by a small but measurable interval my sensory apprehension of them, and this is true not only of events outside me but of the events in my body which I describe as occurring at the present moment. They are all anterior to my apprehension of them. But this is not the deliverance of unsophisticated experience, but a fact which we learn about our process of perceiving external events, and is not given directly in our acquaintance with them. Ask an untrained man whether the events which he sees occur at the same time as his perception of them, and he is merely puzzled by the question. For him the present events are those which he perceives, and he has not asked himself, and does not understand, the question whether they really are simultaneous with the perceiving of them or not. Further experience of a reflective sort, experimental experience of the times of reaction to external objects, shows him that they are not. But equally he may find by reflection or scientific methods that the event he remembers as occurring an hour ago occurred in reality an hour and five minutes ago or longer. Thus the philosophical question of the precise time-relation between our perception and its objects does not arise for us in practice. It remains true that all our mental events stand in some time-relation, whether rightly apprehended or not, to the contemplated physical events. The enjoyed mind is compresent in a time-relation with those objects. This is the whole meaning of a time-relation in which the terms are not both contemplated, as they are when we are dating two physical events with reference to one another in physical Time, but when the one is contemplated and the other enjoyed. That the mental duration or instant stands thus in relation to the contemplated instant in time shows, then, so far as experience directly gives us information, that the times of both terms are parts of one Time.

From this mere vague experience that I who enjoy am in Time along with the event contemplated which is in Time, we may easily pass to a more definite statement. We may date the physical event with reference to the physical events going on in my body at the 'present' moment. Then I am contemplating a stretch of time between the event and me (I may even, as we shall see later when we come to discuss our memory of the past, *enjoy* the interval between me and my apprehension of the physical event). At a later stage in my experience, when I have learnt that my mental act occurs really at the same time as a certain physiological process which corresponds to it, I may contemplate the time-interval between that process and the cognised physical event, and then we have a still exacter notion of the time-relation, but clearly one which is only possible for more advanced experience and not given in the mere cognition, in the mere memory, for example, of my friend's visit an hour ago.

Turning to Space, we find that mind enjoys itself spatially, or is extensive, in the same sense as it is successive and endures in enjoyed time. . . .

Thus just as we enjoy a time filled with mental events, so we enjoy an extension or space filled with mental events. Further, as with time, so here the deliverance of experience is that in apprehending physical extension, say a physical object in space, we are aware in our act of enjoyment of an enjoyed space as related to the extension of the physical object within the one Space. Our mental space and our contemplated space belong experientially to one Space, which is in part contemplated, in part enjoyed. . . .

But this vague experience of interval in space between myself and the place of physical objects in space becomes more definite when I ask where is my mind and its enjoyed space in the whole of Space. I cannot ask where I am in enjoyed space, for the space and time I enjoy are the whole of enjoyed space and time. I can date a mental event in my past; and I can dimly localise a mental event in my space: I can distinguish the outstanding point, or if it is a connected event, the streak in my space which it occupies. But when I ask when I occur in Time as a whole I answer by reference to some physical event in my body with which I am simultaneous. . . .

Chapter IV

Mental Space-Time

Let us for the sake of clearness begin, not with the memory of ourselves, but the memory of objects, that is to say, of things or events which we have experienced before, and in remembering are aware that we have so experienced them. This is the fully developed kind of memory, to which other acts of so-called memory are only approximations. James writes:[6] "The object of memory is only an object imagined in the past (usually very completely imagined there) to which the emotion of belief adheres," and in substance there is little to add to this statement. I prefer to say the object of memory (what I shall call 'the memory' as dintinguished from the mental apprehension of it, which is 'remembering') is an object imagined or thought of in my past. I say 'my past,' for I may believe in teh assassination of Julius Caesar as a past event without being able to remember it. The object, then, is before my mind, bearing on its face the mark of pastness, of being earlier than those objects which I call present. In the mind there corresponds to it the act of imagining or conceiving it, and there is in addition the act of remembering it, the consciousness that I have had it before.

The pastness of the object is a datum of experience, directly apprehended. The object is compresent with me *as past*. The act of remembering is the process whereby this object becomes attached to or appropriated by myself, that is, by my present consciousness of myself which has been already described, in which may be distinguished a subjective and a bodily element unified in the person. The past object is earlier than my present act of mind in remembering, or my equivalent bodily state, whichever may happen to be more predominant in my mind. When the past object is thus appropriated by myself, I am aware of it as belonging to me, as mind, as occurring in *my* past. This is the consciousness that the object is remembered. In precisely the same sense as I am aware of a perceived object when I have before me an experience of the past and appropriate it to my

6. *Psychology,* vol. i. p. 652. [See selection no. 20.—C. M. S.]

personality. The object is then not only past but belongs to a past in which I contemplate myself (that is my body) as having been existent also and related to the object.

In this as in many other psychological inquiries, error may arise from reading into the experience more than is there. The actual past event as we once perceived it is remembered as the memory of it which has been described. I may not say that in memory I am aware of the memory as referred in thought, or in some way, to the actual object I once perceived. It is true that I can in reflection, in a sophisticated modd, so speak. But this is not the deliverance of the experience itself called having a memory. For example, I may see a man and remember that I heard his conversation yesterday. Here I have the actual man before me; but my memory of his conversation is not first taken by itself and then referred to him as I heard him yesterday. The memory-object is itself the object, and the only one I have, of the consciousness that I heard him yesterday. So far as I remember that, there is no reference to any former perception of the man, even though he is now also present in perception. The percept of him and the memory of him are two different appearances which in their connection reveal the one thing, the man, whom we now know to be to-day by perceiving, and to have been yesterday by remembrance. Moreover the memory is as much a physical objects as the percept. He is physical in so far as, in Mr. Russell's happy phrase, he behaves according to the laws of physics. The remembered man does not speak now, but he is remembered as speaking, or, to vary the example, the memory-object is the physical man cutting physical trees yesterday.

Thus we have not in memory itself any reference to the perceived. The memory itself is the only knowledge we have that there ever was something perceived. But there is a real truth misrepresented by the erroneous statement. Like a single perception, a single memory is incomplete. The particular percept is full of movement towards other aspects of the thing perceived, and the memory in like manner throws out feelers to other memories. These memories through their internal coherence and continuity build up for us our memory of the whole thing of which they are partial representations, and, as in the case just given, may blend in turn with fresh perceptions, or, again, with expectations of the future. It is then that in our unsophisticated experience (as distinct from the sophisticated deliverances of the reflective psychologist or philosopher) we can think of our friend as the same thing compresent with us in more than one memory. Even then we only introduce into our experience of him the element of his having once been perceived, through familiarity with the blending of perceptions with memory-ideas of the same thing.[7] For this reason it is that, as has often been observed by psychologists, we learn so much more directly about the nature of Time from expectation of the future than from remembrance of the past. Expectation is precisely like remembering except that the object has the mark of future, that is of later than our present, instead of past or earlier. Now we are practical creatures and look forward to the satisfaction of our needs, and the past interests us only theoretically or, if practically, as a practical guidance for the future. But in expectation the anticipated object is, or very often is, replaced continuously and coherently by the percept, and the expected object may now become a memory. I remember now how the object appeared to me an hour past when I expected it. But whereas the expectation is in the ordinary course succeeded by fulfilment in perception, memory need not be so succeeded and most often is not.

Now it is not the whole thing which we need have before us in memory, but only its appearance altered by the lapse of time, seen through the haze of Time, as things distant in Space are coloured by their remoteness. The lapse of Time may distort, and when to Time is added the subjective prejudices of the experient the memory of the thing may be highly distorted. But it remains what it declares itself to be when supplemented by similar appearances, nothing but the revelation of the thing through that mist of intervening Time, and the thing itself is only given in the actual memory through the mere reaching out of any experience to other experiences of the like sort.

Thus we avoid the first error of interpreting memory to mean more than it contains. No wonder memory is regarded as so mysterious if it

7. For the synthesis of many objects or appearances of a thing into a thing see the discussion later, Bk. III. ch. vii. vol. ii., where it will be seen that the unity of a thing which underlies its various appearances, the objects of perception or memory, is the volume of space-time which it occupies. That volume is filled by each of its appearances, and that is why a single percept or a single memory can be the appearance or 'presentation' of the thing.

is supposed also to inform us of the perceived past, as if that perceived past could be thought of except through some idea other than the memory. A second error is to suppose that the memory is in some sense present and that it is referred to the past through certain indications of a subjective or personal kind. In one form or another this doctrine is very common. Our ideas come to us in succession it is said, but the succession of ideas is not the idea of their succession. To be distinguished as past or future from the present they must all be present together. "All we immediately know of succession is but an interpretation . . . of what is really simultaneous or coexistent." [8] Even for James the feeling of the past is a present feeling. How far this is true of the *feelings* of past, present, and future we shall inquire presently. . . .

So far, we have been concerned with memory proper, where the object is an image or thought with a date, however imperfectly the date is apprehended. It is not necessary for our purpose to describe how we come to be aware of the accurate date, which involves conceptual processes, even in dating a past event five minutes ago. This is a question of the measuring of Time. But partly because of the intrinsic importance of the subject and partly for future use we may make certain observations about the time-characters of ideas in experiences which are not proper memory, but are often loosely called so. In every contemplation we enjoy ourselves, as we have seen, in a time-relation with our object. But the object may have no date. It has its internal time-character, as when I call up in my mind a picture of a man running, or even a thing like a landscape where there is no movement but where the spatial extension involves Time in its intrinsic character. In such a case the image, being a time-saturated object, is contemplated as somewhere in Time, but the position of it as a whole in Time is not dated. This distinguishes a mere revival without memory, or a mere fancy, from a memory proper. It belongs to Time but has no particular date. . . .

When I apprehend a sound and a light at the same moment, they

8. J. Ward, *Psychological Principles* (Cambridge, 1918), p. 214; Art. *Encycl. Brit.* ed. ix. p. 64b.

are for practical purposes taken as simultaneous though they are not so in fact. But we have mental processes which take place successively where yet the objects are present in sensory form. Such an experience is an example of what is called the 'specious present,' because it is not a 'mathematical' moment but experienced as a duration. The familiar example is that of the path of the meteor where the whole movement is sensory and the path of light is seen at once. We never sense an instant of time, which is, taken by itself, a concept or implies conception. Our sensible[9] experience of Time is primarily that of a duration; and experiment has determined what are the smallest intervals between two stimuli of sound or other kinds which are experienced as durations, and what interval of time filled by intermittent experiences of certain kinds, for example the strokes of a bell, can be held together at once in the mind. The specious present does not mean necessarily a duration which is filled with sensa. It is commonly taken to include also the fringe of past or future objects which have ceased to be sensory, or are not yet so, and approach the state of images, as in the case of a succession of sounds retained together in the mind, for example, in hearing strokes of a metronome, or the words of a connected sentence.[10] . . . It has been compared by James to a saddleback as opposed to the present instant which is a knife edge. But there is no reason in the facts to declare that the present is saddlebacked, except so far as sensory objects are simultaneous and not successive as when we see red and blue at once. In other words we should distinguish the 'broad' from the 'deep' present. The present always has breadth as including many simultaneous objects. But it has not depth, that is, breadth in time. Its depth is a succession within duration. No doubt 'the specious present' is a useful conception if it serves as a reminder that we never sense the present instant or the present object by itself, but that we always apprehend a bit of duration, and as a rough, practical description of the present, as rough as the habitual description of present objects as present when they are really slightly past. But otherwise we are compelled to conclude that what it describes is

9. I use this word for convenience. It does not imply that Time (or Space) is sensed but only that it is apprehended through sense. (For the proper apprehension of Time and Space, see later, Bk. III. ch. vi.)

10. [See William James in selection no. 20.—C. M. S.]

not a fundamental fact of our time experience, and that rather it misinterprets that experience. It describes merely the interesting and important fact that our minds are able to hold together a certain number of objects without having recourse to memory proper, and in particular that a certain number of sensa occupying time in their occurrence can thus be held together in our minds. The length of the time interval so filled varies with the sensory events which occur in it. If the specious present is understood in a different way it is specious in the other sense of deceiving us. Perhaps it may be compared with that other interesting and important fact of the existence of a threshold of sensation below which amounts of stimulation are not felt. This was interpreted by Fechner to mean that the threshold was in some special sense the zero of sensation. Whereas any sensation whatever may be taken as zero if we make it the beginning of our scale.

The 'specious present' is a comparatively considerable time interval of some seconds. The minimal duration which we apprehend as duration was vastly smaller. But if we learn by experience that the first contains succession within its duration we may conclude that the elementary duration is successive too, that in fact there is no duration which is not a duration of succession, though the successive moments in a very small duration may not be and are not distinguishable. Or if this is too much to say, then we must urge at least that succession is not something new and additional to duration, but past, present, and future represent distinctions drawn within duration. There is as much difficulty in conceiving elementary durations succeeding one another within a longer duration as in conceiving any duration to be intrinsically a lapse in time and therefore intrinsically successive. We have yet to see how mathematical or conceptual instants can be real. But that our elementary experience of Time should be extensive and yet admit of succession within it is no more difficult to understand than that a blur of red blood should under the microscope reveal itself as a number of red bodies swimming in a yellow plasma, or that sensations we cannot distinguish from one another may under other conditions be known to be distinct. The conclusion is that our sensibility to succession is not so great as our sensibility to duration. Where both can be apprehended the duration and the succession are seen to be of the same stuff. This is true of contemplated or objective Time. In what sense it can be held that Time as we experience it in ourselves is other than a duration which is intrinsically successive passes my understanding.

We have been dealing hitherto with the time of objects and have found that the past is in no sense present but is revealed as past. We have now to turn to the much more difficult matter of enjoyed time. It may be said: past physical events are presented as past, but when the past is declared to be somehow present, the reference is to the apprehension of it, as when, for instance, the feeling of the past is called by James a present feeling or the immediate apprehension of a short succession of events a specious present. I do not feel sure that this is what is meant in all cases. But we may use the easier analysis of the experience of past and future objects as a clue to understanding our enjoyments of time. We shall find that past enjoyments are not experienced in the present but as past, and future ones as future. . . .

If Time is real, if the past is not a mere invention of the mind, and this is our original hypothesis, the mind at any present moment contains its past as past. Otherwise, to fall back on an argument used *ad nauseam* in respect of physical Time, there would be no mind at all but a continual re-creation of quite independent and molecular mental states, which is contrary to elementary experience.

Thus a remembered mental state is a past enjoyment, as it is enjoyed after the lapse of time, the machinery for such awareness of the past being the process by which for one reason or another the brain is thrown into a corresponding, or a partially identical state with the actual past state of the brain during the past experience. The past is revived in imagination of my mental state just as it is revived in imagination of a past object. I know my own past only through the enjoyment of it as past.

The truth that the renewal in memory of a past state of myself is not merely a fresh excitement of myself in the present may perhaps best be seen in the memory of emotions. . . .

So far as I can trust my own experience I believe we can observe a distinction between a remembered and a present emotion. I remember the feeling of shame felt at a social blunder; and the more vividly I represent the circumstances the more intense the emotional excitement becomes, and the more completely it includes the bodily expression proper to the emotion and invades me. Still all this personal experience is detained in attachment to the past object, and despite the urgency of the feelings I am lost in the past, and the whole experience, object-side and subject-side alike, has the mark of the past. But suddenly I may find myself arrested; I forget the past object and I become aware of the emotion as a present state, in which the object is for the most part the bodily reactions, the flushed face and qualms about the heart. I change from a past enjoyment to a present one. What the difference is I find it hard to say; the pastness of the image seems to draw the feeling after it into the past as well. It may be that the whole difference lies in the compresence with the past object. But the difference is for me palpably there. Thus a new or actual emotion, with its sensorial character, ceases to be a present emotion when it is compresent with a past object; whether it is neurally or mentally slightly different from the emotion roused by a present object or not, it becomes a past enjoyment in this connection. Its *actuality* no more makes it a present emotion than the sensory character of the beginning of the meteor's path in the sky makes it present, when the real present is the end of the path.

Before pursuing further the ideas suggested by these facts let us note briefly that where there is not memory proper but only retention in the mind, the earlier stages of the mental enjoyment are past and not present, and that the specious present, the present with a depth, is not really a present in enjoyment, and that consequently, to sum the whole matter up, we cannot hold that the experience of the past is a present feeling, whether we speak of the past object or the past state of mind.

Whether in the study of past and future objects or in that of past and future states of ourselves, we have thus seen that our consciousness of past and future is direct, and is not the alleged artificial process of first having an experience of the present and then referring it by some method to the past or future. There is no such method given in our experience, and we have therefore no right to assume it just because we start with the fancy that all our experiences must be present. If difficulty is still felt in the unfamiliar notion that we enjoy our past as past and our future as future, the answer must be that in the first place facts, however strange the description of them may be, must be accepted loyally and our theories accommodated to them; secondly, that as to the special explanations suggested above, of how a present neural process may be felt as a past enjoyment, an explanation like this is theoretical and designed to remove a theoretical difficulty. For immediate acquaintance with our past and future tells us nothing about neural processes, and if we confine ourselves to our enjoyment of ourselves, we find that the memory or the expectation of a past or future state is the way in which we enjoy past or future, and that there is no more to be said; just as our memory of the past object is that past object as contemplated now in the act of remembering, and there is no more to be said.

But now that by an appeal to experience we have rid ourselves of the confusions as to our past and future enjoyments which were engendered by a mistaken reading of experience, we can proceed to examine the space and time of the mind in their mutual relations, and we shall see that they do not exist separately but are only elements in the one mental space-time which exhibits to inspection of ourselves in the same features, with such qualifications as may be necessary to note, as the Space-Time of the external world, with a part of which it is identical.

We have in the first place at any moment a mass of enjoyments (that is, experiences of ourselves, or experiencings), part of which is present, part memories or remembered enjoyments, part expected enjoyments with the mark of the future. These enjoyments occupy diverse places in the mental space. Present enjoyments are in different places from past and future ones. The enjoyed space is not full of mental states all occurring at one and the same time, but it is occupied, so far as it is occupied, with mental events of different dates. But, as we have seen, what is now a present enjoyment may at another moment be replaced by a remembered one; and what is now a memory may on another occasion be replaced by a present occupying partially at least the same place; the dates or times being on different occasions differently distributed among the places. Thus enjoyed space is full of time. In the same way enjoyed time is distributed over enjoyed space, and spreads over it so as not to be always in the same space. Thus empirically every point in the space has its date and every date has its point, and there is no mental space wihtout its time nor time without its space. There is one mental space-time. Our mind is spatio-temporal. The easiest way to make ourselves a picture of the situation is to suppose the identification of mental space with the corresponding contemplated neural space completed in details, and to substitute for the enjoyed space, for pictorial purposes, the neural space with which it is identical, that is, to think of specific mental events as occurring in their neural tracts. When we do so, we see mental past, present, and future juxtaposed in this space; or the places of mind succeeding each other in their appropriate times.

Such a picture of mental space-time at any moment is the perspective we enjoy of it at any moment or from the point of view of that moment. But the picture is not complete. The present enjoyment and the remembered one are enjoyed as juxtaposed. But they are not in bare unrelated juxtaposition. For a remembered past state is in remembering linked up with the present. There is a felt continuity between them. The same thing is more obviously true where there is not memory proper but a past condition is experienced as retained in the mind only, being at the fringe of a total experience, as when we retain in our minds at this moment the lingering remnants of our past condition, in going through some complex experience, as, for example, in watching the phases of an incident which stirs our feelings. That there is this transitional relation of movement from the one element to the other, is shown by the familiar fact that when one member of a series of mental states is repeated in experience the others also are revived in their time-order. . . .

In the same way, as we have seen abundantly in dealing with memory of mental states, we have mental space repeated in time; that is, several events of the same sort occurring at different times but belonging to the same space; that is, we have time coming back to its old place. And we may repeat a remark like one made before in Chapter I of physical Space-Time, that the repetition of time in space, which is the fact of the broad (not the deep) present, and of space in time, which is the fact of memory, are of the essence of mind as something with a structure and persistence.

We have thus found from simple inspection of our minds, and bringing to bear on the question the most commonplace kind of psychological observation, that space and time in mind are in experienced fact related in the same way as we have seen them to be related in physical Space and Time. Space and time in the mind are indissolubly one. For myself it is easier to be satisfied of this relation between the two, and all the details which enter into it—repetition, variation in the perspective whereby the contemporary becomes successive, and the like—in the case of mind, than in the case of external Space and Time, and to use this result as a clue to interpreting external or physical Space-time. But I have explained already [11] why I have not adopted what for me is the more natural order.

We may now approach the more difficult question, in what sense it is possible, as it was in the case of physical Space-Time, to make a selection from all the perspectives which the mind enjoys of its own space-time and treat the whole of mental space as occurring at the present and the whole of mental time as occupying one point of mental space. In the case of physical Space-Time we saw that an all-comprehensive observer whom we ourselves follow in thought could make such a selection, and we arrived thus at the ordinary notions of a Space in which at a given moment some event or other was occupying every point—Space as the framework of Time; and of Time as the framework in which Space occurred, that is of a Time the whole of which streamed through every point of Space. In mental space-time such results are obtainable, but only approximately and with a qualification.

At first sight it might appear that there was no difficulty in taking a present 'section' through the whole of our mental space. We have only to identify the neural space, say the brain, with the mental space, and then it would seem that at any instant of our life every point in that space was occupied by some event or other that occurred in our history. But mental space enjoyed in any mental state is not merely neural space, but that neural space which is correlated with the mental action. There may be events going on say in the occipital region which happen there but which are not of that particular sort which is correlated with vision, or, as I shall often express it, which *carry* vision. Though the whole contemplated space within which mental action takes place may be considered by proper selection from all the moments of its history as occupied by some present event or other contemporary with the present, they will not necessarily be mental events. They may be unconscious.

11. Ch. III, pp. 243 ff. above.

When we consider mental events as such and neural processes only so far as they carry mind, we cannot find a section of mental space-time which is either the whole mental space occupied by contemporary events or the whole mental time streaming through one point. In our own experience it is clear we get no such thing. Our enjoyed space in a moment of experience may on occasion be so limited that it contains neither memory nor expectation but is wholly present. For example, we may be absorbed in perception. We are then entirely present, for the ideal features in perception are, as we have seen, not expectations or memories, but are merely qualifications of the present, which are there indeed as the result of past experience, but have not the mark of the past nor even of the future. But though in such a perspective our mental space is all present, it is not the whole of our mental space, but only a part of it. Or again I may be seeing a man and also remembering something about him. The one place in the mental space which is common to the perception and the memory may belong both to the perception and to the memory. But it does not belong to them both at the same time, and is alternatly part of the present perceiving and the past remembering.

This is as far as I can get by actual acquaintance. Even the angelic outsider, though he will go farther, and though we may anticipate him by thought, will not get a complete section. He cannot see the whole of our mental space occupied by the present moment. He can realise that any neural process which at this moment of my mind is for me a memory might have been occupied by a mental event contemporary with my present. But it is not certain that he can find such events. Potentially the places now occupied in my perspective by memories or expectations may be occupied in other perspectives by perceptions of the present date. But the selection is only a possibility and nothing more. In the same way he may think that the place of my present mental act may potentially be the scene of some mental event at every moment of my history. But again this is only a remote possibility.

The reason for this difference between mental space-time and physical Space-Time is that the second is infinite and the first finite. We are finite beings, and part of that finitude is that our neural space performs only specific functions. Hearing does not occur in the occipital, nor vision in the temporal region. But in physical Space-Time the reason why in the summation of perspectives a selection could be made of events filling the whole of Space at one moment, or the whole of Time at one place, is that the quality of events was indifferent in the infinite whole. In one perspective a point of space is past, but in some other perspective a quite different sort of event might occupy that point in the present, that is at a moment identical with the point of reference. I see in front of me a point in a tree where a bird alighted a quarter of a second past. But a quarter of a second later, that is at the same moment as my act of seeing the bird, a bud sprouts on the tree at that point. That event is future for me, but for you, the onlooker, it occurs at the same moment as my act of sight, that is, you see tham both as contemporary. There is thus always in some perspective or other some event or other at any point of space contemporary with my present. but places in mental space-time are, because of the specific character of the events which happen there, only occupied when there are events of the same sort. Now I am not every moment using my eyes, still less seeing a particular colour, such as red, at every moment of my history. We cannot, therefore, have a section of our whole history in which our whole space is occupied at each moment with some event or other; nor one in which each point is occupied by some event or other through the whole of our time.

Except for this failure to find corresponding artificial sections in mind, the microcosm, to infinite Space-Time, the macrocosm, a failure founded on the finitude of mental space-time, the relations of Space and Time to one another are identical in the two. It is obvious that the exception would apply equally to any limited piece of Space-Time which is occupied by the life of a finite thing, whether that thing is mental or not, provided it has specific qualities.

V.

The Analysis of Temporal Concepts

Just because thinking is conceptual in character, it has been argued that we ought to clarify our concepts and conceptual structures so that we might eliminate confusions in our thinking. Thereby, we might hopefully resolve many of the debates in which we perhaps needlessly indulge. If we are to come to understand the nature and meaning of the time of human experience, it might then seem that we should give priority to a rigorous examination of the concepts we use to think about time. Such an examination might enable us to clarify misconceptions, avoid confusions, pinpoint those real issues which deserve further investigation or merit extended discussion, and at least enable us to limit our inquiries to those questions which appear to be truly meaningful. Saving ourselves from linguistic and logical confusions and needless conundrums might, indeed, point us to the heart of what is involved in our debates, shrouded as they may be in the jumbled thinking of ordinary utterances in our everyday experience.

By analyzing the concepts we use to discourse about time, John Ellis McTaggart, Bertrand Russell, and Alfred North Whitehead have each attempted to unveil something of the nature of temporal experience. They carried their investigations in different directions but nevertheless shared much in common. Each of these three preeminent Englishmen came out of Trinity College, Cambridge, during the same period; each claimed to start from the factuality of our temporal experience; each demonstrated a real sense of logical rigor and proceeded by means of a conceptual analysis. Each saw time and change as inextricably related and drew important consequences from this relation. Each drew a radical distinction between two different kinds of temporal discourse—between those time concepts which are applicable to the content of experience and those appropriate to description of the experiencing itself. Just which of these is to be regarded as primary is to become a bone of contention, as are the conceptual definitions which are proposed for our use. However similar they were in background and in many aspects of their think-ing, the divergent directions in which they proceeded did indicate that the conceptual clarifications they achieved could not alone remove the deeper differences between them. Following an essentially similar method of conceptual analysis, they propounded quite diverse ques-tions for further thought.

Coming into this discussion from the logical positivism of Berlin, Hans Reichenbach's questions arose out of his attempt to limit philosophic thought to the development of the logic of science. Ar-guing that the many paradoxes which had historically emerged in discussions of time actually arose out of a prescientific sense of human frustration with the necessary mortality of human life, he urged that we ignore our own emotions and turn to contemporary physics for a definitive answer to questions concerning the true nature of time as such. Once a close collaborator of Carnap, who had sought to develop logical formalism beyond Russell, Reichenbach effectively carried forward the thrust of the time analysis Russell had offered. In doing so, he has, for all practical purposes, brought into the discussion a challenge to the basic approach which is being pursued in this volume.

In many ways, McTaggart was as paradoxical a figure as is his place in the development of a philosophy of time. A personal idealist who was also an avowed atheist professing belief in immortality, he regarded himself as an opponent of the Christian religion and yet retained membership in the established church. A critic of Lotze, he saw himself as Hegelian in spirit and wrote several books critically expounding Hegelian views, but he seems to have taken much from Leibniz in the construction of his own systematic metaphysic.

In the developing dialectic of discussion, it might have been ex-pected that the post-Kantian attempts to objectify the ontological status of time would be followed by an outright attack on any kind of temporal realism. In the wake of Bradley's famous attempt to relegate time to the realm of illusionary appearance, because of the con-tradictory predicates normally ascribed to it, McTaggart set out to demonstrate, once and for all, that time is "unreal." Perhaps it was because of this quixotic philosophic project that McTaggart attracted so much attention; for, of him it has been remarked that "no other contemporary philosopher has been graced by so extensive a commentary." [1]

Indeed, McTaggart set out to construct a completely deductive metaphysical system but is best remembered for his preoccupation with this thesis of the unreality of time. The last twenty years of his life were largely dominated by it. Propounded in a paper published at the beginning of this century, it was later incorporated into a two-volume metaphysical study that was largely built around it (and which was posthumously published).

Although he urged that time is not real, he did insist that it is a "well-founded phenomenon," [2] and showed more insight into its, at least phenomenal, nature than many other philosophers. His argument against temporal realism works from the acknowledged pervasive temporality of human experience and builds up from a crucial distinction between two kinds of temporal discourse which have almost always been confused. Ironically, McTaggart's distinction, developed as a prime step in the argument against temporal realism, has proven to be an invaluable conceptual tool for better comprehending earlier temporalist discussions and for avoiding ambiguity in our own.

Sometimes, he pointed out, we say that the object or event of reference is 'past' or 'present' or 'future'. But the appropriateness of one of these temporal predicates depends not on the object itself, but on the relationship it has with the person describing it. In order to ascertain the truth of such a tensed statement, we have to know the who or the when of the speaker. The same event is future to its prophet, and past to its historian; it is progressively future, present and past to one participating in it. The truth of such a statement by prophet, historian, participant, or distinterested observer necessarily depends on the tense which is used at the time that the statement is uttered—and the truth value of the statement changes with the tense. Just because the locus of truth is primarily in the changing perception of the speaking subject, this is a 'subjective' mode of temporal expression; McTaggart dubbed events described in this way as being in an "A series."

But we can also describe the same sequences in a tenseless way wherein the truth does not change and does not depend on either who is making the statement or when it is made. We can describe any sequence of events, or any procedure that is regarded as invariable, in this way: 'Roosevelt died before Churchill', 'the father is born before the son', 'the meat was raw before it was cooked', 'the second eclipse will take place later than the first one'. Such statements, dubbed as "B series" descriptions, are object-oriented, do not depend upon the relationship to the describing subject and are, therefore, 'objective'. Their truth, one might say, is 'permanent' and does not change with time; once true, always true, regardless of the 'when' of the utterance.

Having made this distinction, McTaggart proceeded in clearly marked stages, by a somewhat ingenious argumentation, to establish his conclusion—that time is unreal. First, he proclaimed the complete priority of the A series. It is, he argued, fundamental to the experience of time just because it recognizes the dynamic quality of temporal change and reflects it in utterance. Without the A type of temporal discrimination, we could not even experience the process of change. We say that 'Greek civilization was earlier than ours'—but we can only recognize the truth of this statement, and the process of transition and change it urges by transposing this B description into the past-present-future of A series discourse. The B series, he argued, is immediately seen, when we think about it, to be obviously derived from the authenticity of temporal experience which is always cast in terms of past or present or future.

But, in the second and crucial stage of the argument, he subjected A series discourse to logical analysis and found that the series is inherently contradictory—because the three tense predicates, applied to the same event, are logically incompatible with each other. To the obvious rejoinder that these predicates are correctly supplied at different times, McTaggart argued that each of these different times is itself subject to past-present-future discrimination and to this kind of analysis—which leads us to an infinite regress, that is, again, a reason for dismissing its claim to reality. Finally, he rooted time as a perceptual illusion of an invariable and non-temporal cosmic order which he termed the 'C series' which seems to have been his name for Reality—except that he later rooted this again in a 'D series' in a way that is not altogether clear.[3]

In essential thrust, then, McTaggart followed Hegel in subsuming temporal experience, exhibited in the A series, as a defective perception, under a timeless logically consistent order. But, while disavowing the ultimate reality of time, he yet, like Hegel, displayed a keen appreciation of its central experiential role. With special reference to the "practical consequences," he argued the importance of the B series and pointed to the fundamental import of the concept of the future which is inherent in the A series.[4]

McTaggart's argument for the unreality of time obviously rests on his logicism, his subsumption of time under the sovereignty of a presumably timeless logic. Kant had already warned against this kind of attempt to "ever deduce and explain the concept of time by the help of reason" because the fundamental law of formal logic itself *depends upon* time (see p. 149) and cannot be extended beneath its own grounding. Ignoring this admonition, McTaggart has effectively taken the law of contradiction as non-temporal, thereby as an eternal facet of reality, and effectively presumed it as a non-temporal criterion of the real—although it is not clear just what the evidence for the truth of such a presumption might be.[5]

However we may judge the issue between Kant's grounding of logic on time and McTaggart's subsumption of time under logic, his logicism and the use for which he developed his insight into the dual nature of temporal discourse, the fact remains that his distinction is crucial and we can ill afford not to use it. He has given us an exceedingly useful conceptual tool for clarifying the discussions about the nature and the significance of our temporal experience. It opens up an insight into at least one essential issue implicitly at the heart of many old debates. We can see, for example, that Aristotle was using a B kind of discrimination, that Plotinus' attack implicitly employed, in part, an A criterion to attack a B kind of theory, and that Augustine's perplexities arose from problems yet to be faced in an A kind of inquiry.

But we also face the question for our own thinking of whether McTaggart's announced priority of the A type of series has been securely established. Much of the disputation about the nature of time or temporal experience in contemporary analytic philosophy revolves around this second question.[6] The attack on this A priority was carried forward by McTaggart's student, Bertrand Russell, and most analytic thinkers have chosen up sides in their dispute. To keep McTaggart's distinction in mind is not only a means of keeping one's bearings as one explores that conflict of opinion, whether argued in terms of logic or of language. His priority of the A series, and its implicit significance, is also a key to the discussions of temporal experience in both pragmatism and phenomenology whose acceptance and development of its essential thrust is one of the bonds that draws them together.

In effect, Russell's 1915 article, "On the Experience of Time," was an opening dissent from McTaggart's thesis—which had been announced in his 1908 paper—that the dynamic A series is fundamental to experience and that the logically static B series is derivative from it. In the end, McTaggart did not seem to have taken this priority as seriously as his rather brilliant argumentation seemed to suggest. Although he accorded the A series full priority on an experiential level, he had rooted it metaphysically in a transcendent and non-temporal C series, which looks very much like an hypostatization of the derived B series itself. It is then no surprise that Russell, a young logician who had turned away from his early attachment to McTaggart's brand of Hegelianism, and was suspicious of any transcendent metaphysic, should argue that the C series is really unnecessary because we already have a logically amenable time series in the B series.

Russell, arguing that the A series is "merely" psychological and derived, urged that the B series is fundamental. A transcendent priority of logical categories was thus made mundane, and the B series, just because it is readily adaptable to rules of logical consistency and implication, was taken as the fundamental form of temporal experience. If we accept McTaggart's description of the B series as essentially non-temporal, it is easy to understand this move—as neither the logician's concept of implication nor that of truth is conceived in temporal terms. Clearly in accord with his overall view that "a certain emancipation from slavery to time is essential to philosophical thought,[7] Russell's strategy in his tightly written article was to analyze some concepts of mental and physical time and then to connect them together.

The outline of his argument is important to keep in mind. Seeking to elucidate what we know by acquaintance, he first presented summary definitions of the nine concepts he regarded as crucial, making it clear that he was starting with the objects in perception, whether internal or external. Expanding upon the first five presented—'sensation', 'present', 'simultaneity', 'now', and 'the present time'—he built the definition of each upon the one preceding. The expanded analysis of these five concepts, he made it explicitly clear, "completes our theory of the knowledge of the present" (see p. 305).

The notion of 'immediate memory' was then subjected to a long discussion which concluded that, without considering its extent, it does provide the notion of a time-series. Russell then turned to the concept of 'succession' which "is a relation which is given between objects, and belongs to physical time, where it plays a part analogous to that played by memory in the construction of mental time" (p. 309), and may be immediately experienced. The notion of 'succession' involves those of 'earlier' and 'later' which were next examined, and we are reminded that they do not necessarily imply the notions of 'past' and 'present' just because "there is no logical reason why the relations of earlier and later should not subsist in a world wholly devoid of consciousness" (309). Finally, he turned to the notions of 'past' and 'future': "An event is said to be *past* when it is earlier than the whole of the present, and is said to be *future* when it is later than the whole of the present" (p. 309). The future, he argued, is not directly experienced and thus can only be known in terms of inferential description.

Finally, he proceeded to elucidate the four propositions which tie these definitional discussions together: (a) arguing that both simultaneity and succession give rise to transitive relations, but that the first only does so if the 'instant' can be specified, the term 'instant' was defined, and its existence, as the beginning of an event, maintained; (b) reminding us that 'pastness' is defined in terms of 'earlier', the content of memory was described as what is past; (c) immediately experienced change was declared experiential in sensation because part of the present is earlier than other elements of sensations; (d) two presents may overlap without requiring that they be taken as coinciding.

It is to be noted that although Russell defined with some precision the concepts on which he built, the whole argument rests on the notion of the present which has been left ambiguous throughout, only described in terms of the few seconds of the admittedly 'specious present', and left resting on an almost atomic notion of sensation which was not really defended. To recognize one's own experience in these terms—without memory, much less anticipation, without reference to even the 'train of ideas' in consciousness, and without any consideration of the mental activity involved in perception—seems somewhat grotesque. To claim that this "completes" a "theory of the knowledge of the present" is either audacious or naive. Memory, Russell has only invoked to explain the notion of pastness. Succession has been relegated to the 'external world' while we are told that it is "analogous" to memory, although the nature of the analogy is not spelled out. And hovering in the background is the unacknowledged presence of a notion of Newtonian time, clicking its way as a celestial clock into the past.

Perhaps most crucial is Russell's relegation of 'past' and 'present' to the psychological—"whereas 'earlier' and 'later' can be known by an experience of non-mental objects" (p. 309), which are presumably better because of the implicit invocation of a scheme of value not even alluded to. But this experience itself, according to Russell's own account of sensation, is in terms of the 'present' which is to say that what he has disparaged as the 'merely psychological' is experientially prior, and the only way in which we could have any knowledge of non-mental objects. On his own account, then, and despite his own argument, the notions of 'earlier' and 'later' are derived from within the psychological present, however that 'present' be defined. 'Earlier' and 'later' may indeed "subsist" in a world without consciousness, but they cannot enter into the content of consciousness except in the way in which consciousness is itself structured or functionally operative.

If Russell was, indeed, serious about the title of his article, it would seem that he was working with a rather poverty-stricken notion of experience. Presuming something of a passive receptor theory of sensation–reception, and, in a way, Locke's unempirical notion that 'things' have a determining causal influence or 'power' on the sensations by which we become aware of them, Russell's unspoken thesis is that experience is ultimately derivable from the 'things' that affect us. He had conceded the logical acceptability of constructing the time-series from the Bergsonian thesis of the uniqueness of memory-moments but has given us no reason other than his personal prejudice against whatever it is that he terms 'metaphysics' for his insistence on basing experiential time on the so-called 'external world' instead.[8]

But the continuity which we experience in the 'external world' does not appear to be reflected in his very atomic view of sensation and time experience; his explanation of our experience of time, left as it has been, creates all sorts of problems for the notion of continuity as it appears in our knowledge of the 'external world' as well as in the awarness of our own selves. Russell does not really appear to have been so concerned with the experience of time, claimed by the title of his paper, as much as with ways to discriminate particular times in the 'external' content of our experiences; but it is not readily apparent that he has provided a satisfactory guide to this even more limited goal.

If we push on, however, to understand the full significance of such a B series priority, we find ourselves led to two somewhat different approaches which are not as dissimilar as they may at first seem to be. Having determined to give priority of explanation to the changeless B series, which records changes in an unchangeable way, one is free to propose a complete methodological priority of logical formalism for its study. As Quine's work has clearly shown, the reliance on a timeless logic quickly takes us to a dismissal of the import of time itself and the prescription to translate it into terms more commensurate with its negation.

In our ordinary language, Quine has noted the fact that "relations of date are exalted grammatically as relations of position, weight, and color are not." Without pausing to consider just *why* natural language should give this expression of tensed experience such a privileged position, he has complained of this "tiresome bias in its treatment of time . . . [which] is of itself an inelegance, or breach of theoretical simplicity." [9] Turning our back on the authenticity of that experience which generates a science of nature, he has urged that we follow the precedent of the physical sciences and treat time "on a par with space." [10]

If we are to look to the physical sciences for method, there is no obvious reason why we should not look there for content as well. We are asked to turn to the subject matter of the physical sciences, generally regarded as amenable to conceptual control by formalized systems. Emerging out of a study of the causal links marking the patterns of change in natural phenomena, rather than out of the human experience which discerns them, time is to be understood as little more than their sequential distancings. Concerned not only with the methodology, but also the findings, of modern physics, we are urged to turn to its reports in order to finally find out just what time really is in itself and as the source of our temporal experience.

It is to a position such as this that Reichenbach would have us repair. In accord with the positivism he brings into the discussion, the history of metaphysics is dismissed in a series of caricatures, and legitimate philosophic thought is hopefully to be restricted to the logic of the scientific investigation of the physical aspects of the world which we inhabit.

It would have been helpful if Reichenbach had attended to his philosophic history with that same regard for fact and detail which is so intrinsic to the scientific methodology he claims to revere. Rather than work from a textbook version of Heraclitus, for example, and dismiss his focus on change as "trivial," emotive, and useless regarding "the logical structure of time" (p. 321), he might have considered the import of the *Logos* not only to Heraclitus but to those who have since sought to take the temporality of all experience with the seriousness it might seem to deserve. It is indeed revealing of the lack of empathy for the temporal, that Reichenbach finds more in the eleaticism of Zeno, who was concerned to demonstrate the logical impossibility of change (and, thereby, of time),[11] than in Heraclitus who recognized that it must be subject to a law that pervades it.

The fundamental irrelevance of philosophic thought, at least as it bears on questions of time, is effectively urged. Dismissed as "mere[ly] psychological" (p. 323), Reichenbach sees most of the traditional philosophic thinking in the long prescientific age to have been dominated by an understandable but subjective emotive response to the prospect of imminent death. As the tradition emerged from Parmenides and Plato, it looked to a timeless eternity as its source of emotive stability and sustenance. Having dismissed the Heraclitean insistence that "flux is everything" (p. 329) as "trivial," his unsophisticated overview of philosophic history leads to his essential theme: "that the study of time is a problem of physics;" the concept of time is then reduced to "the time order that connects physical events" (p. 322). And, for some reason, which is wholly unexplicated, "physical events" are taken as synonomous with 'reality' and 'truth'. Kant is credited with having seen the fundamental connection between time and causality in classical mechanics (although Kant's rooting of causality in time is incongruously reversed!); in any event, it is to the further development of this relationship, especially as it appears in contemporary quantum physics, that Reichenbach looks for a solution of the problem of determinism which classical mechanics had apparently imposed on subsequent thought. Working on the assumption that the memory of the human mind is merely a "registering instrument" [12] recording external physical experienced phenomena, he looks to indeterminacy in physical nature in order to find the justification for finding it in man.

Reichenbach dismisses the relevance of the philosophic tradition on the ground that it has produced a nest of puzzles and paradoxes which have so often "been projected into logic" (p. 318); these are, of course, dismissed in perjorative and essentially *ad hominem* pronouncements without explanation as "dissatisfied emotion . . . [in] defense mechanism." (p. 318). Rather, he has turned to the temporal order of nature as seen and understood by different stages of theoretical physics. But, unless such theoretical development is understood as a progressive revelation of divine truth, it would seem to represent, not pure descriptions of physical reality itself, but a growing set of interpretive structures constructed by human thought seeking to comprehend that external reality which appears and responds to increasingly sophisticated systems of human questioning. This is to say that the world which our theoretical physicists are describing is some kind of synthetic conceptual structure which not only involves that independent reality which appears to us but also the structure of the human thinking about it. Therefore, we cannot disengage the structure of the human point of view from what is reported as even the most sophisticated content within its scope. To claim to do so is no service to that science which one claims to venerate or to that scientific tradition which, properly understood, is one of the great testimonies to the reaches of the human spirit.

Seemingly presumed, and never argued, in Reichenbach's dismissal of what he has termed the "emotive significance" of philosophical temporalism is a four-fold set of metaphysical judgments: (a) that human mental activity is nothing more than a passive reflection of what is 'out there', (b) that such 'external' things have some kind of mysterious, yet presumably quantifiable, power to 'cause' the ideas found in human consciousness, (c) that the structure of human thinking processes, whatever it may be, makes no intrinsic contribution to the ideas by means of which thinking is done, (d) that what we are and how we are, in our emotions and our deeper thoughts, can be nothing more than a passive and somewhat irrelevant reflection of "physical fact," whatever that may be. 'Science' is equated with theoretical physics, which is concerned to study the 'dead' matter of which the world is supposedly made. In looking to science for content, no reference is made to the science of living things, to biology which could speak of 'biological clocks' and of the temporal cast of all, even unconscious, life. Just why the time of the living, much less of the psyche, should necessarily be derivative from that of the inert, is not made clear; the question is not even by implication given the honor of acknowledgement.

"Science," Heisenberg has tried to remind us, "is made by men, a self-evident fact that is too often forgotten." As a highly sophisticated expression of the human outlook, it is yet "rooted in conversations," and is "quite inseparable" from "more general questions ... [concerning] human, philosophical or political problems." [13] The meaning of the scientific endeavor itself, as of its reports to us, are then grounded in the human capacities for learning about and dealing with the world. If we are to integrate scientific findings into our philosophic investigations in a truly authentic manner, without distortion or misrepresentation, we are then required not to neglect or overlook the human component or context necessarily foundational to any scientific report. The human concern with time may, indeed, arise, as Reichenbach has argued, from the feeling of finitude compelled by the prospect of eventual death; but this is no excuse for dismissing the human perspective on the science it develops in response to its own questionings.

It would seem that any fundamental defense of a B series priority, whether in terms of a logicism or a physical scientism, must first come to terms with the justification of its priority and of the possibility of our knowing it in a way that squares with the dynamics of the human experiencing within which it comes to be known. This means that, when we begin with the human experiencing within which our knowledge arises, we are starting with something akin to what McTaggart had dubbed the A series. For, in a way that is certainly far from worked out, we see it announcing a fundamental intuition even if that intuition has not yet succeeded in comprehending the dynamic character of the experiential time which it has seen to be foundational to its own outlook.

Whatever may be said of a view such as Reichenbach's, it forces one prime consideration upon our attention—the factuality of our being *in* physical nature. But this is not to say that our minds, functioning within physical nature, are reducible to physical nature, any more than functioning within biological structures, they are necessarily wholly reducible to biological explanation. It is precisely to this kind of point that Whitehead, seeking to reconstruct the philosophy of the physical sciences, directed his attention while appraising our time-experience with a keen sensitivity to many of the issues involved.

His early essay on time in *The Concept of Nature* developed a sharp distinction between the abstract concept of time used in a scientific investigation of natural phenomena and the concrete manner in which time appears in our experiences. Eschewing the vocabulary of A and B series, his analysis of the whole concept of time-relatedness offers a cogent explanation of the real differences between them and indicates how the abstract time of the theoretical physicist arises from his concrete human experience.

Whitehead's discussion is within the context of his fundamental thesis that nature is a unitary process, that the traditional view of physical nature—triply atomic in that it has posited three unreconciled irreducibles: instants of time, points of space, and particles of matter—is fundamentally erroneous and philosophically obsolete. Borrowing freely from both Lotze and Bergson [14] and sharing a high mutuality of respect with Alexander, Whitehead saw the relationship of particulars to time, as to space, to be the central problem of philosophic thought. Using a descriptive method of philosophical reasoning, he worked from the insights of later absolute idealism and sought to transform them onto a realist foundation. This is to say that every meaningful proposition refers to the universe as exhibiting, in its process, some general metaphysical characteristic and what is, therefore, needed is the explanation of the particular fact in terms of the general character of the universe.

What is experienced is the quality of passage. It is experienced in mind, in our sense-awarenesses, and in the procedures of knowing; it is measured in nature by means of spatial extension. Without techniques for measuring time, we would have neither science nor civilized life generally. But 'passage', which is equivalent to Bergson's concept of time, is essential for sense-awareness and is reduced to a serial measurable character from the notion of duration or simultaneous systems. What is experientially primary is the notion of duration or simultaneity which Whitehead conceived as exhibiting within itself the fact of passage and 'temporal thickness' beyond any 'specious present'.

In contrast, he urged that the concept of 'instant' is merely a logical one; without temporal extension, it is non-experiential; science uses it as a conceptual knife-edged boundary to differentiate the 'before' and the 'after'. But because it is a merely conceptual device that is not experiential, its use to describe the real requires that it abandon any claim to being founded on observation. Likewise, a 'moment' is an abstract term designating all nature in an instant. Like Aristotle's nows, it is a boundary-concept lying outside of any given duration. It is useful for the definition of serial relations of temporal order, and it is fundamental to the notion of a time-series which is thus based on an intellectual abstractive process.

The notion of a time-series is necessary for measurement—but it raises the question of just *what* is being measured; this 'whatness' is so "fundamental in experience" that we cannot see it externally. The question, then, is whether time is in nature or nature is in time. If the latter, we have a "metaphysical enigma." Reiterating views shared with his Cambridge colleagues, Whitehead urged that time is, because there are happenings and changes; that apart from dynamic events, there can be no time. So it would seem that time is in nature—but in some sense, it also seems to extend beyond nature. For time is not only an aspect of physical nature; it is also a crucial aspect of sense-awareness, of thinking, of the awareness of the self.

Useful as the notion of a time-series may be, temporal passage is experienced, not in a series, but in terms of the passage of events. We experience passage in sense-awareness as a relationship to nature, but memory is also present to mind; how then may we differentiate between what is present and what is past? We posit memory, as in serial order, while we also posit the thesis that something is going on in the 'external world'. But our sense-experiences exhibit our own minds to us, as in passage; and the fact of memory exhibits the mind's ability to disengage itself from the passage of nature in a somewhat autonomous way.

Just as the distinction between memory and the immediate experiential present is not clear, so it is essentially ambiguous just what can be meant by 'immediate duration'. The experiential present fades into present memory and present anticipation; it points to the essential continuity of experiencing in a moving present which is separate from nature although somehow allied to it. We may, indeed, "speculate, if we like, that this alliance of the passage of mind with the passage of nature arises from their both sharing in some ultimate character of passage which dominates all being." But this speculation is not a matter for present concern.

In looking 'out' to nature, we find that the word 'present' has no meaning; we consider nature in terms of the abstract concept of the 'instant' which divides the temporal field, without any intermediary breadth, into a 'past' and a 'future'. But, when we look into ourselves, we find that what is immediate in our experience is precisely a spread of the present which we call 'duration'; it is internally structured in terms of an indefinite extension into the past and into the future, like a line with an arrowhead at both ends. But this forward-and-backward looking, in terms of which we understand the 'spread' of the present, would seem to be an act of mind which is absent from physical nature.

In our experience, the center is the present which is "ill-defined" but within it "the past and the future meet and mingle." The matrix of experience, then, is the experienced event and the event of experiencing itself; what we find in examining any event is that it is necessarily conceived in terms of past, present, and future, which somehow come together into the unity of a process that essentially incorporates temporal passage.

But if what Whitehead has said here makes any sense, he seems to be suggesting something more. In contrast to Russell's dismissal of the future as a merely conceptual inference, Whitehead has pointed out that it is somehow presently involved with the somehow presently involved past in the experiential present. This suggests that, in contrast to what McTaggart had said, the past-present-future is *not* to be conceived as a series at all; doing so is implicitly reading our experience of these modes in terms of the sequence of before-and-after moments. No wonder that McTaggart could not come to terms with his own announced priority! Somehow, in a way left by Whitehead unexplained, the past-and-future intermingle in the experiential present—in a way that cannot be explained in terms of mere sequentiality or serial order. If Whitehead's thesis of a spread-present holds, then it would seem that *it* is experientially primary and that only within its spread can the whole notion of sequence emerge.

At the outset of this discussion, McTaggart had recognized that we must start with human experientiality and thus accord experiential priority to what he termed the A series. But, it quickly became apparent that he regarded this as a somewhat mistaken way in which finite human reason apprehends the nature of what he assumed to be the non-temporal real. Defining reality as a supervening system of unalterable order, its non-dynamic nature is more accurately reflected in the non-dynamic B series which we derive from our own illusionary mode of apprehension. Russell seems to have taken this forward as he saw the objectivity of the B series as more fundamental and as a key to that continuity which bridges the separate moments in terms of which he defined the temporal aspect of experience. Avoiding McTaggart's speculative metaphysic of a transcendent order, he effectively argued the priority of the B series while dismissing the A series as "merely psychological." Reichenbach apparently picked this up and urged the obvious consequent, the seeming irrelevance of the human in the human understanding of physical time and the authority of today's science as the revealer of time's true nature.

Whitehead is really closer to the insight of McTaggart's original intuition; but he has carried it forward without reverting here to any notion of timelessness. Taking human experience as authentic and not illusionary, he has effectively argued the priority of something like an A orientation and offered an explanation of why and how we derive the notion of a B series from it. He thus pointed up the crucial distinction between (a) how we experience temporal passage and (b) how we seek to describe it with the kind of precision which scientific inquiry into the physical object-world requires. Taking the present as the locus of experience, he saw that it could be understood neither as a point nor as specious, that it is experienced as a 'spread' which somehow encompasses an involvement with past and with future.

We are then pointed to the question of just how this spread of the living present, which defies precise measurement but in which all experience is to be had, is structured. But another inquiry must first be explored. If Whitehead was correct, the heart of the distinction between the A orientation and the B series is the distinction between our experience of time and the way in which we transform it in order to use it for the measuring essential to our understanding and manipulation of nature. And it is this that is at the heart of the pragmatic preoccupation with the pervasiveness of the temporal which provides, in the experiential present, the link between man and nature. For, if our experience is, in any sense, integral, the way we use our concepts of time not only reflects the nature of our experience, it is also reflected back into the experience we do have. In the inquiries which pragmatism undertook into this mutual temporal bond between man's internal experience and the experience of 'external' nature, there is to be found a prefiguring of many elements of the phenomenological concern with the structuring of time experience; we also find many penetrating suggestions for the significance of temporal concepts in temporal experience, and the significance of the temporal for the understanding of that human experience out of which our temporal concepts, as all our other concepts, arise.

16.

McTaggart:

"THE UNREALITY OF TIME" *

John McTaggart Ellis McTaggart (1866–1925) propounded in this essay the theme for which he became famous; originally published in 1908, it reappeared virtually unchanged except for minor stylistic changes and a rejoinder to some critics, as the first chapter on time in the second volume of his major work, *The Nature of Existence*. Published posthumously in 1927 under Broad's editorship, his attack on temporal realism was incorporated into what is an heroic attempt to construct a completely deductive metaphysic which yet accords with human experience. Essential to his argument is a crucial delineation of two very different concepts usually confused in temporal discourse. Although McTaggart did not believe that time is real, his distinction might well be taken to heart by those who do.

It doubtless seems highly paradoxical to assert that Time is unreal, and that all statements which involve its reality are erroneous. Such an assertion involves a far greater departure from the natural position of mankind than is involved in the assertion of the unreality of Space or of the unreality of Matter. So decisive a breach with that natural position is not to be lightly accepted. And yet in all ages the belief in the unreality of time has proved singularly attractive.

In the philosophy and religion of the East we find that this doctrine is of cardinal importance. And in the West, where philosophy and religion are less closely connected, we find that the same doctrine continually recurs, both among philosophers and among the theologians. Theology never holds itself apart from mysticism for any long period, and almost all mysticism denies the reality of time. In philosophy, again, time is treated as unreal by Spinoza, by Kant, by Hegel, and by Schopenhauer. In the philosophy of the present day the

* *Mind,* New Series, no. 68 (October 1908), pp. 457–474.

two most important movements (excluding those which are as yet merely critical) are those which look to Hegel and to Mr. Bradley. And both of these schools deny the reality of time. Such a concurrence of opinion cannot be denied to be highly significant—and is not the less significant because the doctrine takes such different forms, and is supported by such different arguments.

I believe that time is unreal. But I do so for reasons which are not, I think, employed by any of the philosophers whom I have mentioned, and I propose to explain my reasons in this paper.

Positions in time, as time appears to us *prima facie,* are distinguished in two ways. Each position is Earlier than some, and Later than some, of the other positions. And each position is either Past, Present, or Future. The distinctions of the former class are permanent, while those of the latter are not. If M is ever earlier than N, it is always earlier. But an event, which is now present, was future and will be past.

Since distinctions of the first class are permanent, they might be held to be more objective, and to be more essential to the nature of the mind. I believe, however, that this would be a mistake, and that the distinction of past, present and future is as *essential* to time as the distinction of earlier and later, while in a certain sense, as we shall see, it may be regarded as more *fundamental* than the distinction of earlier and later. And it is because the distinctions of past, present and future seem to me to be essential for time, that I regard time as unreal.

For the sake of brevity I shall speak of the series of positions running from the far past through the near past to the present, and then from the present to the near future and the far future, as the A series. The series of positions which runs from earlier to later I shall call the B series. The contents of a position in time are called events. The contents of a single position are admitted to be properly called a plurality of events. (I believe, however, that they can *as* truly, though not *more* truly, be called a single event. This view is not universally accepted, and it is not necessary for my argument.) A position in time is called a moment.

The first question which we must consider is whether it is essential to the reality of time that its events should form an A series as well as a B series. And it is clear, to begin with, that we never *observe* time except as forming both these series. We perceive events in time as being present, and those are the only events which we perceive directly. And all other events in time which, by memory or inference, we believe to be real, are regarded as past or future—those earlier than the present being past, and those later than the present being future. Thus the events of time, as observed by us, form an A series as well as a B series.

It is possible, however, that this is merely subjective. It may be the case that the distinction introduced among positions in time by the A series—the distinction of past, present and future—is simply a constant illusion of our minds, and that the real nature of time only contains the distinction of the B series—the distinction of earlier and later. In that case we could not *perceive* time as it really is but we might be able to *think* of it as it really is.

This is not a very common view, but it has found able supporters. I believe it to be untenable, because, as I said above, it seems to me that the A series is essential to the nature of time, and that any difficulty in the way of regarding the A series as real is equally a difficulty in the way of regarding time as real.

It would, I suppose, be universally admitted that time involves change. A particular thing, indeed, may exist unchanged through any amount of time. But when we ask what we mean by saying that there were different moments of time, or a certain duration of time, through which the thing was the same, we find that we mean that it remained the same while other things were changing. A universe in which nothing whatever changed (including the thoughts of the conscious beings in it) would be a timeless universe.

If, then, a B series without an A series can constitute time, change must be possible without an A series. Let us suppose that the distinction of past, present and future does not apply to reality. Can change apply to reality? What is it that changes?

Could we say that, in a time which formed a B series but not an A series, the change consisted in the fact that an event ceased to be an event, while another event began to be an event? If this were the case, we should certainly have got a change.

But this is impossible. An event can never cease to be an event. It can never get out of any time series in which it once is. If N is ever earlier than O and later than M, it will always be, and has always been, earlier than O and later than M, since the relations of earlier and later are permanent. And as, by our present hypothesis, time is constituted by a B series alone, N will always have a position in a time series, and has always had one.[1] That is, it will always be, and has always been, an event, and cannot begin or cease to be an event.

Or shall we say that one event M merges itself into another event N, while preserving a certain identity by means of an unchanged element, so that we can say, not merely that M has ceased and N begun, but that it is M which has become N? Still the same difficulty recurs. M and N may have a common element, but they are not the same event, or there would be no change. If therefore M changes into N at a certain moment, then, at that moment, M has ceased to be M, and N has begun to be N. But we have seen that no event can cease to be, or begin to be, itself, since it never ceases to have a place as itself in the B series. Thus one event cannot change into another.

Neither can the change be looked for in the numerically different moments of absolute time, supposing such moments to exist. For the same arguments will apply here. Each such moment would have its own place in the B series, since each would be earlier or later than each of the others. And as the B series indicate permanent relations, no moment could ever cease to be, nor could it become another moment.

Since, therefore, what occurs in time never begins or ceases to be, or to be itself, and since, again, if there is to be change it must be change of what occurs in time (for the timeless never changes), I submit that only one alternative remains. Changes must happen to the events of such a nature that the occurrence of these changes does not hinder the events from being events, and the same events, both before and after the change.

Now what characteristics of an event are there which can change and yet leave the event the same event? (I use the word characteristic as a general term to include both the qualities which the event possesses, and the relations of which it is a term—or rather the fact that

1. It is equally true, though it does not concern us on the hypothesis which we are now considering, that whatever is once in a A series is always in one. If one of the determinations past, present, and future can ever be applied to N, then one of them always has been and always will be applicable, though of course not always the same one.

the event is a term of these relations.) It seems to me that there is only one class of such characteristics—namely, the determination of the event in question by the terms of the A series.

Taken any event—the death of Queen Anne, for example—and consider what change can take place in its characteristics. That it is a death, that it is the death of Anne Stuart, that it has such causes, that it has such effects—every characteristic of this sort never changes. "Before the stars saw one another plain" the event in question was a death of an English Queen. At the last moment of time—if time has a last moment—the event in question will still be a death of an English Queen. And in every respect but one it is equally devoid of change. But in one respect it does change. It began by being a future event. It became every moment an event in the nearer future. At last it was present. Then it became past, and will always remain so, though every moment it becomes further and further past.

Thus we seem forced to the conclusion that all change is only a change of the characteristics imparted to events by their presence in the A series, whether those characteristics are qualities or relations.

If these characteristics are qualities, then the events, we must admit, would not be always the same, since an event whose qualities alter is, of course, not completely the same. And, even if the characteristics are relations, the events would not be completely the same, if—as I believe to be the case—the relation of X to Y involves the existence in X of a quality of relationship to Y.[2] Then there would be two alternatives before us. We might admit that events did really change their nature, in respect of these characteristics, though not in respect of any others. I see no difficulty in admitting this. It would place the determinations of the A series in a very unique position among the characteristics of the event, but on any theory they would be very unique characteristics. It is usual, for example, to say that a past event never changes, but

2. I am not asserting, as Lotze did, that a relation between X and Y *consists* of a quality in X and a quality in Y—a view which I regard as quite indefensible. I assert that a relation Z between X and Y *involves* the existence in X of the quality "having the relation Z to Y" so that a difference of relations always involves a difference in quality, and a change in relations always involves a change of quality.

I do not see why we should not say, instead of this, "a past event changes only in one respect—that every moment it is further from the present than it was before". But although I see no intrinsic difficulty in this view, it is not the alternative I regard as ultimately true. For if, as I believe, time is unreal, the admission that an event in time would change in respect of its position in the A series would not involve that anything really did change.

Without the A series then, there would be no change, and consequently the B series by itself is not sufficient for time, since time involves change.

The B series, however, cannot exist except as temporal, since earlier and later, which are the distinctions of which it consists, are clearly time-determinations. So it follows that there can be no B series where there is no A series, since where there is no A series there is no time.

But it does not follow that, if we subtract the determinations of the A series from time, we shall have no series left at all. There is a series—a series of the permanent relations to one another of those realities which in time are events—and it is the combination of this series with the A determinations which gives time. But this other series—let us call it the C series—is not temporal, for it involves no change, but only an order. Events have an order. They are, let us say, in the order M, N, O, P. And they are therefore *not* in the order M, O, N, P, or O, N, M, P, or in any other possible order. But that they have this order no more implies that there is any change than the order of the letters of the alphabet, or of the Peers on the Parliament Roll, implies any change. And thus those realities which appear to us as events might form such a series without being entitled to the name of events, since that name is only given to realities which are in a time series. It is only when change and time come in that the relations of this C series become relations of earlier and later, and so it becomes a B series.

More is wanted, however, for the genesis of a B series and of time than simply the C series and the fact of change. For the change must be in a particular direction. And the C series, while it determines the order, does not determine the direction. If the C series runs M, N, O, P, then the B series from earlier to later cannot run M, O, N, P, or M, P, O, N, or in any way but two. But it can run either M, N, O, P (so that

M is earliest and P latest) or else P, O, N, M (so that P is earliest and M latest). And there is nothing either in the C series or in the fact of change to determine which it will be.

A series which is not temporal has no direction of its own, though it has an order. If we keep to the series of the natural numbers, we cannot put 17 between 21 and 26. But we keep to the series, whether we go from 17, through 21, to 26, or whether we go from 26, through 21, to 17. The first direction seems the more natural to us, because this series has only one end, and it is generally more convenient to have that end as a beginning than as a termination. But we equally keep to the series in counting backward.

Again, in the series of categories in Hegel's dialectic, the series prevents us from putting the Absolute Idea between Being and Causality. But it permits us either to go from Being, through Causality, to the Absolute Idea, or from the Absolute Idea, through Causality, to Being. The first is, according to Hegel, the direction of proof, and is thus generally the most convenient order of enumeration. But if we found it convenient to enumerate in the reverse direction, we should still be observing the series.

A non-temporal series, then, has no direction in itself, though a person considering it may *take* the terms in one direction or in the other, according to his own convenience. And in the same way a person who contemplates a time-order may contemplate it in either direction. I may trace the order of events from the Great Charter to the Reform Bill, or from the Reform Bill to the Great Charter. But in dealing with the time series we have not to do merely with a change in an external contemplation of it, but with a change which belongs to the series itself. And this change has a direction of its own. The Great Charter came before the Reform Bill, and the Reform Bill did not come before the Great Charter.

Therefore, besides the C series and the fact of change there must be given—in order to get time—the fact that the change is in one direction and not in the other. We can now see that the A series, together with the C series, is sufficient to give us time. For in order to get change, and change in a given direction, it is sufficient that one position in the C series should be Present, to the exclusion of all others, and that this characteristic of presentness should pass along the series in such a way that all positions on the one side of the Present have been present, and all positions on the other side of it will be present. That which has been present is Past, that which will be present is Future.[3] Thus to our previous conclusion that there can be no time unless the A series is true of reality, we can add the further conclusion that no other elements are required to constitute a time-series except an A series and a C series.

We may sum up the relations of the three series to time as follows: The A and B series are equally essential to time, which must be distinguished as past, present and future, and must likewise be distinguished as earlier and later. But the two series are not equally fundamental. The distinctions of the A series are ultimate. We cannot explain what is meant by past, present and future. We can, to some extent, describe them, but they cannot be defined. We can only show their meaning by examples. "Your breakfast this morning," we can say to an inquirer, "is past: this conversation is present; your dinner this evening is future." We can do no more.

The B series, on the other hand, is not ultimate. For, given a C series of permanent relations of terms, which is not in itself temporal, and therefore is not a B series, and given the further fact that the terms of this C series also from an A series, and it results that the terms of the C series become a B series, those which are placed first, in the direction from past to future, being earlier than those whose places are further in the direction of the future.

The C series, however, is as ultimate as the A series. We cannot get it out of anything else. That the units of time do form a series, the relations of which are permanent, is as ultimate as the fact that each of them is present, past, or future. And this ultimate fact is essential to time. For it is admitted that it is essential to time that each moment of it shall either be earlier or later than any other moment; and these relations are permanent. And this—the B series—cannot be got out of the A series alone. It is only when the A series, which gives change and direction, is combined with the C series, which gives permanence, that the B series can arise.

3. This account of the nature of the A series is not valid, for it involves a vicious circle, since it uses "has been" and "will be" to explain Past and Future. But, as I shall endeavour to show later on, this vicious circle is inevitable when we deal with the A series, and forms the ground on which we must reject it.

Only part of the conclusion which I have now reached is required for the general purpose of this paper. I am endeavouring to base the unreality of time, not on the fact that the A series is more fundamental than the B series, but on the fact that it is as essential as the B series—that the distinctions of past, present and future are essential to time, and that, if the distinctions are never true of reality, then no reality is in time.

This view, whether it is true or false, has nothing surprising in it. It was pointed out above that time, as we perceive it, always presents these distinctions. And it has generally been held that this is a real characteristic of time, and not an illusion due to the way in which we perceive it. Most philosophers whether they did or did not believe time to be true of reality, have regarded the distinctions of the A series as essential to time.

When the opposite view has been maintained, it has generally been, I believe, because it was held (rightly, as I shall try to show later on) that the distinctions of present, past and future cannot be true of reality, and that consequently, if the reality of time is to be saved, the distinction in question must be shown to be unessential to time. The presumption, it was held, was for the reality of time, and this would give us a reason for rejecting the A series as unessential to time. But of course this could only give a presumption. If the analysis of the notion of time showed that, by removing the A series, time was destroyed, this line of argument would be no longer open, and the unreality of the A series would involve the unreality of time.

I have endeavoured to show that the removal of the A series *does* destroy time. But there are two objections to this theory, which we must now consider.

The first deals with those time-series which are not really existent, but which are falsely believed to be existent, or which are imagined as existent. Take, for example, the adventures of Don Quixote. This series, it is said, is not an A series. I cannot at this moment judge it to be either past, present or future. Indeed I know that it is none of the three. Yet, it is said, it is certainly a B series. The adventure of the galley-slaves, for example, is later than the adventure of the wind-mills. And a B series involves time. The conclusion drawn is that an A series is not essential to time.

The answer to this objection I hold to be as follows. Time only belongs to the existent. If any reality is in time, that involves that the reality in question exists. This, I imagine, would be universally admitted. It may be questioned whether all of what exists is in time, or even whether anything really existent is in time, but it would not be denied that, if anything is in time, it must exist.

Now what is existent in the adventures of Don Quixote? Nothing. For the story is imaginary. The acts of Cervantes' mind when he invented the story, the acts of my mind when I think of the story—these exist. But then these form part of an A series. Cervantes' invention of the story is in the past. My thought of the story is in the past, the present, and—I trust—the future.

But the adventures of Don Quixote may be believed by a child to be historical. And in reading them I may by an effort of the imagination contemplate them as if they really happened. In this case, the adventures are believed to be existent or imagined as existent. But then they are believed to be in the A series, or imagined as in the A series. The child who believes them historical will believe that they happened in the past. If I imagine them as existent, I shall imagine them as happening in the past. In the same way, if any one believed the events recorded in Morris's *News from Nowhere* to exist, or imagined them as existent, he would believe them to exist in the future or imagine them as existent in the future. Whether we place the object of our belief or our imagination in the present, the past, or the future, will depend upon the characteristics of that object. But somewhere in our A series it will be placed.

Thus the answer to the objection is that, just as a thing is in time, it is in the A series. If it is really in time, it is really in the A series. If it is believed to be in time, it is believed to be in the A series. If it is imagined as in time, it is imagined as in the A series.

The second objection is based on the possibility, discussed by Mr. Bradley, that there might be several independent time-series in reality. For Mr. Bradley, indeed, time is only appearance. There is no real time at all, and therefore there are not several real series of time. But the hypothesis here is that there should be within reality several real and independent time-series.

The objection, I imagine, is that the time-series would be all real, while the distinction of past, present, and future would only have meaning within each series, and could not, therefore, be taken as ultimately real. There would be, for example, many presents. Now, of course, many points of time can be present (each point in each time-series is a present once), but they must be present successively. And the presents of the different time-series would not be successive, since they are not in the same time. (Neither would they be simultaneous, since that equally involves being in the same time. They would have no time-relation whatever.) And different presents, unless they are successive, cannot be real. So the different time-series, which are real, must be able to exist independently of the distinction between past, present, and future.

I cannot, however, regard this objection as valid. No doubt, in such a case, no present would be *the* present—it would only be the present of a certain aspect of the universe. But then no time would be *the* time—it would only be the time of a certain aspect of the universe. It would, no doubt, be a real time-series, but I do not see that the present would be less real than the time.

I am not, of course, asserting that there is no contradiction in the eixstence of several distinct A series. My main thesis is that the existence of *any* A series involves a contradiction. What I assert here is merely that, supposing that there could be any A series, I see no extra difficulty involved in there being several such series independent of one another, and that therefore there is no incompatibility between the essentiality of an A series for time and the existence of several distinct times.

Moreover, we must remember that the theory of a plurality of time-series is a mere hypothesis. No reason has ever been given why we should believe in their existence. It has only been said that there is no reason why we should disbelieve in their existence, and that therefore they may exist. But if their existence should be incompatible with something else, for which there is positive evidence, then there would be a reason why we should disbelieve in their existence. Now there is, as I have tried to show, positive evidence for believing that an A series is essential to time. Supposing therefore that it were the case (which, for the reasons given above, I deny) that the existence of a plurality of time-series was incompatible with the essentiality for time of the A series, it would be the hypothesis of a plurality of times which should be rejected, and not our conclusion as to the A series.

I now pass to the second part of my task. Having, as it seems to me, succeeded in proving that there can be no time without an A series, it remains to prove that an A series cannot exist, and that therefore time cannot exist. This would involve that time is not real at all, since it is admitted that, the only way in which time can be real is by existing.

The terms of the A series are characteristic of events. We say of events that they are either past, present, or future. If moments of time are taken as separate realities, we say of them also that they are past, present, or future. A characteristic may be either a relation or a quality. Whether we taken the terms of the A series as relations of events (which seems the more reasonable view) or whether we take them as qualities of events, it seems to me that they involve a contradiction.

Let us first examine the supposition that they are relations. In that case only one term of each relation can be an event or a moment. The other term must be something outside the time-series.[4] For the relations of the A series are changing relations, and the relation of terms of the time-series to one another do not change. Two events are exactly in the same places in the time-series, relatively to one another, a million years before they take place, while each of them is taking place, and when they are a million years in the past. The same is true of the relation of moments to each other. Again, if the moments of time are to be distinguished as separate realities from the events which happen in them, the relation between an event and a moment is unvarying. Each event is in the same moment in the future, in the present, and in the past.

The relations which form the A series then must be relations of events and moments to something not itself in the time-series. What this something is might be difficult to say. But, waiving this point, a more positive difficulty presents itself.

Past, present, and future are incompatible determinations. Every event must be one or the other, but no event can be more than one. This is essential to the meaning of the terms. And, if it were not so, the

4. It has been maintained that the present is whatever is simultaneous with the assertion of its presentness, the future whatever is later than the assertion of its futurity, and the past whatever is earlier than the assertion of its pastness. But this theory involves that time exists independently of the A series, and is incompatible with the results we have already reached.

A series would be insufficient to give us, in combination with the C series, the result of time. For time, as we have seen, involves change, and the only change we can get is from future to present, and from present to past.

The characteristics, therefore, are incompatible. But every event has them all. If M is past, it has been present and future. If it is future, it will be present and past. If it is present, it has been future and will be past. Thus all the three incompatible terms are predicable of each event, which is obviously inconsistent with their being incompatible, and inconsistent with their producing change.

It may seem that this can easily be explained. Indeed it has been impossible to state the difficulty without almost giving the explanation, since our language has verb-forms for the past, present, and future, but no form that is common to all three. It is never true, the answer will run, that M *is* present, past and future. It *is* present, *will be* past, and *has been* future. Or it *is* past, and *has been* future and present, or again *is* future and *will be* present and past. The characteristics are only incompatible when they are simultaneous, and there is no contradiction to this in the fact that each term has all of them successively.

But this explanation involves a vicious circle. For it assumes the existence of time in order to account for the way in which moments are past, present and future. Time then must be pre-supposed to account for the A series. But we have already seen that the A series has to be assumed in order to account for time. Accordingly the A series has to be pre-supposed in order to account for the A series. And this is clearly a vicious circle.

What we have done is this—to meet the difficulty that my writing of this article has the characteristics of past, present and future, we say that it is present, has been future, and will be past. But "has been" is only distinguished from "is" by being existence in the past and not in the present, and "will be" is only distinguished from both by being existence in the future. Thus our statement comes to this—that the event in question is present in the present, future in the past, past in the future. And it is clear that there is a vicious circle if we endeavour to assign the characteristics of present, future and past by the criterion of the characteristics of present, past and future.

The difficulty may be put in another way, in which the fallacy will exhibit itself rather as a vicious infinite series than as a vicious circle. If we avoid the incompatibility of the three characteristics by asserting that M is present, has been future, and will be past, we are constructing a second A series, within which the first falls, in the same way in which events fall within the first. It may be doubted whether any intelligible meaning can be given to the assertion that time is in time. But, in any case, the second A series will suffer from the same difficulty as the first, which can only be removed by placing it inside a third A series. The same principle will place the third inside a fourth, and so on without end. You can never get rid of the contradiction, for, by the act of removing it from what is to be explained, you produce it over again in the explanation. And so the explanation is invalid.

Thus a contradiction arises if the A series is asserted of reality when the A series is taken as a series of relations. Could it be taken as a series of qualities, and would this give us a better result? Are there three qualities—futurity, presentness, and pastness, and are events continually changing the first for the second, and the second for the third?

It seems to me that there is very little to be said for the view that the changes of the A series are changes of qualities. No doubt my anticipation of an experience M, the experience itself, and the memory of the experience are three states which have different qualities. But it is not the future M, the present M, and the past M, which have these three different qualities. The qualities are possessed by three distinct events—the anticipation of M, the experience M itself, and the memory of M, each of which is in turn future, present, and past. Thus this gives no support to the view that the changes of the A series are changes of qualities.

But we need not go further into this question. If the characteristics of the A series were qualities, the same difficulty would arise as if they were relations. For, as before, they are not compatible, and, as before, every event has all of them. This can only be explained, as before, by saying that each event has them successively. And thus the same fallacy would have been committed as in the previous case.[5]

---

5. It is very usual to present Time under the metaphor of a spatial movement. But is it to be a movement from the past to future, or from future to past? If the A series is taken as one of the qualities, it will naturally be taken as a movement from past to future, since the quality of presentness has belonged to the past states and will belong to the

We have come then to the conclusion that the application of the A series to reality involves a contradiction, and that consequently the A series cannot be true to reality. And, since time involves the A series, it follows that time cannot be true of reality. Whenever we judge anything to exist in time, we are in error. And whenever we perceive anything as existing in time—which is the only way in which we ever do perceive things—we are perceiving it more or less as it really is not.

We must consider a possible objection. Our ground for rejecting time, it may be said, is that time cannot be explained without assuming time. But may this not prove—not that time is invalid, but rather that time is ultimate? It is impossible to explain, for example, goodness or truth unless by bringing in the term to be explained as part of the explanation, and we therefore reject the explanation as invalid. But we do not therefore reject the notion as erroneous, but accept it as something ultimate, which, while it does not admit of explanation, does not require it.

But this does not apply here. An idea may be valid of reality though it does not admit of a valid explanation. But it cannot be valid of reality if its application to reality involves a contradiction. Now we began by pointing out that there was such a contradiction in the case of time—that the characteristics of the A series are mutually incompatible and yet all true of every term. Unless this contradiction is removed, the idea of time must be rejected as invalid. It was to remove this contradiction that the explanation was suggested that the char-

---

future states. If the A series is taken as one of relations, it is possible to take the movement either way, since either of the two related terms can be taken as the one which moves. If the events are taken as moving by a fixed point of presentness, the movement is from future to past, since the future events are those which have not yet passed the point, and the past are those which have. If presentness is taken as a moving point successively related to each of a series of events, the movement is from past to future. Thus we say that events come out of the future, but we say that we ourselves move towards the future. For each man identifies himself especially with his present state, as against his future or his past, since the present is the only one of which he has direct experience. And thus the self, if it is pictured as moving at all, is pictured as moving with the point of presentness along the stream of events from past to future.

acteristics belong to the terms successively. When this explanation failed as being circular, the contradiction remained unremoved, and the idea of time must be rejected, not because it cannot be explained, but because the contradiction cannot be removed.

What has been said already, if valid, is an adequate ground for rejecting time. But we may add another consideration. Time, as we have seen, stands and falls with the A series. Now, even if we ignore the contradiction which we have just discovered in the application of the A series to reality, was there ever any positive reason why we should suppose that the A series *was* valid of reality?

Why do we believe that events are to be distinguished as past, present and future? I conceive that the belief arises from distinctions in our own experience.

At any moment I have certain perceptions, I have also the memory of certain other perceptions, and the anticipation of others again. The direct perception itself is a mental state qualitatively different from the memory or the anticipation of perceptions. On this is based the belief that the perception itself has a certain characteristic when I have it, which is replaced by other characteristics when I have the memory or the anticipation or it—which characteristics are called presentness, pastness, and futurity. Having got the idea of these characteristics we apply them to other events. Everything simultaneous with the direct perception which I have now is called present, and it is even held that there would be a present if no one had a direct perception at all. In the same way acts simultaneous with remembered perceptions or anticipated perceptions are held to be past or future, and this again is extended to events to which none of the perceptions I now remember or anticipate are simultaneous. But the origin of our belief in the whole distinction lies in the distinction between perceptions and anticipations or memories of perceptions.

A direct perception is present when I have it, and so is what is simultaneous with it. In the first place this definition involves a circle, for the words "when I have it," can only mean "when it is present". But if we left out these words, the definition would be false, for I have many direct presentations which are at different times, and which cannot, therefore, all be present, except successively. This, however, is the fundamental contradiction of the A series, which has been already considered. The point I wish to consider here is different.

The direct perceptions which I now have are those which now fall within my "specious present". Of those which are beyond it, I can only have memory or anticipation. Now the "specious present" varies in length according to circumstances, and may be different for two people at the same period. The event M may be simultaneous both with X's perception Q and Y's perception R. At a certain moment Q may have ceased to be part of X's specious present. M, therefore, will at that moment be past. But at the same moment R may still be part of Y's specious present. And, therefore, M will be present, at the same moment at which it is past.

This is impossible. If, indeed, the A series was something purely subjective, there would be no difficulty. We could say that M was past for X and present for Y, just as we could say that it was pleasant for X and painful for Y. But we are considering attempts to take time as real, as something which belongs to the reality itself, and not only to our beliefs about it, and this can only be so if the A series also applies to the reality itself. And if it does this, then at any moment M must be present or past. It cannot be both.

The present through which events really pass, therefore, cannot be determined as simultaneous with the specious present. It must have a duration fixed as an ultimate fact. This duration cannot be the same as the duration of all specious presents, since all specious presents have not the same duration. And thus an event may be past when I am experiencing it as present, or present when I am experiencing it as past. The duration of the objective present may be the thousandth part of a second. Or it may be a century, and the accessions of George IV. and Edward VII. may form part of the same present. What reason can we have to believe in the existence of such a present, which we certainly do not observe to be a present, and which has no relation to what we do observe to be a present?

If we escape from these difficulties by taking the view, which has sometimes been held, that the present in the A series is not a finite duration, but a mere point, separating future from past, we shall find other difficulties as serious. For then the objective time in which events are will be something utterly different from the time in which we perceive them. The time in which we perceive them has a present of varying infinite duration, and therefore, with the future and the past, is divided into three durations. The objective time has only two durations, separated by a present which has nothing but the name in common with the present of experience, since it is not a duration but a point. What is there in our experience which gives us the least reason to believe in such a time as this?

And so it would seem that the denial of the reality of time is not so very paradoxical after all. It was called paradoxical because it seemed to contradict our experience so violently—to compel us to treat so much as illusion which appears *primâ facie* to give knowledge of reality. But we now see that our experience of time—centring as it does about the specious present—would be no less illusory if there were a real time in which the realities we experience existed. The specious present of our observations—varying as it does from you to me—cannot correspond to the present of the events observed. And consequently the past and future of our observations could not correspond to the past and future of the events observed. On either hypothesis—whether we take time as real or as unreal—everything is observed in a specious present, but nothing, not even the observations themselves, can ever *be* in a specious present. And in that case I do not see that we treat experience as much more illusory when we say that nothing is ever in a present at all, than when we say that everything passes through some entirely different present.

Our conclusion, then, is that neither time as a whole, nor the A series and B series, really exist. But this leaves it possible that the C series does really exist. The A series was rejected for its inconsistency. And its rejection involved the rejection of the B series. But we have found no such contradiction in the C series, and its invalidity does not follow from the invalidity of the A series.

It is, therefore, possible that the realities which we perceive as events in a time-series do really form a non-temporal series. It is also possible, so far as we have yet gone, that they do *not* form such a series, and that they are in reality no more a series than they are temporal. But I think—though I have no room to go into the question here—that the former view, according to which they really do form a C series, is the more probable.

Should it be true, it will follow that in our perception of these realities as events in time, there will be some truth as well as some error. Through the deceptive form of time, we shall grasp some of their true relations. If we say that the events M and N are simultaneous, we say that they occupy the same position in the time-series. And there will be some truth in this, for the realities, which we perceive as the events M and N, no really occupy the same position in a series, though it is not a temporal series. Again, if we assert that the events M, N, O are all at different times, and are in that order, we assert that they occupy different positions in the time-series, and that the position of N is between the positions of M and O. And it will be true that the realities which we see as these events will be in a series, though not in a temporal series, and that their positions in it will be different, and that the position of the reality which we perceive as the event N will be between the positions of the realities which we perceive as the events M and O.

If this view is adopted, the result will so far resemble those reached by Hegel rather than those of Kant. For Hegel regarded the order of the time-series as a reflexion, though a distorted reflexion, of something in the real nature of the timeless reality, while Kant does not seem to have contemplated the possibility that anything in the nature of the noumenon should correspond to the time order which appears in the phenomenon.

But the question whether such an objective C series does exist, must remain for future discussion. And many other questions press upon us which inevitably arise if the reality of time is denied. If there is such a C series, are positions in it simply ultimate facts, or are they determined by the varying amounts, in the objects which hold those positions, of some quality which is common to all of them? And, if so, what is that quality, and is it a greater amount of it which determines things to appear as later, and a lesser amount which determines them to appear as earlier, or is the reverse true? On the solution of these questions it may be that our hopes and fears for the universe depend for their confirmation or rejection.

And, again, is the series of appearances in time a series which is infinite or finite in length? And how are we to deal with the appearance itself? If we reduce time and change to appearance, must it not be an appearance which changes and which is in time, and is not time, then, shown to be real after all? This is doubtless a serious question, but I hope to show hereafter that it can be answered in a satisfactory way.

17.

Russell:

"ON THE EXPERIENCE OF TIME"

Coming to philosophy from training in mathematics, Bertrand Russell (1872–1970) earned preeminence as one of the great logicians. The godson of John Stuart Mill and a contemporary of McTaggart, he turned from an early enthusiasm for that brand of Hegelian rationalism to the logic of Peano and Frege; he was a colleague of Moore, at one point a collaborator of Whitehead, teacher of Wittgenstein, and a prolific and controversial writer on philosophic and social issues who became, in 1950, the second philosopher to be honored with a Nobel Prize. The concept of time has not played a crucial role in his philosophy, but, in 1915, he published an important article on time which constitutes the selection.*

In the present article, we shall be concerned with all those immediate experiences upon which our knowledge of time is based. Broadly speaking, two pairs of relations have to be considered, namely, (a) sensation and memory, which give time-relations between object and subject, (b) simultaneity and succession, which give time-relations among objects. It is of the utmost importance not to confuse time-relations of subject and object with time-relations of object and object; in fact, many of the worst difficulties in the psychology and metaphysics of time have arisen from this confusion. It will be seen that past, present, and future arise from time-relations of subject and object, while earlier and later arise from time-relations of object and object. In a world in which there was no experience there would be no past, present, or future, but there might well be earlier and later. Let us give the name of *mental time* to the time which arises through

* *The Monist* (Chicago: The Open Court Publishing Company), vol. XXV, no. 2, April 1915, pp. 212–233. By permission.

relations of subject and object, and the name *physical time* to the time which arises through relations of object and object. We have to consider what are the elements in immediate experience which lead to our knowledge of these two sorts of time, or rather of time-relations.

Although, in the finished logical theory of time, physical time is simpler than mental time, yet in the analysis of experience it would seem that mental time must come first. The essence of physical time is succession; but the experience of succession will be very different according as the objects concerned are both remembered, one remembered and one given in sense, or both given in sense. Thus the analysis of sensation and memory must precede the discussion of physical time.

Before entering upon any detail, it may be well to state in summary form the theory which is to be advocated.

1. *Sensation* (including the apprehension of present mental facts by introspection) is a certain relation of subject and object, involving acquaintance, but recognizably different from any other experienced relation of subject and object.

2. Objects of sensation are said to be *present* to their subject in the experience in which they are objects.

3. Simultaneity is a relation among entities, which is given in experience as sometimes holding between objects present to a given subject in a single experience.

4. An entity is said to be *now* if it is simultaneous with what is present to me, i.e., with *this*, where "this" is the proper name of an object of sensation of which I am aware.

5. *The present time* may be defined as a class of all entities that are *now*. [This definition may require modification; it will be discussed later.] [1]

6. *Immediate Memory* is a certain relation of subject to object, involving acquaintance, but recognizably different from any other experienced relation of subject and object.

7. *Succession* is a relation which may hold between two parts of one sensation, for instance between parts of a swift movement which is the object of one sensation; it may then, and perhaps also when one or both objects are objects of immediate memory, be immediately ex-

---

1 [Appears in square brackets in the original publication.—C.M.S.]

perienced, and extended by inference to cases where one or both of the terms of the relation are not present.

8. When one event is succeeded by another, the first is called *earlier* and the second *later.*

9. An event which is earlier than the whole of the present is called *past,* and an event which is later than the whole of the present is called *future.*

This ends our definitions, but we still need certain propositions constituting and connecting the mental and physical time-series. The chief of these are:

a. Simultaneity and succession both give rise to transitive relations; simultaneity is symmetrical, while succession is asymmetrical, or at least gives rise to an asymmetrical relation defined in terms of it.

b. What is remembered is past.

c. Whenever a change is immediately experienced in sensation, parts of the present are earlier than other parts. (This follows logically from the definitions.)

d. It may happen that A and B form part of one sensation, and likewise B and C, but when C is an object of sensation A is an object of memory. Thus the relation "belonging to the same present" is not transitive, and two presents may overlap without coinciding.

The above definitions and propositions must now be explained and amplified.

1. *Sensation,* from the point of view of psychophysics, will be concerned only with objects not involving introspection. But from the point of view of theory of knowledge, all acquaintance with the present may advantageously be combined under one head, and therefore, if there is introspective knowledge of the present, we will include this with sensation. It is sometimes said that all introspective knowledge is of the nature of memory; we will not now consider this opinion, but will merely say that *if* introspection ever gives acquaintance with present mental entities in the way in which the sense give acquaintance with present physical entities, then such acquaintance with mental entities is, *for our purposes,* to be included under the head of sensation. Sensation, then, is that kind of acquaintance with particulars which enables us to know that they are at the present time. The object of a sensation we will call a *sense-datum.* Thus to a given subject sense-data are those of its objects which can be known, from

the nature of their relation to the subject, to be at the present time.

The question naturally arises: how do we know whether an object is present or past or without position in time? Mere acquaintance, as we decided in considering imagination, does not necessarily involve any given temporal relation to the subject. How, then, is the temporal relation given? Since there can be no *intrinsic* difference between present and past objects, and yet we can distinguish by inspection between objects given as present and those given as past, it follows from the criterion set forth at the beginning of the preceding chapter that the relation of subject to object must be different, and recognizably different, according as the object is present or past. Thus sensation must be a special relation of subject to object, different from any relation which does not show that the object is at the present time. Having come to this result, it is natural to accept "sensation" as an ultimate, and define the present time in terms of it; for otherwise we should have to use some such phrase as "given as at the present time," which would demand further analysis, and would almost inevitably lead us back to the relation of sensation as what is meant by the phrase "given as at the present time." For this reason, we accept sensation as one of the ultimates by means of which time-relations are to be defined.

2. Our theory of time requires a definition, without presupposing time, of what is meant by "one (momentary) total experience." This question has been already considered in a previous article, where we decided that "being experienced together" is an ultimate relation among objects, which is itself sometimes immediately experienced as holding between two objects. We cannot analyze this into "being experienced by the same subject," because A and B may be experienced together, and likewise B and C, while A and C are not experienced together: this will happen if A and B form part of one "specious present," and likewise B and C, but A is already past when C is experienced. Thus "being experienced together" is best taken as a simple relation. Although its relation is sometimes perceived, it may of course also hold when it not perceived. Thus "one (momentary) total experience" will be the experience of all that group of objects which are experienced together with a given object. This, however, still contains a difficulty, when viewed as a definition, namely that it assumes that no object is experienced twice, or throughout a longer

time than one specious present. This difficulty must be solved before we can proceed.

Two opposite dangers confront any theory on this point. (*a*) If we say that no one object can be experienced twice, or rather, to avoid what would be *obviously* false, that no one object can be twice an object of sense, we have to ask what is meant by "twice." If a time intervenes between the two occasions, we can say that the object is not numerically the same on the two occasions; or, if that is thought false, we can say at least that the experience is not numerically the same on the two occasions. We can then define "one (momentary) total experience" as everything experienced together with "this," where "this" is an experience, not merely the object of an experience. By this means, we shall avoid the difficulty in the case when "twice" means "at two times separated by an interval when the experience in question is absent." But when what seems to be the same experience persists through a longer continuous period than one specious present, the overlapping of successive specious presents introduces a new difficulty. Suppose, to fix our ideas, that I look steadily at a motionless object while I hear a succession of sounds. The sounds A and B, though successive, may be experienced together, and therefore my seeing of the object while I hear these sounds need not be supposed to constitute two different experiences. But the same applies to what I see while I hear the sounds B and C. Thus the experience of seeing the given object will be the same at the time of the sound A and at the time of the sound C, although these two times may not well be parts of one specious present. Thus our definition will show that the hearing of A and the hearing of C form parts of one experience, which is plainly contrary to what we mean by one experience. Suppose, to escape this conclusion, we say that my seeing the object is a different experience while I am hearing A from what it is while I am hearing B. Then we shall be forced to deny that the hearing of A and the hearing of B form parts of one experience. In that case, the perception of change will become inexplicable, and we shall be driven to greater and greater subdivision, owing to the fact that changes are constantly occurring. We shall thus be forced to conclude that one experience cannot last for more than one mathematical instant, which is absurd.

b. Having been thus forced to reject the view that the existence of one experience must be confined within one specious present, we have now to consider how we can define "one (momentary) total experience" on the hypothesis that a numerically identical experience may persist throughout a longer period than one specious present. It is obvious that no one experience will now suffice for definition. All that falls within one (momentary) total experience must belong to one specious present, but what is experienced together with a given experience need not, on our present hypothesis, fall within one specious present. We can, however, avoid all difficulties by defining "one (momentary) total experience" as a group of objects such that *any two* are experienced together, and nothing outside the group is experienced together with all of them. Thus, for example, if A and B, though not simultaneous, are experienced together, and if B and C likewise are experienced together, C will not belong to one experience with A and B unless A and C also are experienced together. And given any larger group of objects, any two of which are experienced together, there is some one (momentary) total experience to which they all belong; but a new object x cannot be pronounced a member of this total experience until it has been found to be experienced together with all the members of the group. A given object will, in general, belong to many different (momentary) total experiences. Suppose, for example, the sounds A, B, C, D, E occur in succession, and three of them can be experienced together. Then C will belong to a total experience containing A, B, C, to one containing B, C, D, and to one containing C, D, E. In this way, in spite of the fact that the specious present lasts for a certain length of time, experience permits us to assign the temporal position of an object much more accurately than merely within one specious present. In the above instance, C is at the end of the specious present of A, B, C, in the middle of that of B, C, D, and at the beginning of that of C, D, E. And by introducing less discrete changes the temporal position of C can be assigned even more accurately.

We may thus make the following definitions:

"One (momentary) total experience," is a group of experiences such that the objects of any two of them are experienced together, and anything experienced together with all members of the group is a member of the group.

The "specious present" of a momentary total experience is the period of time within which an object must lie in order to be a sense-datum in that experience.

This second definition needs some amplification. If an object has ceased to exist just before a given instant,[2] it may still be an object of sense at that instant. We may suppose that, of all the present objects of sense which have already ceased to exist, there is one which ceased to exist longest ago; at any rate a certain stretch of time is defined from the present instant back through the various moments when present objects of sense ceased to exist. This stretch is the "specious present." It will be observed that this is a complicated notion, involving mathematical time as well as psychological presence. The purely psychological notion which underlies it is the notion of one (momentary) total experience.

Sense-data belonging to one (momentary) total experience are said to be *present* in that experience. This is a merely verbal definition.

The above definitions still involve a certain difficultty, though perhaps not an insuperable one. We have admitted provisionally that a given particular may exist at different times. If it should happen that the whole group of particulars consituting one (momentary) total experience should recur, all our definitions of "the present time" and allied notions would become ambiguous. It is no answer to say that such recurrence is improbable: "the present time" is plainly not ambiguous, and would not be so if such recurrence took place. In order to avoid the difficulty, one of two things is necessary. Either we must show that such complete recurrence is *impossible,* not merely improbable; or we must admit absolute time, i. e., admit that there is an entity called a "moment" (or a "period of time" possibly) which is not a mere relation between events, and is involved in assigning the temporal position of an object. The problem thus raised is serious; but it belongs rather to the physical than to the psychological analysis of time. Within our experience, complete recurrence does not occur. So long, therefore, as we are considering merely the psychological genesis of our knowledge of time, objections derived from the possibility of recurrence may be temporarily put aside. We shall return to this question at a later stage of this article.

2. The word "instant" has a meaning defined later in the present article.

3. *Simultaneity.* This is a relation belonging to "physical" time, i. e., it is a relation between objects primarily, rather than between object and subject. By inference, we may conclude that sense-data are simultaneous with their subjects, i. e., that when an object is present to a subject, it is simultaneous with it. But the relation of simultaneity which is here intended is one which is primarily given in experience only as holding among objects. It does not mean simply "both present together." There are two reasons against such a definition. First, we wish to be able to speak of two entities as simultaneous when they are not both parts of one experience, i. e., when one or both are only known by description; thus we must have a meaning of simultaneity which does not introduce a subject. Secondly, in all cases where there is a change within what is present in one experience, there will be succession, and therefore absence of simultaneity, between two objects which are both present. When two objects form part of one present, they *may* be simultaneous, and their simultaneity *may* be immediately experienced. It is however by no means necessary that they should be simultaneous in this case, nor that, if they are in fact simultaneous, they should form part of what is present in one experience. The only point of connection, so far as knowledge is concerned, between simultaneity and presence, is that simultaneity can only be *experienced* between objects which are both present in one experience.

4. *The definition of "now."* We saw that both "I" and "now" are to be defined in terms of "this," where "this" is the object of attention. In order to define "now," it is necessary that "this" should be a sense-datum. Then "now" means "simultaneous with this." Since the sense-datum may lie anywhere within the specious present, "now" is to that extent ambiguous; to avoid this ambiguity, we may define "now" as meaning "simultaneous with some part of the specious present." This definition avoids ambiguity, but loses the essential simplicity which makes "now" important. When nothing is said to be contrary, we shall adopt the first definition; thus "now" will mean "simultaneous with this," where "this" is a sense-datum.

5. *The present time* is the time of entities which are present, i. e., of all entities simultaneous with some part of the specious present, i. e., of all entites which are "now" in our second, unambiguous sense. If we adopt a relational theory of time, we may define a time simply as

the class of all entities which are commonly said to be at that time, i. e., of all entities simultaneous with a given entity, or with a given set of entities if we do not wish to define a mathematical instant. Thus with a relational theory of time, "the present time" will be simply all entities simultaneous with some part of the specious present. With an absolute theory of time, "the present time" will be the time occupied by the specious present. We shall not at present attempt to decide between the absolute and relative theories of time.

This completes our theory of the knowledge of the present. Although knowledge of *succession* is possible without passing outside the present, because the present is a finite interval of time within which changes can occur, yet knowledge of the *past* is not thus obtainable. For this purpose, we have to consider a new relation to objects, namely *memory*. The analysis of memory is a difficult problem, to which we must now turn our attention.

6. *Immediate memory*. Without, as yet, asserting that there is such a thing as immediate memory, we may define it as "a two-term relation of subject and object, involving acquaintance, and such as to give rise ot the knowledge that the object is in the past." This is not intended as a satisfactory definition, but merely as a means of pointing out what is to be discussed. It is indubitable that we have knowledge of the past, and it would seem, though this is not logically demonstrable, that such knowledge arises from acquaintance with past objects in a way enabling us to know that they are past. The existence, extent, and nature of such immediate knowledge of the past is now to be investigated.

There are two questions to be considered, here as in theory of knowledge generally. First, there is the question: What sort of data would be logically capable of giving rise to the knowledge we possess? And secondly, there is the question: How far does introspection or other observation decide which of the logically possible systems of data is actually realized? We will deal with the first question first.

We certainly know what we mean by saying "such-and-such an event occurred in the past." I do not mean that we know this *analytically,* because that will only be the case with those (if any) who have an adequate philosophy of time; I mean only that we know it in the sense that the phrase expresses a thought recognizably different from other thoughts. Thus we must understand complexes into which "past," or whatever is the essential constituent of "past," enters as a constituent. Again it is obvious that "past" expressses a relation to "present," i. e., a thing is "past" when it has a certain relation to the present, or to a constituent of the present. At first sight, we should naturally say that what is past cannot also be present; but this would be to assume that no particular can exist at two different times, or endure throughout a finite period of time. It would be a mistake to make such an assumption and therefore we shall not say that what is past cannot also be present. If there is a sense in which this is true, it will emerge later, but ought not to be part of what is originally taken as obvious.

The question now arises whether "past" can be defined by relation to some one constituent of the present, or whether it involves the whole present experience. This question is bound up with another question, namely, can "past" be defined as "earlier than the present"? We have seen that *succession* may occur within the present; and when A is succeeded by B, we say that A is earlier than B. Thus "earlier" can be understood without passing outside the present. We cannot say, however, that the past is whatever is earlier than this or that constituent of the present, because the present has no sharp boundaries, and no constituent of it can be picked out as certainly the earliest. Thus if we choose any one constituent of the present, there may be earlier entities which are present and not past. If, therefore, "past" is to be defined in terms of "earlier," it must be defined as "earlier than the whole of the present." This definition would not be open to any *logical* objection, but I think it cannot represent the epistemological analysis of our knowledge of the past, since it is quite obvious that, in order to know that a given entity is in the past, it is not necessary to review the whole present and find that it is all later than the given entity. This argument seems to show that the past must be definable without explicit reference to the whole present, and must therefore not be defined in terms of "earlier."

Another question, by no means easy to answer, is this: Does our knowledge of the past involve *acquaintance* with past objects, or can it be accounted for on the supposition that only knowledge by description is involved in our knowledge of the past? That is, must our knowledge of the past be derived from such propositions as "*This* is past," where *this* is an object of present acquaintance, or can it be wholly derived from propositions of the form: "An entity with such-

and-such characteristics existed in the past"? The latter view might be maintained, for example, by introducing images: it might be said that we have images which we know to be more or less like objects of past experience, but that the simplest knowledge we have concerning such objects is their resemblance to images. In this case, the simplest *cognition* upon which our knowledge of the past is built will be perception of the fact "this-resembles-something-in-the-past," where *this* is an image, and "something" is an "apparent variable." I do not believe that such a view is tenable. No doubt, in cases of remembering something not very recent, we have often only acquaintance with an image, combined with the *judgment* that something like the image occurred in the past. But such memory is liable to error, and therefore does not involve *perception* of a fact of which "past" is a constituent. Since, however, the word "past" has significance for us, there must be perception of facts in which it occurs, and in such cases memory must be not liable to error. I conclude that, though other complications are logically possible, there must, in some cases, be immediate acquaintance with past objects given in a way which enables us to know that they are past, though such acquaintance may be confined to the very recent past.

Coming now to what psychology has to say as to the empirical facts, we find three phenomena which it is important to distinguish. There is first what may be called "physiological" memory, which is simply the persistence of a sensation for a short time after the stimulus is removed. The time during which we see a flash of lightning is longer than the time during which the flash of lightning, as a physical object, exists. This fact is irrelevant to us, since it has nothing to do with anything discoverable by introspection alone. Throughout the period of "physiological memory," the sense-datum is actually *present;* it is only the inferred physical object which has ceased.

Secondly, there is our awareness of the *immediate* past, the short period during which the warmth of sensation gradually dies out of receding objects, as if we saw them under a fading light. The sound we heard a few seconds ago, but are not nearing now, may still be an object of acquaintance, but is given in a different way from that in which it was given when it was a sense-datum. James[3] seems to

3.   [See selection no. 20, pp. 375.—C. M. S.]

include what is thus still given in the "specious present," but however we may choose to define the "specious present," it is certain that the object thus given, but not given in sense, is given in the way which makes us call it *past;* and James[4] rightly states that it is this experience which is "the *original* of our experience of pastness, from whence we get the meaning of the term."

Thirdly, there is our knowledge concerning more remote portions of the past. Such knowledge is more difficult to analyze, and is no doubt derivative and complicated, as well as liable to error. It does not, therefore, belong to the elementary constituents of our acquaintance with the world, which are what concern us at present. Or, if it does contain some elementary constituent, it must be one which is not essential to our having a knowledge of time, though it may increase the extent of our knowledge concerning past events.

Thus of the three phenomena which we have been considering, only the second seems directly relevant to our present problem. We will give the name "immediate memory" to the relation which we have to an object which has recently been a sense-datum, but is now felt as past, though still given in acquaintance. It is essential that the object of immediate memory should be, at least in part, identical with the object previously given in sense, since otherwise immediate memory would not give acquaintance with what is past, and would not serve to account for our knowledge of the past. Hence, by our usual criterion, since immediate memory is intrinsically distinguishable from sensation, it follows that it is a different relation between subject and object. We shall take it as a primitive constituent of experience. We may define one entity as "past" with respect to another when it has to the other that relation which is experienced, in the consciousness of immediate memory, as existing between object and subject. This relation, of course, will come to be known to hold in a vast number of cases in which it is not experienced; the epistemological need of the immediate experience is to make us know what is meant by "past," and to give us data upon which our subsequent knowledge can be built. It will be observed that in order to know a past object we only need immediate memory, but in order to know what is meant by "past," an immediate remembering must be

4. [See selection no 20, p. 368–C. M. S.]

itself made an object of experience. Thus introspection is necessary in order to understand the meaning of "past," because the only cases in which this relation is immediately given are cases in which one term is the subject. Thus "past," like "present," is a notion derived from psychology, whereas "earlier" and "later" can be known by an experience of non-mental objects.

The extent of immediate memory, important as it is for other problems, need not now concern us; nor is it necessary to discuss what is meant by memory of objects with which we are no longer acquainted. The bare materials for the knowledge that there is a time-series can, I think, be provided without considering any form of memory beyond immediate memory.

7. *Succession* is a relation which is given between objects, and belongs to physical time, where it plays a part analogous to that played by memory in the construction of mental time. Succession may be immediately experienced between parts of one sense-datum, for example in the case of a swift movement; in this case, the two objects of which one is succeeded by the other are both parts of the present. It would seem that succession may also be immediately experienced between an object of immediate memory and a sense-datum, or between two objects of immediate memory. The extensions of our knowledge of succession by inference need not now concern us.

8. We say that A is *earlier* than B if A is succeeded by B; and in the same case we say B is *later* than A. These are purely verbal definitions. It should be observed that *earlier* and *later* are relations given as between objects, and not in any way implying past and present. There is no logical reason why the relations of earlier and later should not subsist in a world wholly devoid of consciousness.

9. An event is said to be *past* when it is earlier than the whole of the present, and is said to be *future* when it is later than the whole of the present. It is necessary to include the *whole* of the present, since an event may be earlier than *part* of the present and yet be itself present, in cases where there is succession within the present. It is also necessary to define the past by means of *earlier* rather than by means of memory, since there may be things in the past which are neither themselves remembered nor simultaneous with anything remembered. It should be noted that there is no experience of the future. I do not mean that no particulars which are future are or have been

experienced, because if a particular recurs or endures it may be experienced at the earlier time. What I mean is that there is no experience of anything *as* future, in the way in which sensation experiences a thing as present and memory experiences it as past. Thus the future is only known by inference, and is only known *descriptively*, as "what succeeds the present."

Having now ended our definitions, we must proceed to the propositions constructing and connecting the physical and mental time-series.

a. Simultaneity and succession both give rise to transitive relations, while simultaneity is symmetrical, and succession asymettrical, or at least gives rise to an asymmetrical relation defined in terms of it.

This proposition is required for the construction of the physical time-series. At first sight, it might seem to raise no difficulties, but as a matter of fact it raises great difficulties, if we admit the possibility of recurrence. These difficulties are so great that they seem to make either the denial of recurrence of particulars or the admission of absolute time almost unavoidable.

Let us begin with simultaneity. Suppose that I see a given object A continuously while I am hearing two successive sounds B and C. Then B is simultaneous with A and A with C, but B is not simultaneous with C. Thus it would seem to follow that simultaneity, in the sense in which we have been using the word, is not transitive. We might escape this conclusion by denying that any numerically identical particular ever exists at two different instants: thus instead of the one A, we shall have a series of A's, not differing as to predicates, one for each instant during which we had thought that A endures. Such a view would not be logically untenable, but it seems incredible, and almost any other tenable theory would seem preferable.

In the same way as we defined one (momentary) total experience, we may, if we wish to avoid absolute time, define an "instant" as a group of events *any two* of which are simultaneous with each other, and not all of which are simultaneous with anything outside the group. Then an event is "at" an instant when it is a member of the class which is that instant. When a number of events are all at the same instant, they are related in the way which we have in mind when we think that simultaneity is transitive. It must be observed that we do not thus obtain a transitive two-term relation unless the instant is *specified:* "A and B are at the instant *t*" is transitive, but "there is an instant at which A and B are" holds whenever A and B are simultaneous, and is thus not transitive. In spite of this, however, the above definition of an "instant" provides formally what is required, so far as simultaneity is concerned. It is only so far as succession is concerned that this definition will be found inadequate.

Succession, if the time-series is to be constituted, must give rise to an asymmetrical transitive relation. Now if recurrence or persistence is possible, succession itself will have neither of these properties. If A occurs before B, and again after B, we have a case where succession is not asymmetrical. If B occurs both before A and after C, while A occurs before C but never occurs after C, A will succeed B and B will succeed C, but A will not succeed C; thus succession will not be transitive. Let us consider how this is affected if we pass on to "instants" in the sense above defined. We may say that one instant is *posterior* to another, and the other *anterior* to the one, if every member of the one succeeds every member of the other. But now we are faced with the possibility of *repetition,* i. e., of an instant being posterior to itself. If everything in the universe at one instant were to occur again after a certain interval, so as again to constitute an instant, the anterior and posterior instants would be *identical* according to our present definition. This result cannot be avoided by altering the definition of *anterior* and *posterior.* It can only be avoided by finding some set of entities of which we know that they cannot recur. If we took Bergson's view, according to which our mental life at each moment is intrinsically different, owing to memory, from that of a moment preceded by different experiences, then the experience of each moment of life is unique, and can be used to define an instant. In this way, if the whole universe may be taken as one experience, the time-series can be constructed by means of memory. There is no *logical* error in such a procedure, but there is a greater accumulation of questionable metaphysics than is suitable for our purposes. We must, therefore, seek for some other way of constructing the time-series.

It is no answer to our difficulty to reply that the complete recurrence of the whole momentary state of the universe is *improbable.* The point of our difficulty is this: If the whole state of the universe did recur, it is obvious that there would be *something* not numerically identical in the two occurrences, something in fact, which leads us to speak of "two occurrences." It would be contrary to what is self-evident to say that there was strictly *one* occurrence, which was anterior and posterior to itself. Without taking account of the whole universe, if a thing A exists at one time, then ceases, and then exists again at a later time, it seems obvious that there is *some* numerical diversity involved, even if A is numerically the same. In this case, in fact, where A reappears after an absence, it woudl seem strained to say that the *same* particular had reappeared: we should more naturally say that a new precisely similar particular had appeared. This is by no means so obvious in the case of a thing which persists unchanged throughout a continuous period. Before going further, we must consider whether there can be any substantial difference between persistence and recurrence.

The view which I wish to advocate is the following. An entity may persist unchanged throughout a continuous portion of time, without any numerical diversity corresponding to the different instants during which it exists; but if an entity ceases to exist, any entity existing at a subsequent time must be numerically diverse from the one that has ceased. The object of this hypothesis is to preserve, if possible, a relational theory of time; therefore the first thing to be done is to re-state it in terms which do not even verbally imply absolute time. For this purpose, we may adopt the following definitions. We shall say that a thing *exists at several times* if it is simultaneous with things which are not simultaneous with each other. We shall say that it *exists throughout a continuous time* when, if it is simultaneous with two things which are not simultaneous with each other, it is also simultaneous with any thing which comes after the earlier and before the later of the two things. The assumption that two things which are separated by an interval of time cannot be numerically identical is presupposed in the above definition. This assumption, in relational language, may be stated as follows: *If A precedes B and is not simultaneous with it, while B precedes C and is not simultaneous with it, then A and C are numerically diverse.*[5] We have to inquire whether a logically tenable theory of the time-series can be constructed on this basis.

The difficulty of possible recurrence of the whole state of the

5. Another form of the same axiom is: *If A both precedes and succeeds B, then A is simultaneous with B.*

universe, which troubled us before, is now obviated. It is now possible to define an *instant* as a class of entities of which any two are simultaneous with each other and not all are simultaneous with any entity outside the class. It will follow that it is meaningless to suppose the universe to persist unchanged throughout a finite time. This is perhaps an objection; on the other hand, it may be said that, when we suppose that such persistence is possible, we are imagining ourselves as spectators watching the unusual immobility with continually increasing astonishment; and in this case, our own feelings, at least, are in a state of change. Let us, then, suppose that it is logically impossible, as our present theory requires, for the universe to persist unchanged throughout a finite time. Then if two times are different, something must have changed meanwhile; and if this something has changed back so far as its character goes, yet what has reappeared is, in virtue of our assumption, numerically different from what has disappeared. Thus it is impossible that the world should be composed of numerically the same particulars at two different times.

We may now define an *instant* as a class which is identical with all the terms that are simultaneous with every member of itself. We will say that one event "wholly precedes" another when it precedes it without being simultaneous with it; and we will say that one instant is "anterior" to another when there is at least one member of the one instant which wholly precedes at least one member of the other instant. We shall assume that simultaneity is symmetrical, and that every event is simultaneous with itself, so that nothing can wholly precede itself. We will also assume that "wholly preceding" is transitive. These two assumptions together imply our previous assumption, which was that "wholly preceding" is asymmetrical, i. e., that if A wholly precedes B, then B does not wholly *precede* A. Finally, we will assume that of any two events which are not simultaneous one must wholly precede the other. Then we can prove that "anterior" is a serial relation, so that the instants of time form a series. The only remaining thing that needs to be proved is that there are instants, and that every event belongs to some instant. For this purpose let us call one event an "early part" of another when everything simultaneous with the one is simultaneous with the other, and nothing wholly preceding the one is simultaneous with the other. Let us define the "beginning" of an event as the class of events simultaneous with all its early parts. Then it will be found that, if we assume that any event wholly after something simultaneous with a given event is wholly after some early part of the given event, then the beginning of an event is an instant of which the event in question is a member.[6]

It would seem, therefore, that the physical time-series can be constructed by means of the relations considered in the earlier part of this article. Our few remaining propositions, which are chiefly concerned wiht mental time, offer less difficulty.

b. What is remembered is past. It should be noted that the past was defined as "what is earlier than the whole of the present," so that it cannot be supposed that whatever is passed is remembered, nor does memory enter into the *definition* of the past.

c. When a change is immediately experienced in sensation, parts of the present are earlier than other parts. This follows, because, since the change, by hypothesis, lies within sensation, it follows that the earlier and the later state of things are both present according to the definition.

d. If A, B, and C succeed each other rapidly, A and B may be parts of one sensation, and likewise B and C, while A and C are not parts of one sensation, but A is remembered when C is present in sensation. In such a case, A and B belong to the same present, and likewise B and C, but not A and C; thus the relation "belonging to the same present" is not transitive. This has nothing to do with the question of persistence or recurrence which we considered under (*a*), but is an independent fact concerned with mental time, and due to the fact that the present is not an instant. It follows that, apart from any question of duration in objects, two presents may overlap without coinciding.

---

6. In symbols, the above theory, with certain logical simplifications, has been set forth by Dr. Norbert Wiener in his "Contributions to the Theory of Relative Position," *Proc. Camb. Phil. Soc.,* Vol. XVII. Part 5, (1914).

18.

Reichenbach:

THE PRIMACY OF PHYSICAL TIME

Hans Reichenbach (1891–1953), though coming to America from Berlin rather than Vienna, was a prominent spokesman for the philosophy of logical positivism. Closely associated with Rudolf Carnap, he did work in logic and the philosophy of science. Concerned with the logical structure of modern physics as well as with scientific views bearing on some traditional philosophic questions, Reichenbach regarded questions concerning the nature of time as of prime import. The following selection is taken from the first chapter, "The Emotive Significance of Time," in *The Direction of Time*,* posthumously published under the editorship of his widow. He urged here that the primary reference to our temporal concepts must be to the ordered sequences in physical nature from which they emanate and to which they ultimately refer. Once we look to the developments of modern physics for clarification, he believed, the many paradoxes in our thinking of time will be cleared and many philosophic problems associated with it such as that of free will might be open to resolution.

Undisturbed by our query, the flow of time goes on. Already our present is filled with other experiences which, at the earlier time, we could not completely anticipate. Though in part predictable, the present experience contains many unexpected and previously unknowable features. What was uncertain is now determined. Possibilities which we feared, or hoped for, have now become realities; others, which we never had thought of, have intervened. And even the

* Hans Reichenbach, *The Direction of Time*, ed. Maria Reichenbach (Berkeley, Los Angeles, and London: University of California Press, 1971). Copyright, 1956, by the Regents of the University of California. By permission.

familiar daily experiences, though highly predictable, reveal in their actual occurrences some specific characteristics that could not have been foreseen. What else awaits us in the future? Will there be a war, or some other political catastrophe? Shall we get the long-hoped-for salary raise? Will a letter arrive that tells us about the death of a friend whom we believed to be in good health? Or will a letter announce that some distant relative bequeathed a fortune to us? And what will the little things which we expect be like? Will the car start right away? Shall we get through the intersection before the traffic light turns red? What will Fred say when I tell him that Doris is going to marry John?

All these things are in the future. What is it, this future? Does it keep events in stock, so to speak, and distribute them according to a plan? Or do events grow from chance? Growing means becoming. What is Becoming? How can something unreal become real? And as soon as it is real, it slides into the past, only to become unreal again, leaving nothing but a shadow in our memory. The present is the only reality. While it slips away, we enter into a new present, thus always remaining in the eternal Now. What is time, if all we have of it is this Now, this one moment gliding with us through the current of events that flows from the unchangeable past to the unknowable future?

Questions of this kind reveal the highly emotional content associated with the experience of time. They tempt us to look for answers that satisfy emotions rather than clarify meanings. I do not wish to say that such questions are unreasonable. But the answers to them may look very different from what we expect; and we may even be unable to find the answers, unless we first revise the questions and make precise what, at this stage, is mere groping for meanings. Human thought processes do not follow the pattern of calculating machines, which have an answer to any question, provided the question is asked correctly. We cannot answer every correct question—but we can often answer questions which are not correctly asked, by first giving them a form in which they have meaning. Often the process of reformulating the question and giving the answer is the same process. Looking for answers, we discover new meanings and find out what it was that we were asking for.

This is the scientific approach. Do not expect answers before you have found clear meanings. But do not throw away unclear questions. Keep them on file until you have the means at the same time to clarify and to answer them. Often these means result from developments in other fields, which at first sight appear to have nothing to do with the question.

The history of philosophy offers many illustrations of this process of clarification of meanings. Thales of Miletus believed that water is the substance of which all things are made. Heraclitus argued that, instead, this mysterious substance was fire. But neither of them knew precisely what it means to say that a piece of matter is composed of several substances. Modern chemistry has made this meaning precise by its methods of chemical analysis and has shown that neither water nor fire is a chemical element. Another illustration is found in Plato's philosophy. Plato believed that geometrical relations are known through visions of ideas, a reminiscence of experiences which our souls had in a world beyond the heavens long before their terrestrial lives began. Modern mathematics has shown that the act of visualizing geometrical figures can be understood in a this-worldly way: it is a recollection of everyday experiences with objects of our environment. It is the meaning of the term "visualization" that was clarified in this answer to a question. And only with the modern answer did the question assume a distinct meaning.

The inquiry into the nature of time has a similar history. It greatly puzzled the ancients, remained unsolved for two thousand years, and found an answer in developments of modern physics which were not directly concerned with the problem of time, but with that of causality. Before turning to these developments, it may be appropriate to examine more closely the conception of time contained within older philosophical systems, since they reveal the emotional reactions and formulate the logical puzzles which every one of us encounters in the experience of time.

Our emotional response to the flow of time is largely determined by the irresistibility of its passing away. The flow of time is not under our control. We cannot stop it; we cannot turn it back; we have the feeling of being carried away by it, helplessly, like a piece of lumber in the current of a river. We can know the past, but we cannot change it. Our activity can be directed toward the future only. But the future is incompletely known, and unexpected events may turn up which make our plans break down. It is true, the future may also have favorable turns in store. Yet we know that they are limited in number and that

318 THE ANALYSIS OF TEMPORAL CONCEPTS

adjusting ourselves to what the future may bring cannot help us too much—there is only a limited stretch of time ahead of us, and the end of all this striving and responding to new situations is death. The coming of death is the inescapable result of the irreversible flow of time. If we could stop time, we could escape death—the fact that we cannot makes us ultimately impotent, makes us equals of the piece of lumber drifting in the river current. The fear of death is thus transformed into a fear of time, the flow of time appearing as the expression of superhuman forces from which there is no escape. The phrase "passing away", by means of which we evasively speak of death without using its name, reveals our emotional identification of time flow with death.

Dissatisfied emotion has frequently been projected into logic. In theories of the universe it often appears in the guise of logical queries and pseudo-logical constructions. A philosopher argues that he has discovered a puzzle of Being which logic cannot solve—he might as well say that he has discovered a fact that arouses his emotional resistance. The fear of death has greatly influenced the logical analysis which philosophers have given of the problem of time. The belief that they had discovered paradoxes in the flow of time is called a "projection" in modern psychological terminology. It functions as a defense mechanism; the paradoxes are intended to discredit physical laws that have aroused deeply rooted emotional antagonism.

Religious philosophers have maintained that the happenings in time do not constitute the sum total of reality. They insist that there is another reality, a higher reality, which is exempt from time flow. Only the inferior reality of human experience is bound to time. The assumed superior reality, strangely enough, has been called *eternal,* which is a term referring to time. But in the language of these philosophers the term no longer pertains to permanent duration, but rather to something existing beyond time, not subject to time flow. Its opposite in the teminology of the church is *secular,* a term originally referring to the time span of a human life (in an extended meaning, of a century), but having assumed the meaning of something subject to time flow and thus something earthly, displaying the inferior nature of physical reality. The desire to survive death and to live eternally, in the sense of an unlimited time, a desire obviously incompatible with physical facts, has thus led to a conception in which eternal life is not life in time, but in a different reality. In order to escape the "passing away" with time, a timeless reality was invented.

Among the ancients, Parmenides and Plato developed such concepts of reality, though in different forms. Parmenides tells us that the higher reality does not come into being and does not pass out of being. "It is uncreated and indestructible; for it is complete, immovable, and without end. Nor was it ever, nor will it be; for now it *is,* all at once, a continuous *one.*" And Plato explains that "time is the moving image of eternity". Here "eternity" does not mean "infinite time". It is supposed to denote a reality not controlled by time flow, which, however, is reflected, so to speak, in the river of time. The happenings in time are, at best, an inferior form of reality; for Parmenides, it seems, they are not real at all, but illusions.

Such philosophies are documents of emotional dissatisfactions. They make use of metaphors invented to appease the desire to escape the flow of time and to allay the fear of death. They cannot be brought into a logically consistent form. Yet, strangely enough, they are often presented as the results of logical analysis. The grounds offered for them are the alleged paradoxes of Becoming. Parmenides argues that if there were Becoming, a thing must grow from nothing into something, which he regards as logically impossible. And his successor in the Eleatic school, Zeno, has supplied us with a number of famous paradoxes which, he thought, demonstrate the impossibility of motion and the truth of Parmenides' conception of Being as timeless. . . .

Concerning the arrow paradox [of Zeno], we answer today that rest at one point and motion at one point can be distinguished. "Motion" is defined, more precisely speaking, as "travel from one point to another in a finite and nonvanishing stretch of time"; likewise, "rest" is defined as "absence of travel from one point to another in a finite and non-vanishing stretch of time". The term "rest at one point at one moment" is not defined by the preceding definitions. In order to define it, we define "velocity" by a limiting process of the kind used for a differential quotient; then "rest at one point" is defined as the value zero of the velocity. This logical procedure leads to the conclusion that the flying arrow, at each point, possesses a velocity greater than zero and therefore is not at rest. Furthermore, it is not permissible to ask how the arrow can get to the next point, because in a continuum there is no next point. Whereas for every integer there exists a next integer, it is different with a continuum of points: between any two points there is another point. Concerning the other paradox, we argue that Achilles can catch up with the tortoise because an infinite number of nonvanishing distances converging to zero can have a finite sum and can be traversed in a finite time.

These answers, in order to be given in all detail, require a theory of infinity and of limiting processes which was not elaborated until the nineteenth century.[1] In the history of of logic and mathematics, therefore, Zeno's paradoxes occupy an important place; they have drawn attention to the fact that the logical theory of the ordered totality of points on a line—the continuum—cannot be given unless the assumption of certain simple regularities displayed by the series of integers is abandoned. In the course of such investigations, mathematicians have discovered that the concept of infinity is capable of a logically consistent treatment, that the infinity of points on a line differs from that of the integers, and that Zeno's paradoxes are not restricted to temporal flow, since they can likewise be formulated and solved for a purely spatial continuum.

What makes Zeno's paradoxes psychologically interesting, however, is the fact that they were discovered, not as part of the pursuit of a mathematical theory of the continuum, but through a process of rationalization; that they were found because the Eleatic school wanted to prove the unreality of time. Had Zeno not constructed his paradoxes under the spell of this preoccupation wih a "metaphysical" aim, he would have come to a different solution. He would have argued that, since arrows do fly and a fast runner does overtake a tortoise, there must be something wrong with his conception of logic, but not with physical reality. But he did not want to come to this conclusion. He wanted to show that change and Becoming are illusory, and he wanted to show that Reality has a timeless existence exempt from the shortcomings of time-controlled human experience —from passing away and from death.

1. For a modern discussion of Zeno's paradoxes, see Bertrand Russell, *Our Knowledge of the External World* (Chicago, Open Court Publ. Co., 1914, chap. vi; and Adolf Grünbaum, "Relativity and the Atomicity of Becoming", *Rev. Metaphys.*, Vol. 4 (1950), p. 176.

The time theory of Parmenides has become the historical symbol of a negative emotional attitude toward the flow of time. But the actual structure of time is compatible with different emotive reactions; and there has always existed a positive attitude toward time flow, an alternative emotional response to change and Becoming, for which the future is an inexhaustible source of new experience and a challenge to our abilities to make the best of emerging opportunities. The historical symbol of this positive emotional attitude toward time flow was created in the philosophy of Parmenides' contemporary and opponent Heraclitus.

"All things are in flux" is the formula in which Heraclitus' philosophy has been summed up. Becoming is for him the very essence of life. "The sun is new every day"—this means, for him, that it is good that every day produces something new. We need not cling to what has been; we can get along very well in a world of continuous change. "You cannot step twice into the same river, for fresh waters are ever flowing in upon you." This seeming paradox is not as profound as Heraclitus believed, for we can very well call the river *the same* even though its waters change. But Heraclitus' aphorism draws our attention to the logical nature of the physical thing as a series of different states in time; it is not necessary for physical identity that these states be exactly alike. A human being is the same, identical person all the time, although the body grows and changes its chemical building blocks. A physics of things does not require a denial of time flow. Common sense, as well as science, agrees with this conception of Heraclitus.

Yet the prophet of time flow has not been able to tell us very much about the logical structure of time. Heraclitus' aphorisms supply an emotive commentary rather than a logical analysis of time. Logic has not profited from his insistence on change, whereas it did profit from Zeno's queries about change. Heraclitus' attempt to show that opposites are the same, obviously springing from his recognition that different states in time can constitute the same thing, is one of those over-simplified generalizations which are in part truistic, in part obviously false. "The way up and the way down is one and the same" is merely a formulation of the trivial fact that a relation and its converse can be used equally well to make equivalent statements: "taller" can express the same fact as "shorter," through a reversal of the order of the terms they connect. But to say that the statement "Peter is taller than Paul" means the same as the statement "Peter is shorter than Paul" would be a contradiction. In other examples of so-called identical opposites, Heraclitus merely cites instances in which extremes are at opposite ends of the same scale, like hot and cold. In others, again, he illustrates the trivial fact that a thing can have opposite relations to different other things, as in his instance of sea water, which is drinkable for fishes but undrinkable for men. An alleged logic of opposites cannot solve the problem of time and Becoming.

Heraclitus' approach to the problem of time order is naive; it is the attempt to understand time by mere reflection on meanings derived from everyday experiences. Unfortunately, this kind of approach has been regarded by many, even in our day, as the truly philosophical approach, particularly if the conclusions arrived at are formulated in an obscure and oracular language, like that of Heraclitus. But darkness of language has too often been the guise of a philosophy of trivialities mingled with falsehood and nonsense—whether it teaches the identity of opposites, the doctrine that contradiction is the root of motion and life, or the conception that the nothing is something. A clarification of the meaning of time and Becoming can be expected only if questions raised by common sense are answered with the help of scientific method. The precision of the scientific formulation, its testability by observation in combination with logic, and its far-reaching power of connecting facts from very different domains combine to form an instrument of research capable of shedding new light on problems emerging with everyday experience. The analysis of time had to be connected with the analysis of science in order to become accessible to logical clarification.

A brief consideration shows that the study of time is a problem of physics. Emotive reaction to time flow cannot determine the answer to the question: What *is* time? Subjective experience of time, though giving rise to emotional attitudes, cannot give us sufficient information about the time order that connects physical events. We know that subjective judgment about the speed of time flow is deceptive; that on some occasions, time seems to pass quickly, on others, it seems to drag, depending, for instance, on whether we are fascinated or bored. Psychologists have shown that what we call the present is not a time *point*, strictly speaking, but a short *interval* of time, the length of which characterizes the psychological threshold of time perception. An optical impression, for example, takes time to "build up"; this fact explains the perception of motion in a moving picture, which consists of static pictures shown in rapid succession. But the existence of a temporal threshold appears irrelevant to the study of time as an objective process, in the same sense as the existence of perceptual thresholds is irrelevant to the investigation of geometrical length or of sound intensity. In fact, if Zeno's criticism of the continuum is to be applicable to time, the physical process of time flow must be assumed to be independent of the psychological experience of time, which does not have the structure of a mathematical continuum but is "atomistic" in nature.[2] What matters is the structure of that time which controls physical events; what we wish to know is whether our emotional reaction to time is justified, whether there is a time flow, objectively speaking, which makes events slide into the past and prevents them from ever again returning to the present. In what sense does the future differ from the past? For the answer, we must turn to physics, if we wish to understand time itself, rather than mere psychological reactions to it.

It was said above that the past is distinguished from the future as the unchangeable from the unknowable. Is this distinction to mean that the future is still changeable? We would be inclined to answer in the affirmative, because of our simple daily experiences. Our control of the future, though certainly limited in extent, is often sufficient to satisfy our needs. Planned action, based on anticipation of what the future will bring us, has enabled us to turn many of its gifts to practical use. We sow and reap; we provide ourselves with shelter; we organize human society; we build machines that facilitate our daily work.

The scientist, however, might be inclined to question the belief that the future is changeable. Being unknowable, he might argue, does not imply being undetermined; perhaps the future is as determined as the past and the difference between past and future is merely a difference between knowing and not knowing. The apparent asymmetry of time would then be only a matter of knowledge and ignorance; time itself would be symmetrical, its objective nature would be the same in the direction of the past as in the direction of the future. Such conceptions

2. This has been correctly emphasized by Grünbaum, *op. cit.*, p. 162.

suggest themselves within the scientific approach, because science has accepted the universal validity of causality.

Causal laws govern the past as well as the future. We see them at work in past facts; but we also see them confirmed by future facts which we have correctly predicted and which have later become reality. The future is not entirely unknowable. Quite a few occurrences can be predicted. Among these are the motions of the stars, the seasons, the growth of plants, animals, human beings; and certainly death is predictable. What led philosophers to question the reality of time is the fact that some undesirable future facts—in particular, death—are predictable. Why not assume that the future is as determined as the past? . . .

It is no wonder that, with the progress of modern science, determinism became an increasingly influential doctrine. Newton's physics had unveiled the physical laws governing both celestial and terrestrial bodies; and the same laws were supposed to reign in the realm of the atom and control atomic motion. The mathematician Laplace did not hesitate to assume that the precision of astronomical laws also holds within the atomic domain. Since even human thinking and feeling merely reflect the constellations of atoms within the brain, he concluded that every future occurrence is as determined as the past. Only human ignorance prevents us from foretelling the future. In his famous remark about a logically possible superman he has formulated the complete symmetry between past and future:

> We must consider the present state of the universe as the effect of its former state and as the cause of the state which will follow it. An intelligence which for a given moment knew all the forces controlling nature, and in addition, the relative situations of all the entities of which nature is composed—if it were great enough to carry out the mathematical analysis of these data—would hold, in the same formula, the motions of the largest bodies of the universe and those of the lightest atom: nothing would be uncertain for this intelligence, and the future as well as the past would be present to its eyes.[3]

3.   Pierre S. Laplace, *Essai philosophique sur les probabilités* (first published in Paris, 1814), section "De la probabilité".

If this passage, which has become the classical formulation of determinism, were true, it would spell the breakdown of a realistic interpretation of time flow. If the future is as determined as the past, the present cannot create anything new; the causal structure which constitutes the physical world can then be held to extend either from negative infinity of time to positive infinity, or, if you like, from positive endlessness to negative endlessness. And the human mind then appears to be a mere spectator who happens to see this structure from a vantage point, the present, that slides along without being a scene of Becoming. For causal determinism, as for Parmenides, there is no Becoming.

The determinism of Newton's mechanics has received an even more explicit and precise formulation in Einstein's and Minkowski's four-dimensional space-time continuum. The three dimensions of space and the one dimension of time constitute its four axes, and physical happenings are represented as "world lines", like the lines of a diagram. The present is merely a cross section in this diagram, and it makes no difference where we put it. It is only a reference point from which we count time distances, like the year *one* from which we count our era. The structure of the space-time manifold is the same everywhere, and in both directions of time; the shape of all world lines in it is determined by mathematical laws. This timeless universe is a four-dimensional Parmenidean Being, in which nothing happens, "complete, immovable, without end . . . ; it *is* all at once, a continuous *one*". Time flow is an illusion, Becoming is an illusion; it is the way we human beings experience time, but there is nothing in nature which corresponds to this experience.

The deterministic conception of time flow may be compared to the happenings seen in a motion picture theater. While we watch a fascinating scene, its future development is already imprinted on the film; Becoming is an illusion, because it makes no difference to the happenings at what point we look at them. What we regard as Becoming is merely our acquisition of knowledge of the future, but it has no relevance to the happenings themselves. The following story, which was told to me as true, may illustrate this conception. In a moving picture version of *Romeo and Juliet*, the dramatic scene was shown in which Juliet, seemingly dead, is lying in the tomb, and Romeo, believing she is dead, raises a cup containing poison. At this moment an outcry from the audience was heard: "Don't do it!" We laugh at the person who, carried away by the emotion of subjective experience, forgets that the time flow of a movie is unreal, is merely the unwinding of a pattern imprinted on a strip of film. Are we more intelligent than this man when we believe that the time flow of our actual life is different? Is the present more than our cognizance of a predetermined pattern of events unfolding itself like an unwinding film?

Now it is true that the person in the audience is outside the happenings of the film, whereas we are part of the happenings in this world. But if our future is as determined as the "future" which a film strip holds in stock, what difference would it make if we knew it in advance? We could not change it: we would be mere spectators of our own actions, and as far as control of the future is concerned, we would not be better off than a person in the audience of a movie theater who wants to keep Romeo from committing suicide.

The paradox of determinism and planned action is a genuine one. I shall later analyze it in more precise terms; at the moment it may be sufficient to point out that if time has no direction, planned action appears incomprehensible. We plan to go to a theater tomorrow; but it seems utterly senseless to plan to go to a theater yesterday. Why do we make this distinction? We answer that we cannot change the past; we can only change the future. If we went to the theater yesterday, we did; and if we did not go, we did not; what we think of it today makes no difference. Why do we not argue in the same way about going to the theater tomorrow? In our behavior we express the conviction that time goes in one direction, because planning presupposes time flow. Can scientific analysis support this conviction?

The criticism of time flow raised by the ancients can be overcome by an improved logic of infinity. The objections resulting from a deterministic physics are much more difficult to refute. Before entering into a detailed analysis of these problems, let us study another historical attempt at a theory of time, which, in contrast to earlier approaches, was undertaken in full knowledge of the results of science and with the intention to find a way out of the dilemma of determinism or freedom. This attempt, in which Parmenides' conception of time as an illusion has been revived in modern form, is given by Kant's theory of time. . . .

Now for Kant causal order is as subjective as time. It is one of the categories by means of which we order our experience, but it does not express a property of things-in-themselves. Therefore physics can only tell us how the world appears to us, not how it is, independent of a human observer. And determinism refers only to the world of appearance—the world in itself is free from the rule of causal laws.

This is a strange doctrine. In order to make time flow subjective, Kant has to make even causality subjective. In order to save freedom and morality, he has to sacrifice physics as a science of objective things; it is for him merely a science of experienced things. He goes beyond Parmenides in exempting Being not only from time flow, but also from causal determination.

It is difficult to understand in what sense Kant could claim that his theory reestablished freedom. When we act, we want to change those occurrences which take place in time; we want to change the future. But temporal happenings, Kant tells us, are subject to causality and determinism. What kind of influence, then, can we have upon them? It does not help us to assume that behind the experienced things there are other things not controlled by physical laws. We want to change those things that *are* subject to causality and *are* perceived in our experience; we even use causal laws to control them, knowing very well that if things did not conform to such laws we would be hopelessly lost in our attempts to plan the future. What is a philosophy good for, if it evades answers to questions about what men can do, by telling us that there is another realm of Being which we certainly cannot control? Kant's philosophy of subjective time and subjective causality is a form of escapism. It does not solve the paradox of freedom and determinism; it does not clarify the experience of time flow; it cannot account for the distinction between past and future, between the unchangeable and the realm of what we hope can be changed.

The distinction between the past and the future, the objective interpretation of time as a process of Becoming and not merely a form of human experience, has strong support in common sense. It would not be easy to acquiesce in a philosophy which regards such conceptions as illusions. Yet the convictions of common sense cannot be accepted without criticism by the philosopher. The problem of time cannot be solved by an appeal to intuitive knowledge, which tells us that there must be a process of Becoming, making planned action possible. Reliance on so-called intuition has too often turned out to be misleading. There is such a thing as an escape into common sense as well as an escape into metaphysical speculation. Neither of them offers answers acceptable to those who look for an unprejudiced approach, guided by logical analysis in combination with observation.

This remark applies to such attempts as were made by Bergson, who claims to have constructed a philosophy for which Becoming is the essence of time. Like his predecessor in antiquity, Heraclitus, Bergson is one of those who react to time flow with a positive emotional attitude, who see in change the very element of life, and whose optimistic approach to the problem of time flow is not troubled by the threat of death. To Bergson, as to Heraclitus, the idea of Becoming offers emotional reward.

Such an attitude may be a good help in starting a logical investigation, but it cannot replace it. Bergson's spirited appeal to intitution, to the "immediate data of consciousness", cannot establish a theory of time. He argues that the physicist has misunderstood time—that he has "spatialized" time by treating it like a dimension of space. Real time, or "duration", he says, can only be understood by immediate awareness, which reveals to us that time is an act of Becoming. Such arguments do not say very much to someone who wants to know whether he can trust his intuitions. If a man is under posthypnotic suggestion, his intuition tells him that he is free; but we know he is not. If intuition tells us that the future is generated by "creative evolution", I should like to know, not only whether this is true, but also what this phrase means. But the answer cannot be found by another appeal to intuition. The meaning of Becoming can be clarified only by logical analysis; and this analysis is to be based on all we know about the physical world, including ourselves. An act of vision is no substitute.

There is no other way to solve the problem of time than the way through physics. More than any other science, physics has been concerned with the nature of time. If time is objective the physicist must have discovered that fact, if there is Becoming the physicist must know it; but if time is merely subjective and Being is timeless, the physicist must have been able to ignore time in his construction of reality and describe the world without the help of time. Parmenides' claim that time is an illusion, Kant's claim that time is subjective, and

Bergson's and Heraclitus' claim that flux is everything, are all insufficiently grounded theories. They do not take into account what physics has to say about time. It is a hopeless enterprise to search for the nature of time without studying physics. If there is a solution to the philosophical problem of time, it is written down in the equations of mathematical physics.

Perhaps it would be more accurate to say that the solution is to be read between the lines of the physicist's writings. Physical equations formulate specific laws, general as they may be; but philosophical analysis is concerned with statements *about* the equations rather than with the content of the equations themselves. It was mentioned that the question of determinism has a great bearing upon the problem of time, but there is no physical equation that formulates determinism. Whether determinism holds is a question *about* physical equations; it is the question whether certain equations supply strict predictions and cover all possible phenomena. For such reasons, the philosophical investigation of physics is not given in the language of physics itself, but in the metalanguage, which speaks *about* the language of physics.

It is well known that the determinism of classical physics has been abandoned in modern quantum physics, and it will be most important to investigate the implications which the turn toward indeterminism entails for the problem of time. But there are other investigations which must precede the study of quantum mechanics. In thermodynamics, physics has been explicitly concerned with the problem of time flow; it is a gross misunderstanding of physics to say that it has "spatialized" time. The specific nature of time, as different from space, has found an expression in certain very fundamental physical equations. It will be seen that the contribution of quantum mechanics to the time problem can be understood only after a study of the thermodynamical approach. And it will turn out that physics can account for time flow and for Becoming, that common sense is right, and that we can change the future. But the proof requires the use of scientific method. Even the meaning of the terms "time" and "becoming" can only be understood by a common sense which has assimilated the results of scientific thought.

19

Whitehead:

TWO KINDS OF TIME RELATEDNESS

One of the truly great and imaginatively synoptic thinkers of our time, Alfred North Whitehead (1861–1947) brought to philosophy a keen sensitivity to issues involved in our thinking about time. A senior lecturer in mathematics at Trinity College, Cambridge, he moved into philosophy, wrote *Principia Mathematica* with Bertrand Russell, and shortly thereafter moved to Cambridge, Massachusetts, where, as a new member of the Harvard faculty, he wrote *Process and Reality*. That work, originally delivered as Gifford lectures, is one of the classics of contemporary metaphysics. The selection is taken from one of his earlier books, from the chapter, "Time," in *The Concept of Nature;* * here he was concerned to work out a new philosophy of the physical sciences emphasizing the category of relatedness, and understanding perception, *not* in terms of isolated atomic sensations, but as an experience within nature which itself is regarded as a connected system of events.

In the first place there is posited for us a general fact: namely, something is going on; there is an occurrence for definition.

This general fact at once yields for our apprehension two factors, which I will name, the 'discerned' and the 'discernible.' The discerned is comprised of those elements of the general fact which are discriminated with their own individual peculiarities. It is the field directly perceived. But the entities of this field have relations to other entities which are not particularly discriminated in this individual way. These other entities are known merely as the relata in relation to the entities of the discerned field. Such an entity is merely a 'some-

---

* Published by Cambridge University Press (Cambridge, England and New York, 1920); reprinted by permission.

thing' which has such-and-such definite relations to some definite entity or entities in the discerned field. As being thus related, they are—owing to the particular character of these relations—known as elements of the general fact which is going on. But we are not aware of them except as entities fulfilling the functions of relata in these relations. . . .

The concept of 'period of time' marks the disclosure in sense-awareness of entities in nature known merely by their temporal relations to discerned entities. Still further, this separation of the ideas of space and time has merely been adopted for the sake of gaining simplicity of exposition by conformity to current language. What we discern is the specific character of a place through a period of time. This is what I mean by an 'event.' We discern some specific character of an event. But in discerning an event we are also aware of its significance as a relatum in the structure of events. This structure of events is the complex of events as related by the two relations of extension and cogredience. The most simple expression of the properties of this structure are to be found in our spatial and temporal relations. A discerned event is known as related in this structure to other events whose specific characters are otherwise not disclosed in that immediate awareness except so far as that they are relata within the structure.

The disclosure in sense-awareness of the structure of events classifies events into those which are discerned in respect to some further individual character and those which are not otherwise disclosed except as elements of the structure. These signified events must include events in the remote past as well as events in the future. We are aware of these as the far off periods of unbounded time. But there is another classification of events which is also inherent in sense-awareness. These are the events which share the immediacy of the immediately present discerned events. These are the events whose characters together with those of the discerned events comprise all nature present for discernment. They form the complete general fact which is all nature now present as disclosed in that sense-awareness. It is in this second classification of events that the differentiation of space from time takes its origin. The germ of space is to be found in the mutual relations of events within the immediate general fact which is all nature now discernible, namely within the one event which is the totality of present nature. The relations of other events to this totality of nature form the texture of time.

The unity of this general present fact is expressed by the concept of simultaneity. The general fact is the whole simultaneous occurrence of nature which is now for sense-awareness. This general fact is what I have called the discernible. But in future I will call it a 'duration,' meaning thereby a certain whole of nature which is limited only by the property of being a simultaneity. Further in obedience to the principle of comprising within nature the whole terminus of sense-awareness, simultaneity must not be conceived as an irrelevant mental concept imposed upon nature. Our sense-awareness posits for immediate discernment a certain whole, here called a 'duration'; thus a duration is a definite natural entity. A duration is discriminated as a complex of partial events, and the natural entities which are components of this complex are thereby said to be 'simultaneous with this duration.' Also in a derivative sense they are simultaneous with each other in respect to this duration. Thus simultaneity is a definite natural relation. The word 'duration' is perhaps unfortunate in so far as it suggests a mere abstract stretch of time. This is not what I mean. A duration is a concrete slab of nature limited by simultaneity which is an essential factor disclosed in sense-awareness.

Nature is a process. As in the case of everything directly exhibited in sense-awareness, there can be no explanation of this characteristic of nature. All that can be done is to use language which may speculatively demonstrate it, and also to express the relation of this factor in nature to other factors.

It is an exhibition of the process of nature that each duration happens and passes. The process of nature can also be termed the passage of nature. I definitely refrain at this stage from using the word 'time,' since the measurable time of science and of civilised life generally merely exhibits some aspects of the more fundamental fact of the passage of nature. I believe that in this doctrine I am in full accord with Bergson, though he uses 'time' for the fundamental fact which I call the 'passage of nature.' Also the passage of nature is exhibited equally in spatial transition as well as in temporal transition. It is in virtue of its passage that nature is always moving on. It is involved in the meaning of this property of 'moving on' that not only is any act of sense-awareness just that act and no other, but the terminus of each act is also unique and is the terminus of no other act. Sense-awareness seizes its only chance and presents for knowledge something which is for it alone.

There are two senses in which the terminus of sense-awareness is unique. It is unique for the sense-awareness of an individual mind and it is unique for the sense-awareness of all minds which are operating under natural conditions. There is an important distinction between the two cases. (i) For one mind not only is the discerned component of the general fact exhibited in any act of sense-awareness distinct from the discerned component of the general fact exhibited in any other act of sense-awareness of that mind, but the two corresponding durations which are respectively related by simultaneity to the two discerned components are necessarily distinct. This is an exhibition of the temporal passage of nature; namely, one duration has passed into the other. Thus not only is the passage of nature an essential character of nature in its *rôle* of the terminus of sense-awareness, but it is also essential for sense-awareness in itself. It is this truth which makes time appear to extend beyond nature. But what extends beyond nature to mind is not the serial and measurable time, which exhibits merely the character of passage in nature, but the quality of passage itself which is in no way measurable except so far as it obtains in nature. That is to say, 'passage' is not measurable except as it occurs in nature in connexion with extension. In passage we reach a connexion of nature with the ultimate metaphysical reality. The quality of passage in durations is a particular exhibition in nature of a quality which extends beyond nature. For example passage is a quality not only of nature, which is the thing known, but also of sense-awareness which is the procedure of knowing. Durations have all the reality that nature has, though what that may be we need not now determine. The measurableness of time is derivative from the properties of durations. So also is the serial character of time. We shall find that there are in nature competing serial time-systems derived from different families of duration. These are a peculiarity of the character of passage as it is found in nature. This character has the reality of nature, but we must not necessarily transfer natural time to extra-natural entities. (ii) For two minds, the discerned components of the general facts exhibited in their respective acts of sense-awareness must be different. For each mind, in its awareness of nature is aware of a certain complex of related natural entities in their relations to the living body as a focus. But the associated durations may be identical. Here we are touching on that character of the passage nature which issues in the spatial relations of simultaneous bodies. This possible identity of the durations in the case of the sense-awareness of distinct minds is what binds into one nature the private experiences of sentient beings. We are here considering the spatial side of the passage of nature. Passage in this aspect of it also seems to extend beyond nature to mind.

It is important to distinguish simultaneity from instantaneousness. I lay no stress on the mere current usage of the two terms. There are two concepts which I want to distinguish, and one I call simultaneity and the other instantaneousness. I hope that the words are judiciously chosen; but it really does not matter so long as I succeed in explaining my meaning. Simultaneity is the property of a group of natural elements which in some sense are components of a duration. A duration can be all nature present as the immediate fact posited by sense-awareness. A duration retains within itself the passage of nature. There are within it antecedents and consequents which are also durations which may be the complete specious presents of quicker consciousnesses. In other words a duration retains temporal thickness. Any concept of all nature as immediately known is always a concept of some duration though it may be enlarged in its temporal thickness beyond the possible specious present of any being known to us as existing within nature. Thus simultaneity is an ultimate factor in nature, immediate for sense-awareness.

Instantaneousness is a complex logical concept of a procedure in thought by which constructed logical entities are produced for the sake of the simple expression in thought of properties of nature. Instantaneousness is the concept of all nature at an instant, where an instant is conceived as deprived of all temporal extension. For example we conceive of the distribution of matter in space at an instant. This is a very useful concept in science especially in applied mathematics; but it is a very complex idea so far as concerns its connexions with the immediate facts of sense-awareness. There is no such thing as nature at an instant posited by sense-awareness. What sense-awareness delivers over for knowledge is nature through a period. Accordingly nature at an instant, since it is not itself a natural entity, must be defined in terms of genuine natural entities. Unless we do so, our science, which employs the concept of instantaneous nature, must abandon all claim to be founded upon observation.

I will use the term 'moment' to mean 'all nature at an instant.' A moment, in the sense in which the term is here used, has no temporal extension, and is in this respect to be contrasted with a duration which has such extension. What is directly yielded to our knowledge by sense-awareness is a duration. Accordingly we have now to explain how moments are derived from durations, and also to explain the purpose served by their introduction.

A moment is a limit to which we approach as we confine attention to durations of minimum extension. Natural relations among the ingredients of a duration gain in complexity as we consider durations of increasing temporal extension. Accordingly there is an approach to ideal simplicity as we approach an ideal diminution of extension.

The word 'limit' has a precise signification in the logic of number and even in the logic of non-numerical one-dimensional series. As used here it is so far a mere metaphor, and it is necessary to explain directly the concept which it is meant to indicate.

Durations can have the two-termed relational property of extending one over the other. Thus the duration which is all nature during a certain minute extends over the duration which is all nature during the 30th second of that minute. This relation of 'extending over' —'extension' as I shall call it—is a fundamental natural relation whose field comprises more than durations. It is a relation which two limited events can have to each other. Furthermore as holding between durations the relation appears to refer to the purely temporal extension. I shall however maintain that the same relation of extension lies at the base both of temporal and spatial extension. This discussion can be postponed; and for the present we are simply concerned with the relation of extension as it occurs in its temporal aspect for the limited field of durations.

The concept of extension exhibits in thought one side of the ultimate passage of nature. This relation holds because of the special character which passage assumes in nature; it is the relation which in the case of durations expresses the properties of 'passing over.' Thus the duration which was one definite minute passed over the duration which was its 30th second. The duration of the 30th second was part of the duration of the minute. I shall use the terms 'whole' and 'part' exclusively in this sense, that the 'part' is an event which is extended over by the other event which is the 'whole.' Thus in my nomenclature 'whole' and 'part' refer exclusively to this fundamental relation of extension; and accordingly in this technical usage only events can be either wholes or parts.

The continuity of nature arises from extension. Every event extends over other events, and every event is extended over by other events. Thus in the special case of durations which are now the only events directly under consideration, every duration is part of other durations; and every duration has other durations which are parts of it. Accordingly there are no maximum durations and no minimum durations. Thus there is no atomic structure of durations, and the perfect definition of a duration, so as to mark out its individuality and distinguish it from highly analogous durations over which it is passing, or which are passing over it, is an arbitrary postulate of thought. Sense-awareness posits durations as factors in nature but does not clearly enable thought to use it as distinguishing the separate individualities of the entities of an allied group of slightly differing durations. This is one instance of the indeterminateness of sense-awareness. Exactness is an ideal of thought, and is only realised in experience by the selection of a route of approximation.

The absence of maximum and minimum durations does not exhaust the properties of nature which make up its continuity. The passage of nature involves the existence of a family of durations. When two durations belong to the same family either one contains the other, or they overlap each other in a subordinate duration without either containing the other; or they are completely separate. The excluded case is that of durations overlapping in infinite events but not containing a third duration as a common part.

It is evident that the relation of extension is transitive; namely as applied to durations, if duration A is part of duration B, and duration B is part of duration C, then A is part of C. Thus the first two cases may be combined into one and we can say that two durations which belong to the same family *either* are such that there are durations which are parts of both *or* are completely separate.

Furthermore the converse of this proposition holds; namely, if two durations have other durations which are parts of both *or* if the two durations are completely separate, then they belong to the same family.

The further characteristics of the continuity of nature—so far as durations are concerned—which has not yet been formulated arises in connexion with a family of durations. It can be stated in this way: There are durations which contain as parts any two durations of the same family. For example a week contains as parts any two of its days. It is evident that a containing duration satisfies the conditions for belonging to the same family as the two contained durations.

We are now prepared to proceed to the definition of a moment of time. Consider a set of durations all taken from the same family. Let it have the following properties: (i) of any two members of the set one contains the other as a part, and (ii) there is no duration which is a common part of every member of the set.

Now the relation of whole and part is asymmetrical; and by this I mean that if A is part of B, then B is not part of A. Also we have already noted that the relation is transitive. Accordingly we can easily see that the durations of any set with the properties just enumerated must be arranged in a one-dimensional serial order in which as we descend the series we progressively reach durations of smaller and smaller temporal extension. The series may start with any arbitrarily assumed duration of any temporal extension, but in descending the series the temporal extension progressively contracts and the successive durations are packed one within the other like the nest of boxes of a Chinese toy. But the set differs from the toy in this particular: the toy has a smallest box which forms the end box of its series; but the set of durations can have no smallest duration nor can it converge towards a duration as its limit. For the parts either of the end duration or of the limit would be parts of all the durations of the set and thus the second condition for the set would be violated.

I will call such a set of durations an 'abstractive set' of durations. It is evident that an abstractive set as we pass along it converges to the ideal of all nature with no temporal extension, namely, to the ideal of all nature at an instant. But this ideal is in fact the ideal of a nonentity. What the abstractive set is in fact doing is to guide thought to the consideration of the progressive simplicity of natural relations as we progressively diminish the temporal extension of the duration considered. Now the whole point of the procedure is that the quantitative expressions of these natural properties do converge to limits though the abstractive set does not converge to any limiting duration. The laws relating these quantitative limits are the laws of nature 'at an instant,' although in truth there is no nature at an instant and there is only the abstractive set. Thus an abstractive set is effectively the entity meant when we consider an instant of time without temporal extension. It subserves all the necessary purposes of giving a definite meaning to the concept of the properties of nature at an instant. I fully agree that this concept is fundamental in the expression of physical science. The difficulty is to express our meaning in terms of the immediate deliverances of sense-awareness, and I offer the above explanation as a complete solution of the problem.

In this explanation a moment is the set of natural properties reached by a route of approximation. An abstractive series is a route of approximation. There are different routes of approximation to the same limiting set of the properties of nature. In other words there are different abstractive sets which are to be regarded as routes of approximation to the same moment. Accordingly there is a certain amount of technical detail necessary in explaining the relations of such abstractive sets with the same convergence and in guarding against possible exceptional cases. Such details are not suitable for exposition in these lectures, and I have dealt with them fully elsewhere.[1]

It is more convenient for technical purposes to look on a moment as being the class of all abstractive sets of durations with the same convergence. With this definition (provided that we can successfully explain what we mean by the 'same convergence' apart from a detailed knowledge of the set of natural properties arrived at by approximation) a moment is merely a class of sets of durations whose relations of extension in respect to each other have certain definite peculiarities. We may term these connexions of the component durations the 'extrinsic' properties of a moment; the 'intrinsic' properties of the moment are the properties of nature arrived at as a limit as we

1. Cf. *An Enquiry concerning the Principles of Natural Knowledge*, Cambridge University Press, 1919.

proceed along any one of its abstractive sets. These are the properties of nature 'at that moment,' or 'at that instant.'

The durations which enter into the composition of a moment all belong to one family. Thus there is one family of moments corresponding to one family of durations. Also if we take two moments of the same family, among the durations which enter into the composition of one moment the smaller durations are completely separated from the smaller durations which enter into the composition of the other moment. Thus the two moments in their intrinsic properties must exhibit the limits of completely different states of nature. In this sense the two moments are completely separated. I will call two moments of the same family 'parallel.'

Corresponding to each duration there are two moments of the associated family of moments which are the boundary moments of that duration. A 'boundary moment' of a duration can be defined in this way. There are durations of the same family as the given duration which overlap it but are not contained in it. Consider an abstractive set of such durations. Such a set defines a moment which is just as much without the duration as within it. Such a moment is a boundary moment of the duration. Also we call upon our sense-awareness of the passage of nature to inform us that there are two such boundary moments, namely the earlier one and the later one. We will call them the initial and the final boundaries.

There are also moments of the same family such that the shorter durations in their composition are entirely separated from the given duration. Such moments will be said to lie 'outside' the given duration. Again other moments of the family are such that the shorter durations in their composition are parts of the given duration. Such moments are said to lie 'within' the given duration or to 'inhere' in it. The whole family of parallel moments is accounted for in this way by reference to any given duration of the associated family of durations. Namely, there are moments of the family which lie without the given duration, there are the two moments which are the boundary moments of the given duration, and the moments which lie within the given duration. Furthermore any two moments of the same family are the boundary moments of some one duration of the associated family of durations.

It is now possible to define the serial relation of temporal order among the moments of a family. For let A and C be any two moments of the family, these moments are the boundary moments of one duration d of the associated family, and any moment B which lies within the duration d will be said to lie between the moments A and C. Thus the three-termed relation of 'lying-between' as relating three moments A, B, and C is completely defined. Also our knowledge of the passage of nature assures us that this relation distributes the moments of the family into a serial order. I abstain from enumerating the definite properties which secure this result, I have enumerated them in my recently published book [2] to which I have already referred. Furthermore the passage of nature enables us to know that one direction along the series corresponds to passage into the future and the other direction corresponds to retrogression towards the past.

Such an ordered series of moments is what we mean by time defined as a series. Each element of the series exhibits an instantaneous state of nature. Evidently this serial time is the result of an intellectual process of abstraction. What I have done is to give precise definitions of the procedure by which the abstraction is effected. This procedure is merely a particular case of the general method which in my book I name the 'method of extensive abstraction.' This serial time is evidently not the very passage of nature itself. It exhibits some of the natural properties which flow from it. The state of nature 'at a moment' has evidently lost this ultimate quality of passage. Also the temporal series of moments only retains it as an extrinsic relation of entities and not as the outcome of the essential being of the terms of the series.

Nothing has yet been said as to the measurement of time. Such measurement does not follow from the mere serial property of time; it requires a theory of congruence which will be considered in a later lecture.

In estimating the adequacy of this definition of the temporal series as a formulation of experience it is necessary to discriminate between the crude deliverance of sense-awareness and our intellectual theories. The lapse of time is a measurable serial quantity. The whole of scientific theory depends on this assumption and any theory of time

2. Cf. *Enquiry*

which fails to provide such a measurable series stands self-condemned as unable to account for the most salient fact in experience. Our difficulties only begin when we ask what it is that is measured. It is evidently something so fundamental in experience that we can hardly stand back from it and hold it apart so as to view it in its own proportions.

We have first to make up our minds whether time is to be found in nature or nature is to be found in time. The difficulty of the latter alternative—namely of making time prior to nature—is that time then becomes a metaphysical enigma. What sort of entities are its instants or its periods? The dissociation of time from events discloses to our immediate inspection that the attempt to set up time as an independent terminus for knowledge is like the effort to find substance in a shadow. There is time because there are happenings, and apart from happenings there is nothing.

It is necessary however to make a distinction. In some sense time extends beyond nature. It is not true that a timeless sense-awareness and a timeless thought combine to contemplate a timeful nature. Sense-awareness and thought are themselves processes as well as their termini in nature. In other words there is a passage of sense-awareness and a passage of thought. Thus the reign of the quality of passage extends beyond nature. But now the distinction arises between passage which is fundamental and the temporal series which is a logical abstraction representing some of the properties of nature. A temporal series, as we have defined it, represents merely certain properties of a family of durations—properties indeed which durations only possess because of their partaking of the character of passage, but on the other hand properties which only durations do possess. Accordingly time in the sense of a measurable temporal series is a character of nature only, and does not extend to the processes of thought and of sense-awareness except by a correlation of these processes with the temporal series implicated in their procedures.

So far the passage of nature has been considered in connexion with the passage of durations; and in this connexion it is peculiarly associated with temporal series. We must remember however that the character of passage is peculiarly associated with the extension of events, and that from this extension spatial transition arises just as much as temporal transition. The discussion of this point is reserved for a later lecture but it is necessary to remember it now that we are proceeding to discuss the application of the concept of passage beyond nature, otherwise we shall have too narrow an idea of the essence of passage.

It is necessary to dwell on the subject of sense-awareness in this connexion as an example of the way in which time concerns mind, although measurable time is a mere abstract from nature and nature is closed to mind.

Consider sense-awareness—not its terminus which is nature, but sense-awareness in itself as a procedure of mind. Sense-awareness is a relation of mind to nature. Accordingly we are now considering mind as a relatum in sense-awareness. For mind there is the immediate sense-awareness and there is memory. The distinction between memory and the present immediacy has a double bearing. On the one hand it discloses that mind is not impartially aware of all those natural durations to which it is related by awareness. Its awareness shares in the passage of nature. We can imagine a being whose awareness, conceived as his private possession, suffers no transition, although the terminus of his awareness is our own transient nature. There is no essential reason why memory should not be raised to the vividness of the present fact; and then from the side of mind. What is the differnce between the present and the past? Yet with this hypothesis we can also suppose that the vivid remembrance and the present fact are posited in awareness as in their temporal serial order. Accordingly we must admit that though we can imagine that mind in the operation of sense-awareness might be free from any character of passage, yet in point of fact our experience of sense-awareness exhibits our minds as partaking in this character.

On the other hand the mere fact of memory is an escape from transience. In memory the past is present. It is not present as over-leaping the temporal succession of nature, but it is present as an immediate fact for the mind. Accordingly memory is a disengagement of the mind from the mere passage of nature; for what has passed for nature has not passed for mind.

Furthermore the distinction between memory and the immediate present is not so clear as it is conventional to suppose. There is an intellectual theory of time as a moving knife-edge, exhibiting a present fact without temporal extension. This theory arises from the concept of an ideal exactitude of observation. Astronomical observations are successively refined to be exact to tenths, to hundredths, and to thousandths of seconds. But the final refinements are arrived at by a system of averaging, and even then present us with a stretch of time as a margin of error. Here error is merely a conventional term to express the fact that the character of experience does not accord with the ideal of thought. I have already explained how the concept of a moment conciliates the observed fact with this ideal; namely, there is a limiting simplicity in the quantitative expression of the properties of durations, which is arrived at by considering any one of the abstractive sets included in the moment. In other words the extrinsic character of the moment as an aggregate of durations has associated with it the intrinsic character of the moment which is the limiting expression of natural properties.

Thus the character of a moment and the ideal of exactness which it enshrines do not in any way weaken the position that the ultimate terminus of awareness is a duration with temporal thickness. This immediate duration is not clearly marked out for our apprehension. Its earlier boundary is blurred by a fading into memory, and its later boundary is blurred by an emergence from anticipation. There is no sharp distinction either between memory and the present immediacy or between the present immediacy and anticipation. The present is a wavering breadth of boundary between the two extremes. Thus our own sense-awareness with its extended present has some of the character of the sense-awareness of the imaginary being whose mind was free from passage and who contemplated all nature as an immediate fact. Our own present has its antecedents and its consequents, and for the imaginary being all nature has its antecedent and its consequent durations. Thus the only difference in this respect between us and the imaginary being is that for him all nature shares in the immediacy of our present duration.

The conclusion of this discussion is that so far as sense-awareness is concerned there is a passage of mind which is distinguishable from the passage of nature though closely allied with it. We may speculate, if we like, that this alliance of the passage of mind with the passage of nature arises from their both sharing in some ultimate character of passage which dominates all being. But this is a speculation in which we have no concern. The immediate deduction which is sufficient for us is that—so far as sense-awareness is concerned—mind is not in time or in space in the same sense in which the events of nature are in time, but that it is derivatively in time and in space by reason of the peculiar alliance of its passage with the passage of nature. Thus mind is in time and in space in a sense peculiar to itself. This has been a long discussion to arrive at a very simple and obvious conclusion. We all feel that in some sense our minds are here in this room and at this time. But it is not quite in the same sense as that in which the events of nature which are the existences of our brains have their spatial and temporal positions. The fundamental distinction to remember is that immediacy for sense-awareness is not the same as instantaneousness for nature. This last conclusion bears on the next discussion with which I terminate this lecture. This question can be formulated thus, Can alternative temporal series be found in nature? . . .

The uniqueness of the temporal series is presupposed in the materialist philosophy of nature. But that philosophy is merely a theory, like the Aristotelian scientific theories so firmly believed in the middle ages. If in this lecture I have in any way succeeded in getting behind the theory to the immediate facts, the answer is not nearly so certain. The question can be transformed into this alternative form, Is there only one family of durations? In this question the meaning of a 'family of durations' has been defined earlier in this lecture. The answer is now not at all obvious. On the materialistic theory the instantaneous present is the only field for the creative activity of nature. The past is gone and the future is not yet. Thus (on this theory) the immediacy of perception is of an instantaneous present, and this unique present is the outcome of the past and the promise of the future. But we deny this immediately given instantaneous present. There is no such thing to be found in nature. As an ultimate fact it is a nonentity. What is immediate for sense-awareness is a duration. Now a duration has within itself a past and a future; and the temporal breadths of the immediate durations of sense-awareness are very indeterminate and dependent on the individual percipient. Accordingly there is no unique factor in nature which for every percipient is preeminently and necessarily the present. The passage of nature leaves nothing between the past and the future. What we perceive as present is the vivid fringe of memory tinged with anticipation. This vividness lights up the discriminated field within a duration. But no assurance can thereby be given that the happenings of nature cannot be assorted into other durations of alternative families. We cannot even know that the series of immediate durations posited by the sense-awareness of one individual mind all necessarily belong to the same family of durations. There is not the slightest reason to believe that this is so. Indeed if my theory of nature be correct, it will not be the case.

The materialistic theory has all the completeness of the thought of the middle ages, which had a complete answer to everything, be it in heaven or in hell or in nature. There is a trimness about it, with its instantaneous present, its vanished past, its non-existent future, and its inert matter. This trimness is very medieval and ill accords with brute fact.

The theory which I am urging admits a greater ultimate mystery and a deeper ignorance. The past and the future meet and mingle in the ill-defined present. The passage of nature which is only another name for the creative force of existence has no narrow ledge of definite instantaneous present within which to operate. Its operative presence which is now urging nature forward must be sought for throughout the whole, in the remotest past as well as in the narrowest breadth of any present duration. Perhaps also in the unrealised future. Perhaps also in the future which might be as well as the actual future which will be. It is impossible to meditate on time and the mystery of the creative passage of nature without an overwhelming emotion at the limitations of human intelligence.

VI.

The Significance of Experiential Time

Whether or not time is 'ultimately real', whatever conceptual distinctions seem most helpful for understanding it, these concerns, inquiries and questionings arise and transpire within a temporal context and are thereby colored by it. Whatever import time may prove to have, it must be reflected in that experience in which it is discovered or comes to be known.

If meaning and significance are regarded as being not so much inherent in things as in the human relationships with them; if truth is regarded not as some transcendent object of contemplation but as in the warp and woof of human activity; if, in short, human experience is seen in any sense as bearing meaning and truth, then experiential time becomes crucial. A philosophic outlook which places a premium on creativity, novelty, openness, deliberate decision, and purposive activity must look to some kind of temporal realism in human experience as the ontological ground of its values and their possible realization. An outlook which centers its attention on lived experience must take the time which provides its content and context with philosophic seriousness.

This has been the general orientation of that distinctively American philosophic outlook which came to maturity in the form of pragmatism. Emerging at about the time when Lotze was replacing Hegel as the prime philosophic figure, it gave expression to the American experience which manifested an intrinsically unique concern with the seriousness of the temporal. Conceived in the religious ethos of puritanism, it saw real moral significance in present activity as it tried to redeem the past while yet looking to the future for vindication. Insofar as membership in the American experience largely derived from a deliberate quest for new beginnings, its natural stance was to shift prime perspective from the past to the future. It saw itself as rooted in history while yet bringing to a respect for history a new attitude—not of mere reverence but of utilization. Emerson epitomized it well when he admonished us to "honor truth by its use." For practical or 'cash' value in the building of the future became the key working concept in the philosophic inquiries into the wisdom of the past and the concerns of the present.

Having been raised by theologians and *philosophes* who looked to time and history for meaning and significance, having been instructed in the Lockean teaching, which included the thesis that the activity of consciousness is marked by the temporal, its thinkers followed Emerson's lead in looking to the idealist movement for nurture. From Kant they learned that time is the pervasive form of all cognitive experience, from Fichte that it is also the form of all moral experience, and from Hegel that it is the form of nature as well. American thought came of age when the idealist concern with organic or contextualistic interrelationships as pervasively dynamic was most triumphant. If our experience of working in the world tells us something of the world within which we work, our experience of the world should also tell us something about ourselves. If this experiential relationship is dynamic and thereby pervasively temporal, then the time of our experience is potentially revelatory of the nature of the self and also of the world within which it is able to find itself.[1]

Meaning and truth, perhaps, may then be found in examining the nature, and thereby the significance, of our temporal experience, of the experiential time upon which experience is based. Working from such a stance, often unvoiced in quite this way, each of the prime figures in the development of the American pragmatic tradition took experiential time with full seriousness as providing a key to understanding the nature and significance of experience itself. Severally and together, they have exercised a profound influence on the further development of those traditions from which they borrowed. However they may have differed in interests or in emphasis, they shared a mutual respect and built upon each other; the divergences between them are not to be belittled, but neither should they be allowed to obscure the many bonds which held them together into a community of development.

Within a strictly chronological perspective, William James should have been paired with Henri Bergson. For James, Bergson's senior by seventeen years, published *The Principles of Psychology* in 1890, a few months after Bergson's first book appeared. (The virtual conjunction of publication was perhaps portentous of the new focus on the systematic study of the psyche, in place of the older science of physics, which was soon to come in the understanding of experiential time.) They came to know and esteem each other and, although they travelled in somewhat different directions, there was a mutuality of interest as well. Together they provided an essential preface to the development of contemporary phenomenology—from the proto-phenomenology, focus on internality and on time which were features of Bergson's thought generally, and from the tremendous influence which James's book was to have on Husserl as well as othjers.[2] Within a few years of publication, James's book had achieved an international reknown; its chapter on time became the standard text and point of departure for many writers; cited by people as diverse as Alexander and Russell, it also seems to have suggested elements in Whitehead's discussions.

James was in this early book already giving expression to the outlook which shaped the development of American pragmatism and it is in the context of its preoccupation with action that the implications of what he has had to say are most clearly seen. Often referred to as the one person who had made pragmatism a household word, his essay on time (in the *Principles)* set out themes which were developed by Peirce, Royce, Santayana, and Dewey.

In facing the question of time, each of them, in his own way, appears to have worked from what James has had to say: the insistence on an *extended* present together with disavowal of the experiential primacy of time as a series of now-points; the focus on the internality of temporal experience with the companion thesis that it is somehow our tie to the world of nature; the protest against substitution of conceptual abstraction for concrete lived experience; the thesis that the understanding of experience itself depends on an authentic understanding of its temporality, that experiential time is essential to the being of the experiencing person.

James saw himself as working from Locke's exploratory essay. Asserting the primacy of duration as the rudimentary notion of time, James pointed out that before we could apprehend a 'train of ideas' we would have to see them together in one perception. Were perception confined to separate moments and deprived of the present-ness of memory, we would not be able to have any sense of temporal continuity. Because we do, in fact, experience temporal continuity—of ourselves and of the objects we perceive—the "constitution of consciousness" must be able to be aware of a 'spread' of moments. The separable moment, James argued, is a non-experiential abstraction from a 'duration-block', from a present that has a 'thickness' to it.

The momentary 'now', as James was to point out later, refers to no true existent. "The literally present moment is a purely verbal supposition, not a position; the only present ever realized concretely being the 'passing moment' in which the dying rearward of time and its dawning future forever mix their lights. Say 'now' and it *was* even while you say it." [3] The present is not a momentary 'now'; it is, he suggested in a famous metaphor, a "saddle-back, with a certain breadth of its own on which we sit perched, and from which we look in two directions into time" (p. 371). Our perception must then be both retrospective and prospective; we must see into the stream of temporal flow in order to discern what is immediately before us. Such a 'duration-block', James insisted, is primary; it is "a synthetic datum", and it is from its internal flow that we are able to derive the notion of succession. With time, as with space, we first see whole units that are given in immediate experience, and it is only within them that we *then* discriminate particular constituent parts.

In contrast to our perception of space—we can see for miles—we can only take in a few seconds of time as separable experiential durations. James (in a long excised section) cited a number of scientific studies by his contemporaries to point out that although the apprehension of a duration may cover a few seconds, its extent—never in excess of twelve seconds—is somewhat dependent upon the emotional mood of the beholder.

Time is linked to change: without some sensible changing content in our awareness, we would have no sense of time. Our attention is focused on what is happening in the experienced durations, but whenever content seems devoid of change—such as a very long pause in a speech—we turn our attention to the time itself. But this is to say that what we experience in time is process; setting out a theme he was to develop later, he argued that we do not experience separate things as much as things in relation, as "wholes already formed." Drawing a sharp distinction between what is experienced and our primary experience of it, he pointed out, "The condition of *being* of the wholes may be the elements; but the condition of our *knowing* the elements is our having already felt the wholes as wholes." These experiences of the processes which are experienced, as the durations in which they are experienced, are immediately known and only later are they analyzed. In abstract terms, they are later set out in the measurable units of experientially fictional moments—in terms of minutes, hours, and dates.

Under the heading, "The Feeling of Past Time Is a Present Feeling," James has taken forth his opening thesis—that memory *is in* the present perception, that in the experiential present we do find perception of something as 'immediately passed' as well as what reproductive memory brings into the present from the more distinct past.

But the focus of our perception is on things 'external' to us. "Why," he asks, "do we perceive them at all?" In reply, he advanced the Lotzean thesis of the uniqueness of time. Unlike all other 'impressions', the *only* way in which our perception truly corresponds with the "outer reality, is that of the *time succession* of phenomena" (p. 380). Events, as our perceptions of them, take time. Time-awareness is our one direct experiential link to the world which we inhabit.

But there is a crucial distinction to be made between the succession of ideas in our minds and our awareness of this succession. A succession of feelings is not in itself a feeling of succession. The 'train of ideas' could not be the source of our idea of time were our experience of separate moments; we need to be able to see them within the "intuited duration" which James called the "specious present." This specious present "of which we are immediately and incessantly sensible"—a duration of a few seconds within which succession is experienced as integral to the whole—is "the original paragon and prototype of all conceived times" (p. 383).

What James has accomplished is a systematic and empirical examination of just what is involved in the experience of time-awareness. Taking forward the Lockean examination from which he worked, he emphatically focused the investigation on *how* we experience time. This experiential examination reveals the fact that the experience of sequence requires our being able to apprehend time-spreads that include sequences. The whole is prior to the parts and the whole of the sequence is prior to the 'before' and 'after' discriminations within it.

But introspection itself takes time; it is not clear that its perception of its own activity is broken down into the few-second bits, which are then in sequential relation themselves. If we start with wholes, it would seem that these duration-bits themselves are parts of larger wholes which are then analyzed into component elements. One might well then say that James's "saddle-back" is a very small one indeed. James had seen that the past and the future are, in some way, elements in the present apprehension, but he neither gave us any real insight into how they are integrated into the structure of the present nor any understanding of how we develop these duration-bits into our notions of real time-spreads. This is to say that James had seen the experiential present to be a spread and not a point—but seems to have substituted dashes for dots on a model that is still essentially linear; he has not really shown us how we get from one 'dash' to the next and how we do, indeed, form the 'line' of which they are components; or, whether the 'dashes' are but distinguishable elements of the experienced 'line'.

Although James has urged that our perception must be prospective as well as retrospective, the focus of his discussion in the *Principles* was on the involvement of the past in the present; but little was really said there about the functions of expectation and anticipation. This seems strange for a pragmatist who, emphasizing results and goal-oriented behavior, should thereby be expected to regard all experience as future-oriented. For this reason, and because of its inherent brevity, the notion of a "specious present" seems like a useful fiction but not really real, analytically derivative, not experientially immediate.

Perhaps the greatest question which James has raised is the old Aristotelian thesis that we 'perceive' time. Because an hour or a day is an abstraction, we can well understand why it has no direct referent —like the green of the leaf in front of me. But, what is the immediate referent of the 'duration-block', of the few seconds of the specious present? I can perceive things, things in process; but can I really perceive their relations as such? For James's thesis of the perception of time seems to be involved with this other doctrine which he will be at pains to develop in later work. On the face of it, it would seem to rest on some kind of ambiguity concerning the distinction between perception and sensation—just what are we to understand as the literal meaning of a statement that time, *qua* short duration, is a perception or a sensation? [4] We should also ask whether or not the notion of a 'specious present' is not itself an abstraction from the experienced continuity we each have.

James has given us a cogent argument that whatever may be involved in our time-awareness, it must be centered on more than an experientially fictional point. He has drawn on some illuminating comparisons—in their similarities and differences—between the facets of temporal and of spatial experience. He has urged the primacy of the temporal facet of our experience as that which binds us to the world we do in fact experience. But he has left us with questions about the structure of the living present—how the past and how the future enter into it, how we may integrate it into our understanding of the lived experience which does not seem to be broken down into the serial dashes of a few seconds. And, if our time is what binds us to the experienced world, we find ourselves asking about the mode of correlation of our temporal experience with the time of those continuities which appear in it as experiential content.

If William James provided the basic text on the nature of experiential time for his generation here and abroad, he is best remembered as the person who had popularized pragmatism while turning in his own development from psychology to philosophy itself. But, it was, of course, Charles Sanders Peirce who pioneered the pragmatic way of philosophic thinking and brought to bear in its development the heritage he took from his studies of Scotus, Leibniz, Kant, and the idealisms of his day.

Although Peirce's mathematical and logical interests were hardly taken up by his successors (and, in fact, have only been accorded real recognition in recent years), he set out essential themes which were to be critically developed in somewhat divergent directions by those who followed: an emphasis on pluralism, concrete individuality, and the significance of community; a focus on relations, with a tendency to emphasize those which are triadic in character; an impatience with deterministic doctrines and the avowal of a future still open, in some degree, to being shaped by human activity; a continuing concern with the genuine possibilities displayed in actual experience together with an insistence on temporal realism in human experience.

This is nowhere more apparent than in the 1905 article in which Peirce set out to differentiate his own pragmatism from James's interpretation of it. In the second half of this very influential paper, he succinctly set forth some of these themes and, most significantly for a philosophic understanding of time, shifted the focus of attention and meaning from the present which James had emphasized, and from the past which had been generally honored by most traditional discussion, to the future. Meaning, significance, even rationality, Peirce argued, are to be understood in terms of the future. Defending the thesis of experimentalism, as a prime mode of knowledge, he emphasized the principle of continuity, the importance of human action, and the pragmatic theme of control. He also brought out the notion of the 'boomerang' effect of the modification of human activity itself in future time by new collective knowledge.

In a very real sense, the significance of what Peirce has done, in the few pages of his paper, for the philosophical understanding of time should not be overlooked. He has, of course, carried forward James's attack on the notion of experiential time as a series of points abstracted from lived experience. But, in taking the whole notion of purpose as a fundamental index to human rational behavior he has made it necessary to rethink the nature of the temporal perspective.

In taking purpose as fundamental, he saw that it involves a new understanding of cognition, of truth, and of meaning. For purpose involves the deliberate use of intelligence as a guide to activity in altering the developing state of the future. It involves planning into the future on the basis of the collectively distilled information accumulated in the past. It turns cognition from a more or less passive contemplation of external happenings revealed in perception to intellectually grounded expectations; the significance of perception itself has been changed from the source of information in the present to the presentation of fulfillment of a present idea in the future. Truth, somehow, is no longer to be seen in particular apprehensions or description–judgments about them but in the degree of fulfillment of expectations, fears, and hopes. But this is to say that cognition is no longer to be seen as a purely mental exercise; cognition, in a very real sense, is (anticipating Heidegger) a special kind of activity, an activity which involves a deliberate and selective involvement with the world of physical entities and other persons.

To take activity as preeminent in human experience, immediately demands, further consideration shows, at least two radical changes in the way we usually consider experiential time. It involves a new focus on the importance of temporal continuity and a new conception of meaning.

Activity requires the essential continuity of time. It requires that the continuity of what are taken as temporal 'moments' be real, that temporal continuity can be *counted on* in the planning that deliberate activity involves. This means that the continuity is fundamentally prior to the 'moments' we select for the 'measuring' which our activity may require, the criterion for which is suitability for the purpose at hand. This temporal continuity must, further, be two-sided: it must be the temporal continuity of the experiencing person who is the actor of his planned action; it must also be the temporal continuity of those aspects of the environment with which he is involved in the course of that activity. The continuity of time, then, is the further justification of James's contention that it is time that binds us to our worlds of experienced entities. Whether metaphysically explained as a dual track, as occasionalism, as a preestablished harmony, as arising out of the essential one-ness of the world, the continuity of time, of the experiencer and the experienced, must functionally be one. The unity of time is requisite for the plurality of activities and of experiences which it 'permits'.

But, if purposive activity is fundamental to understanding it also involves a radical temporalization of the concept of meaning. For the meaning and the significance of present activity is not to be found in the narrow limits of an epistemologically defined 'present perception' or 'present cognition'; it is to be found in the expectations, desires, hopes, and fears for a future outcome which conspire to motivate the present act. It is these, strictly speaking, non-cognitive aspects of human thinking that provide the rationale for the present activity. The source of meaning and significance is, then, not to be found in the present reception of sense-data but in the capacity to transcend the momentary limits of the immediate present—whether defined in terms of 'moments' or 'duration-blocks'—and to take the envisioned future outcome that is hoped for or expected into the judgment of the meaning of the present.[5]

Taking the whole notion of rational purpose as fundamental, Peirce had seen that it provides the source of the meanings we see in our time-consuming activities. It also suggests that, as such activity is fundamental to human experience, we necessarily presuppose the temporal continuity within the experiential world and the capacity for an essential thrust into the future for understanding the meaning and thereby the nature of the action itself.

These very important suggestions for understanding experiential temporality were not fully developed by Peirce, presumably because his interests lay elsewhere. But what immediately became apparent from Peirce's work was the requisite centrality of temporal considerations in the understanding of human experience. This centrality of time was to become something of the pragmatic 'creed'; the suggestion for rethinking the whole nature of experiential time, implicit in Peirce's work, was developed not only in James's later work and by Royce, Santayana, and Dewey in diverse ways; it also was to be brought into the center of philosophic consideration of the nature of experiential time by another generation of thinkers.

The man who probably came closer than most to sharing Peirce's general *weltanschauung* was Josiah Royce.[6] Coming into philosophy by way of an early voluntarism which he never really surrendered, Royce clearly sought to mature and transform it into that 'absolute pragmatism' which he identified with his own statement of absolute idealism. Student of Lotze and of Peirce, protégé of James, teacher of Santayana, Mead, and Lewis, his impact on the development of thought is deeper than the current disfavor of his espoused idealism might at first indicate. The "foremost logician in the country," after Peirce,[7] his range of interest was wide and his insight into philosophic issues often pointed beyond the particular conclusions he was immediately concerned to establish. Indeed, Gabriel Marcel has remarked that Royce "marks a kind of transition between absolute idealism and existentialist thought;"[8] Royce's work certainly displays the roots of many themes of existential phenomenology as rooted in the center of the idealist tradition, but a careful reading also points to the ways in which the perspectives of pragmatism fed into the same stream.

An early and original author of the phenomenological notion of intentionality, one of Royce's prime working concepts was a pragmatic understanding of the notion of 'idea': an idea is *not* a mere representation; it is a plan for action. As such it involves an act of will. Like an act of will, the meaning of an idea is to be seen on two levels: internally, in terms of the purpose which it is a means of actualizing; externally, in terms of the outward condition at which it aims and claims to describe. The first meaning is appreciative, subjective, and teleological; the more fundamental of the two, it sets out the parameters within which the second, as description, is to be judged. A map, for example, accurately describes an area only in terms of the purpose for which it is framed; a geological map, which an oil explorer needs, is of little value to the person who needs a road map of the same area. The descriptive value of either map of the same area obviously depends upon the way in which it enables the purpose for which it is intended to be realized. Just so, Royce argued, an idea is a purpose in the process of fulfillment and its descriptive, or objective, value depends upon how effectively it fulfills its reason for being.

Royce used this two-fold intentional nature of the concept of 'idea' to further his own idealist thesis concerning the relationship between time and eternity. But in so doing, he brought much into the discussion of the nature and significance of experiential time.

Working from James's notion of the duration-bit of the specious present, Royce saw that one cannot get from it to our ordinary working notion of an extended past and future by sensation or immediate awareness alone. We must have a working *notion* of time that is far broader than the few seconds of an immediate perception. Its extension and integration into the spread of lived time requires conceptual interpretation of its content. The 'specious present' does present us with consciousness of change, succession that is irreversible, and the mingling of past and future in the present. But experience is not only of outer facts appearing in awareness; we always experience these facts as signs of meanings which extend far beyond them. Meanings are conceptual links connecting separable specious-presents together as elements in larger experiential wholes.

If James's understanding of time awareness may be likened to dashes, so Royce's may be pictured by a moving chain pulling in one direction: each of the interlocking links necessarily is involved with the one that is pulling it and the one it in turn pulls along, as it connects past and future *into* itself. For Royce has strongly protested taking the notion of the present in anything resembling a mathematical point or using any description that belies its essential dynamicity.

But, if such specious-presents are thus linked together conceptually, this says that any notion of perceptual time needs conceptual time for its understanding. As Kant had already pointed out, perceptions alone "without concepts are blind." [9] Without conceptual organization, we would know neither the 'what' nor the 'what-for' of any perceptual object before us. Without conceptual integration, we would be unable to connect separate perceptions into meaningful continuities. Our understanding of perceptual objects is not momentary; it is by the conceptual interpretation by which we take the objects we see to endure beyond the few seconds of any glance that we are able to understand and to 'see' things and events in broad temporal sweeps that fuse into an historical perspective.

But this is to say that Royce has shifted the focus from perceptual time given in the seconds of a specious-present to an experiential time in which percept and concept are fused in an organically unified experience of lived duration.[10] The specious-present, then, like a Lockean idea, would seem to be, not experiential, but an analytical abstraction from the concrete continuity of actual experience.

If the notion of an 'idea', as Royce has developed it, is essentially a plan of action, it requires no great jump by reason to see that the concept of time is essentially directional, thrusting continually toward the future. It is only because time takes the past into itself as it pulls toward the future, and only because its direction is irreversible, that we can undertake plans of activity, commit ourselves to processes, and pursue purposes in acts of deliberate will. Time, then, like the will, is a continuity of future-oriented activity; it is, as he has urged, "the form of practical activity" and "the form of the will."

We say that time is directly experienced in the present; but such a statement has a two-sided meaning. The present is the present of consciousness, but it is also the present of the world that is being experienced in that consciousness; the meaning of the present is what binds together these two sides of the present in significant activity and points out the continuity of the self with the world in which it operates. The 'present' is but the way in which the continuing reality of the world and of the self are presented to consciousness which is, itself, an interpretive awareness of meaning in patterns of change.

Time is, then, the form of the self. This is Royce's most startling thesis. That time is the form, or principle of order, in the changes of physical nature is a thesis at least as old as Plato. But Royce forcefully tied this principle to the nature of man. Previous discussion had primarily focused on the temporal import of cognitive activity. Royce has here introduced a note that should have been obvious—time is crucial to meaning, to the significance we see in events and in things; the implications drawn out of Peircean insights into the pragmatic nature of the idea have now been made fully explicit. As the "form of the will," time is crucial to directed change, to the meaning of activity, to the being of the person who undertakes the activity and understands himself in terms of it. Time is then the form of active being. It is 'ingredient' not only to the world of nature; it is 'ingredient' to the being of the person whose existence and consciousness are continually involving change. Our consciousness of our own beings, and of the world to which we belong, is marked by change and given meaning by the temporal order by which we and it take the past up into the living present and press on to a future which may not yet be but which somehow pulls us on.

That time is the form of the physical world had already been enunciated by Plato. That it is the prime form of human apprehension was the point made by Kant. What Royce has enunciated, however, had not really been enunciated that explicitly and forcefully before: previous discussion had been primarily concerned with nature—our knowledge of it or the way in which we apprehend its time. What Royce has seen is that the mode of apprehending says much about the apprehender: all human knowledge is found in particular acts of will and expressed in ideas that nevertheless take us beyond the immediate in plans of action that are future-oriented. Time is the necessary form of this activity, and it is, thereby, the form, not only of cognitive, but of all consciousness.

Time in its essential continuity is then fundamental. Early in his philosophic career, in 1880, Royce, who in many ways worked more from Kant than from Hegel, had urged that the question of the relationship "of every conscious moment to every other" is "more fundamental" than even Kant's concern with the relationship of human knowledge to its object. It is, he urged, the foundation from which any true epistemology must proceed; he called for a "new phenomenology" which works from one fundamental fact: "Every man lives in a present, and contemplates a past and a future. In this consists his whole life." [11] If philosophic thought is to throw light on the nature and significance of human experience, it must start from the temporal nature of human consciousness in which the self, the world, and their dynamic nexus of meaning are presented in the form of willful activity.

Royce may not have kept fully to his own announced goal because of his consuming drive to delineate the nature of the transcendent as well as of the transcendental, of the supervening absolute as well as of the necessary presuppositions to be found in human experience. But the key to his drive to the transcendent was his concept of time-filled experience and his notion of the absolute, in a way which does not really concern us here, became, despite himself, interestingly dynamic and patterned after the structure of a temporally delineated human experience. The significance of experience, Royce's completed metaphysic argued, was that it pointed beyond itself to the world in which it had come to be.

But with regard to experiential time, Royce's work helps us to understand it and the questions we must ask even more carefully than he asked them but in the direction to which he pointed. He has not really given us a new time-based epistemology nor a new phenomenology which would start from man's essential temporality. He had seen that time and consciousness are a stream of continuity, but he did not show us how past and future actually do integrate into the formation of the present.

His voluntarism, however, upon which he sought to build a speculative metaphysic, was more directly keyed to the essential thrust of pragmatism than James's psychological study which only hinted at it. A voluntarism necessarily presupposes the future-oriented temporal stream of experience and this may be one reason why Royce saw so deeply into the ramifications of the question of human time. He identified it with will, with selfhood—and in his developed metaphysic suggested another note that has been lacking in discussion and which still remains strangely unexplored in any depth: the import of the social in the understanding, and, indeed, formation of the temporality of experience.

Whatever deficiencies there are in Royce's developments of his own insights, he has taken us far. For, starting from Lotze's insistence on the uniqueness of the temporal, James's insight that the experiential present must be seen as a spread instead of a point, and Peirce's conception of an 'idea', Royce saw the ramifications which come from joining them together. For, joined together, as the pragmatic tradition has found itself bound to do, they insist on the fundamental temporality of the human experiencer in the shaping of his experience in temporal form.

Royce and Santayana, in the usually facile textbook clichés, have been painted as critics, not partisans, of the pragmatism predominant in American thought. Both were, indeed, critical of the apparent direction in which pragmatism seemed to be developing—a positivistic, object-oriented, and science-dominated pre-Kantian kind of sense-empiricism. Pragmatism had, indeed, been developed by Peirce and James, both of whom were recognized scientific thinkers of their time; perhaps it was natural that their initial philosophic direction, however they may have differed, should have been in terms of method and methodology.

But a methodology must refer to content. And Royce, thoroughly at home in the schools of modern idealism, keenly aware of pragmatism's roots in Kant, Fichte, and Hegel, saw it as the developed expression of certain elements in early idealist philosophy and criticized it for ignoring its roots and their implications. Santayana, looking toward the import of the broader aspects of cultural consciousness, criticized pragmatism's actual development for its seeming narrowness of vision and for retaining too much of its ancestral idealism which he was concerned to replace with a new naturalism. Despite his mounted attacks, we can see Santayana's "cultivated, dissenting voice," Aiken pointed out, "like that of Royce, as speaking in one sense for a pragmatism wiser, more liberal, and more comprehensive than that of Peirce or James or even Dewey." [12]

Looking back to the ancient pre-Socratic preoccupation with the dynamics of nature for inspiration, Santayana made a sharp distinction between the physical time of natural processes and its 'echo' in human subjective or "sentimental" time of our experience. Nature, the physical world, has its own time as the field of action in which the notion of the 'present' as a division between past and future is meaningless. But this notion is crucial for human thinking which frames its activity and its science in a perceptual specious-present which is then promptly ignored in practice.

Working from the Jamesian notion of the specious-present as the locus of experience, Santayana did not quantify its extent and seems to have used it more as a wide-open 'now', "a single temporal landscape" (p. 413). Pointing to the future in terms of action, it seems to have anchored itself in the past in terms of contemplation. As such, then, the present has a two-fold meaning: in one sense, it is a synonym for apprehended actuality; in the other, it focuses the area of transition of experience between the source of a process and the result of one. But, in this second sense, it incorporates the essence of change which includes direction of development: for change includes the "notion of the earlier absence of something now given, and of the earlier presence of something now absent. This is the very sense of existence and of time, the key to all intelligence and dominion over reality" (p. 409).

The lesson of the Kantian 'schematism' [13] has been learned well: time is "the key to all intelligence." The pragmatic lesson has been carried forward: if intelligence is to be identified with control of one's environment, then time is "the key ... to dominion over reality." Lotze's teaching has been absorbed: if personal or sentimental time is but a metaphoric version of the natural changes which we experience, it is nevertheless genuine. It is an imaginative synthesis which focuses attention on the 'now' primarily in terms of propensity to action. As a 'spread', our living present is no knife-edge; it brings the future and also the past into an extended present in which we see with confidence the transformations of what has been given. But this is to say that one's confidence in the developing future, without which activity would be pointless if not impossible, is tied to his conviction that he helps to determine that transformation himself.

Although one's own temporal experience is not a direct reading of the material or physical world, it is a genuine, if poetically imaginative, meshing with it. A man's experience is his *own* vantage point onto a world into which he is already closely tied. And it is to this world of real change that one's temporal experience claims to refer, although it arises from his own particular 'here' and 'now'. Although one's past is 'dead', we know that the past is somehow present in the process of imaginative interpretation just as one's imaginative vision of a not-yet future calls him to present activity and understanding.

Pointing out the dimensions of the imaginative extension of the experiential present, Santayana has, indeed, extended the Jamesian insight into the spread of the present and united this with the Peircean and Roycean emphases on the integral unity of understanding and action. Seeing this as essentially an emotive unification, he has seen no need to subject it to rigorous analysis. His insight, although left vague in detail, is nevertheless important. For in his own charming manner, Santayana has forcefully pointed out the fact which only poets, not philosophers, seem to hav recognized as fundamental fact—that temporal meaning is not merely an intellectual affair; time as the form of the self—working from Royce—is the form of the *whole* self and this means that it is a matter of sentiment, of emotion, of internalization, of involvement and commitment, and not merely an aridly intellec-tual, if explanatory, formula. What Santayana has brought out of Royce's temporalization of the human self, if it was not already clear, is this: If conceptualization is to be understood in terms of will, expectation, fear, and hope, as the conjunction of Royce and Peirce would indicate, this must be understood in terms of the integral unity of the person who is epitomized in his emotive stance, in his feelings, and in his sentiments. For it is from these, more than from any purely rationalistic notion of the will, that the individual's impulses to movement and to action come.

It is in a two-fold sense that Dewey has been regarded as the Aristotle of pragmatism. Like his illustrious predecessor, he brought diverse strands of thought together. We find in his work much of James's concern for the personal and the practical together with Peirce's experimentalism, respect for science, and forward-looking stance; although their approaches were markedly different, he shared with Santayana the attempt to mold a new naturalism. He had little interest in Royce's drive to transcendent metaphysics, but he shared with him an Hegelian heritage, a conviction that the individual and the social are inextricably connected in free community, and a passionate concern for the application of philosophic insight to the problems of life. But there is also much of Aristotle explicitly in Dewey: the notion of man's community with nature which is viewed in terms of change and motion, a respect for the principle of potentiality as a key to understanding the continuity of change, a comprehensive sweep of philosophic concern, and a determination to accept the factuality of what is taken as the given.

His long career touched virtually every aspect of philosophic concern and to each he brought a real sense of temporal perspective and historic nexus. Significantly, two of his first books were on Leibniz and Darwin; they set much of the mood that was to pervade the development of his thinking. The first suggests his abiding interests in the fundamental import of individuality, context, dynamicity and temporal quality; the second, his continuing concern for the significance of the scientific experience and his conviction that rationality and understanding of natural processes are to be found in development, evolution, and growth.

Dewey's lecture, "Time and Individuality," which was delivered at the midpoint of his career in the late 1930's, has generally been ignored. This is unfortunate. In a uniquely focused way it provides an incisive view into a central theme of his philosophic enterprise. Demonstrating something of a community of outlook with Royce and Santayana, it delineates the essential thrust of the pragmatic preoccupation with the meaning of time in human experience generally, and with the temporal context of the social individuality it so highly prized.

The lecture disavowed the old tradition that time is the source of frustration instead of hope, the destroyer rather than the creator, that temporal concerns are not central to knowledge, science, and value. Instead, it insisted on the primordiality of individuality which is grounded in "time as fundamental reality."

In *Experience and Nature,* Dewey had made a distinction between temporal quality and temporal order: "Temporal order is a matter of science; temporal quality is an immediate trait of every occurrence whether in or out of consciousness." [14] Dewey's essay carried this forward. If time is real, he has reasoned, then we must disavow a causal metaphysic which seeks to explain the present in terms of the past: causal explanation does not explain the individual suffering change, but merely the changes themselves, and always in terms of prior individual existents whose being is still left unexplained. Seeking the source of explanation, not in the dominion of the past over the present but in the changing entities themselves, Dewey urged the need for reviving and rethinking the Aristotelian thesis of potentiality—as inherent, forward-looking propensities for interaction with other existent entities in an interrelated and reciprocally operative order.

If individuality is conceived as a 'temporal career', then we can begin to understand why it essentially involves uncertainty, indeterminacy, contingency, continuity, novelty, and creativity. If change is a real aspect of the world we experience, then the real question is not whether there is to be change, but which specific changes are to be encouraged or avoided. The reality of individuality, of time, and of change join to focus concern on evaluation of real possibilities, so that change can be selected and directed; the application of temporal intelligence by beings who are constituted in terms of time means a deliberate molding of the future rather than mere acquiesence before the alleged dictation of the past.

Dewey has voiced here, as elsewhere, the continuing awareness of the social implications of his philosophic outlook: The affirmation of time as the ground of individuality necessitates the affirmation of freedom for individual development. It also suggests a limitation on the pertinence of science here insofar as modern science has concerned itself merely with the relations of individual existents to each other and not with the individuals who, as such, constitute the real order. In contrast, we find in art a new significance. Art is not timeless; it achieves a new significance as it epitomizes the necessity and value of an individuality that is, if one may use the words in a Deweyan context, ontologically grounded as temporal being.

Taking up Peirce's shift to the future as the source of cognitive meaning that is necessarily tied to action, Dewey has expanded this forward-looking experimentalism beyond the area of scientific concerns. In many ways, he has taken up Royce's identification of self, will, and time, and urged, in a naturalist setting, the meaning of individuality in a different kind of temporalist ontology. If nature is seen as an interacting system of temporally defined individuals, then the meaning of time is the key to the meaning of the individual self as it is to the meaning of the whole—for both microcosm and macrocosm are to be understood primarily in terms of the time that is the essence of their being.

Going beyond his predecessors, Dewey has provided, in spirit if not in letter, an insight into the fundamental importance accorded to the concept of time throughout their work. Individually and together, all of them were concerned to combat more contemplative traditions, to account for the human activity that is essentially involved in cognitive endeavor, and to exercise the philosophic responsibility for illuminating the ramifications of the problems of men; they each urged, in one way or another, that time is fundamental to meaning, intelligence, value, and truth, and to the import of that socially involved individuality they all deeply prized. They each took seriously Kant's teaching that time is the first form of experience and then set about to broaden the concept of experience itself. They presumed as truth what Fichte developed out of Kant: that practical reason is the encompassing form of the rational. They accepted the Hegelian thesis that time is the form of the existent reality of the world, insisted on the concreteness of its temporal quality—and saw James's thesis, to logically follow from it: that time is the form or medium of our cognitive and practical continuity of contact with the reality of the world. They absorbed Lotze's insistence that time is not authentically representable in spatial terms, and Bergson's that experiential time is more concerned with the duration of events than with their precise measurement. They seemed to take spatiality as given and do not seem to have shared Alexander's worry about its reality—although they generally seemed to approach it under a temporal aegis. When these lessons are put together, it is not difficult to see why Dewey, at the end, pointed out that if time is the form of experience, it is also the form of the experiencer, and—in a way reminiscent of Leibniz—that one's person *qua* person *is* one's continuing history.

But, however they absorbed and developed these lessons and integrated them in their own specific but related ways, however forcefully they enunciated the principle that time is the first form of meaning and significance of our experience and of ourselves, they left open at least one basic question which can be stated in several ways.

Given the thesis that experiential time is some kind of spread and not a point on a line, how do we explain just how this is possible or what this does mean? If the 'specious present' is, indeed, 'specious', just what meaning can we give to the notion of the present that avoids the spatial reduction to the abstraction of a non-experiential point on a line? Given the thesis that cognition is a form of temporal activity, just how does it arise out of, and yet remain part of, the temporal experience to which it belongs? If experience is to be had in the somehow-defined present which somehow brings the past and the future into it, just how are these modes of time integrated into a living present and how do they 'mingle' without losing their temporal identity without being homogenized into some kind of tenseless timeless moment? If the thrust is to be to the future instead of to the past as the source of explanation, meaning, and understanding, what does it mean when we consider that the future does not yet exist and has never existed? In brief, if experiential time, as Augustine pointed out long ago, is some kind of 'extension' of the mind in which past and future somehow join to form the experiential present, just how does this happen, just how is the present, the lived time of living experience, structured and brought to be?

It is to questions such as these that the phenomenologists have turned in considering the time of human experience. For, if the pragmatic thesis is correct that human experience is essentially and pervasively temporal, we must examine this temporality itself in order to understand how it forms our experience. We must, in short, examine our experiencing itself so that we can see experiential time at work and come to know what and how it is.

20

James:

"THE PERCEPTION OF TIME"

William James (1842–1910) was born in New York but spent
most of his life at Harvard where he had been trained in medicine
and where he began his academic career in his thirtieth year as a
lecturer in anatomy. Usually read as the foremost spokesman for
pragmatism, his most widely known book is *The Principles of
Psychology* in which are to be found many themes receiving
further development in his later and more explicitly philosophical works. The selection is taken from his famous essay on
time, chapter XV,* in this early work; it has been referred to and
cited in several selections and provided the common point of
departure for those which follow; it presents an outright attack
on the concept of experiential time as a serial order of separate
moments.

. . . I shall deal with what is sometimes called internal perception, or
the perception of *time,* and of events as occupying a date therein,
especially when the date is a past one, in which case the perception in
question goes by the name of *memory.* To remember a thing as past, it
is necessary that the notion of 'past' should be one of our 'ideas.' We
shall see in the chapter on Memory that many things come to be
thought by us as past, not because of any intrinsic quality of their own,
but rather because they are associated with other things which for us
signify pastness. But how do these things get *their* pastness? What is
the *original* of our experience of pastness, from whence we get the
meaning of the term? It is this question which the reader is invited to
consider in the present chapter. We shall see that we have a constant
feeling *sui generis* of pastness, to which every one of our experiences in

* *The Principles of Psychology,* chap. XV, vol. I (New York: Henry Holt and
Company, Inc., 1890).

turn falls a prey. To think a thing as past is to think it amongst the objects or in the direction of the objects which at the present moment appear affected by this quality. This is the original of our notion of past time, upon which memory and history build their systems. And in this chapter we shall consider this immediate sense of time alone.

If the constitution of consciousness were that of a string of bead-like sensations and images, all separate,

"we never could have any knowledge except that of the present instant. The moment each of our sensations ceased it would be gone for ever; and we should be as if we had never been. . . . We should be wholly incapable of acquiring experience. . . . Even if our ideas were associated in trains, but only as they are in imagination, we should still be without the capacity of acquiring knowledge. One idea, upon this supposition, would follow another. But that would be all. Each of our successive states of consciousness, the moment it ceased, would be gone forever. Each of those momentary states would be our whole being." [1]

We might, nevertheless, under these circumstances, *act* in a rational way, provided the mechanism which produced our trains of images produced them in a rational order. We should make appropriate speeches, though unaware of any word except the one just on our lips; we should decide upon the right policy without ever a glimpse of the total grounds of our choice. Our consciousness would be like a glow-worm spark, illuminating the point it immediately covered, but leaving all beyond in total darkness. Whether a very highly developed practical life be possible under such conditions as these is more than doubtful; it is, however, conceivable.

I make the fanciful hypothesis merely to set off our real nature by the contrast. Our feelings are not thus contracted, and our consciousness never shrinks to the dimensions of a glow-worm spark. *The knowledge of some other part of the stream, past or future, near or remote, is always mixed in with our knowledge of the present thing.*

A simple sensation, as we shall hereafter see, is an abstraction, and all our concrete states of mind are representations of objects with

1. James Mill, Analysis, vol. I. p. 319 (J. S. Mill's Edition).

some amount of complexity. Part of the complexity is the echo of the objects just past, and, in a less degree, perhaps, the foretaste of those just to arrive. Objects fade out of consciousness slowly. If the present thought is of A B C D E F G, the next one will be of B C D E F G H, and the one after that of C D E F G H I—the lingerings of the past dropping successively away, and the incomings of the future making up the loss. These lingerings of old objects, these incomings of new, are the germs of memory and expectation, the retrospective and the prospective sense of time. They give that continuity to consciousness without which it could not be called a stream.

The Sensible Present Has Duration.

Let any one try, I will not say to arrest, but to notice or attend to, the *present* moment of time. One of the most baffling experiences occurs. Where is it, this present? It has melted in our grasp, fled ere we could touch it, gone in the instant of becoming. As a poet, quoted by Mr. Hodgson, says,

"Le moment où je parle est déjà loin de moi,"

and it is only as entering into the living and moving organization of a much wider tract of time that the strict present is apprehended at all. It is, in fact, an altogether ideal abstraction, not only never realized in sense, but probably never even conceived of by those unaccustomed to philosophic meditation. Reflection leads us to the conclusion that it *must* exist, but that it *does* exist can never be a fact of our immediate experience. The only fact of our immediate experience is what Mr. E. R. Clay has well called 'the *specious* present.' His words deserve to be quoted in full:

The relation of experience to time has not been profoundly studied. Its objects are given as being of the present, but the part of time referred to by the datum is a very different thing from the conterminous of the past and future which philosophy denotes by the name Present. The present to which the datum refers is really a part of the past—a recent past—delusively given as being a time that intervenes between the past and the future. Let it be named the specious present, and let the past, that is given as being the past, be known as the obvious past. All the notes of a bar of a song seem to the listener to be contained in the present. All the changes of place of a meteor seem to the beholder to be contained in the present. At the instant of the termination of such series, no part of the time measured by them seems to be a past. Time, then, considered relatively to human apprehension, consists of four parts, viz., the obvious past, the specious present, the real present, and the future. Omitting the specious present, it consists of three . . . nonentities—the past, which does not exist, the future, which does not exist, and their conterminous, the present; the faculty from which it proceeds lies to us in the fiction of the specious present.

In short, the practically cognized present is no knife-edge, but a saddle-back, with a certain breadth of its own on which we sit perched, and from which we look in two directions into time. The unit of composition of our perception of time is a *duration,* with a bow and a stern, as it were—a rearward- and a forward-looking end.[2] It is only as parts of this *duration-block* that the relation of *succession* of one end to the other is perceived. We do not first feel one end and then feel the other after it, and from the perception of the succession infer an interval of time between, but we seem to feel the interval of time as a whole, with its two ends embedded in it. The experience is from the outset a synthetic datum, not a simple one; and to sensible perception its elements are inseparable, although attention looking back may easily decompose the experience, and distinguish its beginning from its end.

When we come to study the perception of Space, we shall find it quite analogous to time in this regard. Date in time corresponds to position in space; and although we now mentally construct large spaces by mentally imagining remoter and remoter positions, just as we now construct great durations by mentally prolonging a series of successive dates, yet the original experience of both space and time is always of something already given as a unit, inside of which attention

2. Locke in his dim way, derived the sense of duration from reflection on the succession of our ideas (*Essay,* book II. chap. XIV. § *3* [See pp. 122 above.] chap. XV. § 12.). . . .

afterward discriminates parts in relation to each other. Without the parts already given as *in* a time and *in* a space, subsequent discrimination of them could hardly do more than perceive them as *different* from each other; it would have no motive for calling the difference temporal order in this instance and spatial position in that.

And just as in certain experiences we may be conscious of an extensive space full of objects, without locating each of them distinctly therein; so, when many impressions follow in excessively rapid succession in time, although we may be distinctly aware that they occupy some duration, and are not simultaneous, we may be quite at a loss to tell which comes first and which last; or we may even invert their real order in our judgment. In complicated reaction-time experiments, where signals and motions, and clicks of the apparatus come in exceedingly rapid order, one is at first much perplexed in deciding what the order is, yet of the fact of its occupancy of time we are never in doubt.

Accuracy of Our Estimate of Short Durations.

We must now proceed to an account of the *facts* of time-perception in detail as preliminary to our speculative conclusion. Many of the facts are matters of patient experimentation, others of common experience.

First of all, we note a marked *difference between the elementary sensations of duration and those of space*. The former have a much narrower range; the time-sense may be called a myopic organ, in comparison with the eye, for example. The eye sees rods, acres, even miles, at a single glance, and these totals it can afterward subdivide into an almost infinite number of distinctly identified parts. The units of duration, on the other hand, which the time-sense is able to take in at a single stroke, are groups of a few seconds, and within these units very few subdivisions—perhaps forty at most, as we shall presently see—can be clearly discerned. The durations we have practically most to deal with—minutes, hours, and days—have to be symbolically conceived, and constructed by mental addition, after the fashion of those extents of hundreds of miles and upward, which in the field of space are beyond the range of most men's practical interests altogether. To 'realize' a quarter of a mile we need only look out of the window and *feel* its length by an act which, though it may in part result from the organized associations, yet seems immediately performed. To realize an hour, we must count 'now!—now!—now!—now!—' indefinitely. Each 'now' is the feeling of a separate *bit* of time, and the exact sum of the bits never makes a very clear impression on our mind.

How many bits can we clearly apprehend at once? Very few if they are long bits, more if they are extremely short, most if they come to us in compound groups, each including smaller bits of its own.

Hearing is the sense by which the subdivision of durations is most sharply made. Almost all the experimental work on the time-sense has been done by means of strokes of sound. How long a series of sounds, then, can we group in the mind so as not to confound it with a longer or a shorter series?

Our spontaneous tendency is to break up any monotonously given series of sounds into some sort of a rhythm. We involuntarily accentuate every second, or third, or fourth beat, or we break the series in still more intricate ways. Whenever we thus grasp the impressions in rhythmic form, we can identify a longer string of them without confusion. . . .

Wundt and his pupil Dietze have both tried to determine experimentally the *maximal extent of our immediate distinct consciousness for successive impressions.*

Wundt found that twelve impressions could be distinguished clearly as a united cluster, provided they were caught in a certain rhythm by the mind, and succeeded each other at intervals not smaller than 0.3 and not larger than 0.5 of a second. This makes the total time distinctly apprehended to be equal to from 3.6 to 6 seconds.

Dietz gives larger figures. The most favorable intervals for clearly catching the strokes were when they came at from 0.3 second to 0.18 second apart. *Forty* strokes might then be remembered as a whole, and identified without error when repeated, provided the mind grasped them in five sub-groups of eight, or in eight sub-groups of five strokes each. When no grouping of the strokes beyond making *couples* of them by the attention was allowed—and practically it was found impossible not to group them in at least this simplest of all ways—16 was the largest number that could be clearly apprehended as a whole.[3] This would make 40 times 0.3 second, or 12 seconds, to be the *maximum filled duration* of which we can be both *distinctly and immediately* aware.

The maximum unfilled, or *vacant duration,* seems to lie within the same objective range. Estel and Mehner, also working in Wundt's laboratory, found it to vary from 5 or 6 to 12 seconds, and perhaps more. The differences seemed due to practice rather than to idiosyncrasy.

These figures may be roughly taken to stand for the most important part of what, with Mr. Clay, we called, a few pages back, the *specious present.* The specious present has, in addition, a vaguely vanishing backward and forward fringe; but its nucleus is probably the dozen seconds or less that have just elapsed. . . .

There is a certain emotional *feeling* accompanying the intervals of time, as is well known in music. *The sense of haste goes with one measure of rapidity, that of delay with another;* and these two feelings harmonize with different mental moods. . . .

We Have No Sense For Empty Time.

Although subdividing the time by beats of sensation aids our accurate knowledge of the amount of it that elapses, such subdivision does not seem at the first glance essential to our perception of its flow. Let one sit with closed eyes and, abstracting entirely from the outer world, attend exclusively to the passage of time, like one who wakes, as the poet says, "to hear time flowing in the middle of the night, and all things moving to a day of doom." There seems under such cir-

---

3. *Counting* was of course not permitted. It would have given a symbolic concept and no intuitive or immediate perception of the totality of the series. With counting we may of course compare together series of any length—series whose beginnings have faded from our mind, and of whose totality we retain no sensible impression at all. To count a series of clicks is an altogether different thing from merely perceiving them as discontinuous. In the latter case we need only be conscious of the bits of empty duration between them; in the former we must perform rapid acts of association between them and as many names of numbers.

cumstances as these no variety in the material content of our thought, and what we notice appears, if anything, to be the pure series of durations budding, as it were, and growing beneath our indrawn gaze. Is this really so or not? The question is important, for, if the experience be what it roughly seems, we have a sort of special sense for pure time—a sense to which empty duration is an adequate stimulus; while if it be an illusion, it must be that our perception of time's flight, in the experiences quoted, is due to the *filling* of the time, and to our *memory* of a content which it had a moment previous, and which we feel to agree or disagree with its content now.

It takes but a small exertion of introspection to show that the latter alternative is the true one, and that *we can no more intuit a duration than we can intuit an extension, devoid of all sensible content.* Just as with closed eyes we preceive a dark visual field in which a curdling play of obscurest luminosity is always going on; so, be we never so abstracted from distinct outward impressions, we are always inwardly immersed in what Wundt has somewhere called the twilight of our general consciousness. Our heart-beats, our breathing, the pulses of our attention, fragments of words or sentences that pass through our imagination, are what people this dim habitat. Now, all these processes are rhythmical, and are apprehended by us, as they occur, in their totality; the breathing and pulses of attention, as coherent successions, each with its rise and fall; the heart-beats similarly, only relatively far more brief; the words not separately, but in connected groups. In short, empty our minds as we may, some form of *changing process* remains for us to feel, and cannot be expelled. And along with the sense of the process and its rhythm goes the sense of the length of time it lasts. Awareness of *change* is thus the condition on which our perception of time's flow depends; but there exists no reason to suppose that empty time's own changes are sufficient for the awareness of change to be aroused. The change must be of some concrete sort—an outward or inward sensible series, or a process of attention or volition.[4]

4. I leave the text just as it was printed in the Journal of Speculative Philosophy (for 'Oct. 1886') in 1887. Since then Münsterberg in his masterly Beiträge zur experimentellen Psychologie (Heft 2, 1889) seems to have made it clear what the sensible changes are by which we measure the lapse of time. . . .

And here again we have an analogy with space. The earliest form of distinct space-perception is undoubtedly that of a movement over some one of our sensitive surfaces, and this movement is originally given as a simple whole of feeling, and is only decomposed into its elements—successive positions successively occupied by the moving body—when our education in discrimination is much advanced. But a movement is a change, a process; so we see that in the time-world and the space-world alike the first known things are not elements, but combinations, not separate units, but wholes already formed. The condition of *being* of the wholes may be the elements; but the condition of our *knowing* the elements is our having already felt the wholes as wholes.

In the experience of watching empty time flow—'empty' to be taken hereafter in the relative sense just set forth—we tell it off in pulses. We say 'now! now! now!' or we count 'more! more! more!' as we feel it bud. This composition out of units of duration is called the law of time's *discrete flow*. The discreteness is, however, merely due to the fact that our successive acts of *recognition* or *apperception* of *what* it is are discrete. The sensation is as continuous as any sensation can be. All continuous sensations are *named* in beats. We notice that a certain finite 'more' of them is passing or already past. To adopt Hodgson's image, the sensation is the measuring-tape, the perception the dividing-engine which stamps its length. As we listen to a steady sound, we *take it in* in discrete pulses of recognition, calling it successively 'the same! the same! the same!' The case stands no otherwise with time.

After a small number of beats our impression of the amount we have told off becomes quite vague. Our only way of knowing it accurately is by counting, or noticing the clock, or through some other symbolic conception.[5] When the times exceed hours or days, the conception is absolutely symbolic. We think of the amount we mean either solely as a *name,* or by running over a few salient *dates* therein,

---

5. "Any one wishing yet further examples of this mental substitution will find one on observing how habitually he thinks of the spaces on the clock-face instead of the periods they stand for; how, on discovering it to be half an hour later than he supposed, he does not represent the half hour in its duration, but scarcely passes beyond the sign of it marked by the finger." (H. Spencer: Psychology, § 336.)

with no pretence of imagining the full durations that lie between them. No one has anything like a *perception* of the greater length of the time between now and the first century than of that between now and the tenth. To an historian, it is true, the longer interval will suggest a host of additional dates and events, and so appear a more *multitudinous* thing. And for the same reason most people will think they directly perceive the length of the past fortnight to exceed that of the past week. But there is properly no comparative time *intuition* in these cases at all. It is but dates and events, *representing* time; their abundance *symbolizing* its length. I am sure that this is so, even where the times compared are no more than an hour or so in length. It is the same with Spaces of many miles, which we always compare with each other by the numbers which measure them.[6]

From this we pass naturally to speak of certain familiar variations in our estimation of lengths of time. *In general, a time filled with varied and interesting experiences seems short in passing, but long as we look back. On the other hand, a tract of time empty of experiences seems long in passing, but in retrospect short.* A week of travel and sight-seeing may subtend an angle more like three weeks in the memory; and a month of sickness hardly yields more memories than a day. The length in retrospect depends obviously on the multitudinousness of the memories which the time affords. Many objects, events, changes, many subdivisions, immediately widen the view as we look back. Emptiness, monotony, familiarity, make it shrivel up. In Von Holtei's 'Vagabonds' one Anton is described as revisiting his native village.

"Seven years," he exclaims, "seven years since I ran away! More like seventy it seems, so much has happened. I cannot think

6. The only objections to this which I can think of are: (1) The accuracy with which some men judge of the hour of day or night without looking at the clock; (2) the faculty some have of waking at a preappointed hour; (3) the accuracy of time-perception reported to exist in certain trance-subjects. It might seem that in these persons some sort of a sub-conscious record was kept of the lapse of time *per se*. But this cannot be admitted until it is proved that there are no physiological processes, the feeling of whose course may serve as a *sign* of how much time has sped, and so lead us to infer the hour. That there are such processes it is hardly possible to doubt. . . .

of it all without becoming dizzy—at any rate not now. And yet again, when I look at the village, at the church-tower, it seems as if I could hardly have been seven days away." . . .

The same space of time seems shorter as we grow older—that is, the days, the months, and the years do so; whether the hours do so is doubtful, and the minutes and seconds to all appearance remain about the same. . . .

[I]t is certain that, in great part at least, the foreshortening of the years as we grow older is due to the monotony of memory's content, and the consequent simplification of the backward-glancing view. In youth we may have an absolutely new experience, subjective or objective, every hour of the day. Apprehension is vivid, retentiveness strong, and our recollections of that time, like those of a time spent in rapid and interesting travel, are of something intricate, multitudinous, and long-drawn-out. But as each passing year converts some of this experience into automatic routine which we hardly note at all, the days and the weeks smooth themselves out in recollection to contentless units, and the years grow hollow and collapse.

So much for the apparent shortening of tracts of time in *retrospect*. They shorten *in passing* whenever we are so fully occupied with thier content as not to note the actual time itself. A day full of excitement, with no pause, is said to pass 'ere we know it.' On the contrary, a day full of waiting, of unsatisfied desire for change, will seem a small eternity. *Tœdium, ennui, Langweile, boredom,* are words for which, probably, every language known to man has its equivalent. It comes about whenever, from the relative emptiness of content of a tract of time, we grow attentive to the passage of the time itself. Expecting, and being ready for, a new impression to succeed; when it fails to come, we get an empty time instead of it; and such experiences, ceaselessly renewed, make us most formidably aware of the extent of the mere time itself.[7] Close your eyes and simply wait to hear somebody tell you that a minute has elapsed. The full length of your

7. "Empty time is most strongly perceived when it comes as a *pause* in music or in speech. Suppose a preacher in the pulpit, a professor at his desk, to stick still in the midst of his discourse; or let a composer (as is sometimes purposely done) make all his instruments stop at once; we await every instant the resumption of the performance, and, in this

leisure with it seems incredible. You engulf yourself into its bowels as into those of that interminable first week of an ocean voyage, and find yourself wondering that history can have overcome many such periods in its course. All because you attend so closely to the mere feeling of the time *per se,* and because your attention to that is susceptible of such fine-grained successive subdivision. The *odiousness* of the whole experience comes from its insipidity; for *stimulation* is the indispensable requisite for pleasure in an experience, and the feeling of bare time is the least stimulating experience we can have.[8] The sensation of tædium is a *protest,* says Volkmann, against the entire present.

Exactly parallel variations occur in our consciousness of space. A road we walk back over, hoping to find at each step an object we have dropped, seems to us longer than when we walked over it the other way. A space we measure by pacing appears longer than one we traverse with no thought of its length. And in general an amount of space attended to in itself leaves with us more impression of spaciousness than one of which we only note the content.

I do not say that *everything* in these fluctuations of estimate can be accounted for by the time's content being crowded and interesting, or simple and tame. Both in the shortening of time by old age and in its lengthening by *ennui* some deeper cause *may* be at work. This cause can only be ascertained, if it exist, by finding out *why we perceive time at all.* To this inquiry let us, though without much hope, proceed.

---

awaiting, perceive, more than in any other possible way, the empty time. To change the example, let, in a piece of polyphonic music—a figure, for instance, in which a tangle of melodies are under way—suddenly a single voice be heard, which sustains a long note, while all else is hushed. . . . This one note will appear very protracted—why? Because we *expect* to hear accompanying it the notes of the other instruments, but they fail to come." (Herbart: Psychol. als W., § 115.)—Compare also Münsterberg, Beiträge, Heft 2, p. 41.

8. A night of pain will seem terribly long; we keep looking forward to a moment which never comes—the moment when it shall cease. But the odiousness of this experience is not named *ennui* or *Langweile,* like the odiousness of time that seems long from its emptiness. The more positive odiousness of the pain, rather, is what tinges our memory of the night. What we fee, as Prof. Lazarus says (Ideale Fragen [1878], p. 202), is the long time of the suffering, not the suffering of the long time *per se.*

The Feeling of Past Time is a Present Feeling.

If asked why we perceive the light of the sun, or the sound of an explosion, we reply, "Because certain outer forces, ether-waves or air-waves, smite upon the brain, awakening therein changes, to which the conscious perceptions, light and sound, respond." But we hasten to add that neither light nor sound *copy* or *mirror* the ether- or air-waves; they represent them only symbolically. The *only* case, says Helmholtz, in which such copying occurs, and in which

"our perceptions can truly correspond with outer reality, is that of the *time succession* of phenomena. Simultaneity, succession, and the regular return of simultaneity or succession, can obtain as well in sensations as in outer events. Events, like our perceptions of them, take place in time, so that the time-relations of the latter can furnish a true copy of those of the former. The sensation of the thunder follows the sensation of the lightning just as the sonorous convulsing of the air by the electric discharge reaches the observer's place later than that of the luminiferous ether."

One experiences an almost instinctive impulse, in pursuing such reflections as these, to follow them to a sort of crude speculative conclusion, and to think that he has at last got the mystery of cognition where, to use a vulgar phrase, 'the wool is short.' What more natural, we say, than that the sequences and durations of things *should* become known? The succession of the outer forces stamps itself as a like succession upon the brain. The brain's successive changes are copied exactly by correspondingly successive pulses of the mental stream. The mental stream, feeling itself, must feel the time-relations of its own states. But as these are copies of the outward time-relations, so must it know them too. That is to say, these latter time-relations arouse their own cognition; or, in other words, the mere existence of time in those changes out of the mind which affect the mind is a sufficient cause why time is perceived by the mind.

This philosophy is unfortunately too crude. Even though we *were* to conceive the outer successions as forces stamping their image on the brain, and the brain's successions as forces stamping their image on the mind,[9] still, between the mind's own changes *being* successive, and *knowing their own succession,* lies as broad a chasm as between the object and subject of any case of cognition in the world. *A succession of feelings, in and of itself, is not a feeling of succession. And since, to our successive feelings, a feeling of their own succession is added, that must be treated as an additional fact requiring its own special elucidation,* which this talk about outer time-relations stamping copies of themselves within, leaves all untouched.

I have shown, at the outset of the article, that what is past, to be known as past, must be known *with* what is present, and *during* the 'present' spot of time. As the clear understanding of this point has some importance, let me, at the risk of repetition, recur to it again. Volkmann has expressed the matter admirably, as follows:

"One might be tempted to answer the question of the origin of the time-idea by simply pointing to the train of ideas, whose various members, starting from the first, successively attain to full clearness. But against this it must be objected that the successive ideas are not yet the idea of succession, because succession *in* thought is not the thought *of* succession. If idea A follows idea B, consciousness simply exchanges one for another. That B *comes after* A is for our consciouness a non-existent fact; for this *after* is given neither in B nor in A; and no third idea has been supposed. The thinking of the sequence of B upon A is another kind of thinking from that which brought forth A and then brought forth B; and this first kind of thinking is absent so long as merely the thinking of A and the thinking of B are there. In short, when we look at the matter sharply, we come to this antithesis, that if A and B are to be represented *as occurring in succession* they must be *simultaneously represented;* if we are to think *of* them as one after the other, we must *think* them both at once." [10]

9. Succession, time *per se, is* no force. Our talk about its devouring tooth, etc., is all elliptical. Its *contents* are what devour. The law of inertia is incompatible with time's being assumed as an efficient cause of anything.

10. Lehrbuch d. Psych., § 87. Compare also H. Lotze, Metaphysik, § 154. [See also selection no. 13, pp. 209-210 above.—C. M. S.]

If we represent the actual time-stream of our thinking by an horizontal line, the thought *of* the stream or of any segment of its length, past, present, or to come, might be figured in a perpendicular raised upon the horizontal at a certain point. The length of this perpendicular stands for a certain object or content, which in this case is the time thought of, and all of which is thought of together at the actual moment of the stream upon which the perpendicular is raised. . . .

There is thus a sort of *perspective projection* of past objects upon present consciousness, similar to that of wide landscapes upon a camera-screen.

And since we saw a while ago that our maximum distinct *intuition* of duration hardly covers more than a dozen seconds (while our maximum vague intuition is probably not more than that of a minute or so), we must suppose that *this amount of duration is pictured fairly steadily in each passing instant of consciousness* by virtue of some fairly constant feature in the brain-process to which the consciousness is tied. *This feature of the brain-process, whatever it be, must be the cause of our perceiving the fact of time at all.*[11] The duration thus steadily perceived is hardly more than the 'specious present,' as it was called a few pages back. Its *content* is in a constant flux, events dawning into its forward end as fast as they fade out of its rearward one, and each of them changing its time-coefficient from 'not yet,' or 'not quite yet,' to 'just gone' or 'gone' as it passes by. Meanwhile, the specious present, the intuited duration, stands permanent, like the rainbow on the waterfall, with its own quality unchanged by the events that stream through it. Each of these, as it slips out, retains the power of being reproduced; and when reproduced, is reproduced with the duration and neighbors which it originally had. Please observe, however, that the reproduction of an event, *after* it has once completely dropped out of the rearward end of the specious present, is an entirely different psychic fact from its direct perception in the specious present as a thing immediately past. A creature might be entirely devoid of *reproductive* memory, and yet have the time-sense; but the latter would be limited, in his case, to the few seconds immediately passing by. Time older than that he would never recall. I assume reproduction in the text, because I am speaking of human beings who notoriously

11. The cause of the perceiving, not the object perceived!

possess it. Thus memory gets strewn with *dated* things—dated in the sense of being before or after each other.[12] The date of a thing is a mere relation of *before* or *after* the present thing or some past or future thing. Some things we date simply by mentally tossing them into the past or future *direction.* So in space we think of England as simply to the eastward, of Charleston as lying south. But, again, we may date an event exactly, by fitting it between two terms of a past or future series explicitly conceived, just as we may accurately think of England or Charleston being just so many miles away.[13]

The things and events thus vaguely or exactly dated become thence forward those signs and symbols of longer time-spaces, of which we previously spoke. According as we think of a multitude of them, or of few, so we imagine the time they represent to be long or short. But *the original paragon and prototype of all conceived times is the specious present, the short duration of which we are immediately and incessantly sensible.*

12. " 'No more' and 'not yet' are the proper time-feelings, and we are aware of time in no other way than through these feelings," says Volkmann (Psychol., § 87). This, which is not strictly true of our feelings of *time per se,* as an elementary bit of duration, is true of our feeling of *date* in its events.

13. We construct the miles just as we construct the years. Travelling in the cars makes a succession of different fields of view pass before our eyes. When those that have passed from present sight revive in memory, they maintain their mutual order because their contents overlap. We think them as having been before or behind each other; and, from the multitude of the views we can recall behind the one now presented, we compute the total space we have passed through.

It is often said that the perception of time develops later than that of space, because children have so vague an idea of all dates before yesterday and after to-morrow. But no vaguer than they have of extensions that exceed as greatly their unit of space-intuition. Recently I heard my child of four tell a visitor that he had been 'as much as one week' in the country. As he had been there three months, the visitor expressed surprise; whereupon the child corrected himself by saying he had been there 'twelve years.' But the child made exactly the same kind of mistake when he asked if Boston was not one hundred miles from Cambridge, the distance being three miles.

21

Peirce:

FUTURITY, MEANING, AND ACTION

Charles Sanders Peirce (1839–1914) was the generally acknowledged pioneering father of pragmatism. In 1905 he published an
article in *The Monist* under the title, "What Pragmatism Is;" * its
intent was to distinguish his own original position, which he
renamed 'pragmaticism', from the less formal and more psychological version being espoused by James. In explicating this
distinctive position, he gave clear expression to the new temporal
outlook implicit in pragmatism from the beginning; the second
half of the article in which he did this constitutes the selection.

Let us now hasten to the exposition of pragmaticism itself. Here it
will be convenient to imagine that somebody to whom the doctrine is
new, but of rather preternatural perspicacity, asks questions of a
pragmaticist. Everything that might give a dramatic illusion must be
stripped off, so that the result will be a sort of cross between a dialogue
and a catechism, but a good deal more like the latter—something
rather painfully reminiscent of Mangnall's *Historical Questions.*
Questioner: I am astounded at your definition of your pragmatism,
because only last year I was assured by a person above all suspicion of
warping the truth—himself a pragmatist—that your doctrine precisely
was "that a conception is to be tested by its practical effects." You
must surely, then, have entirely changed your definition very recently.
Pragmatist: If you will turn to Volumes VI and VII of the *Revue
Philosophique,* or to the *Popular Science Monthly* for November 1877
and January 1878, you will be able to judge for yourself whether the
interpretation you mention was not then clearly excluded. The exact
wording of the English enunciation (changing only the first person
into the second) was: "Consider what effects that might conceivably

* Reprinted by permission of Open Court Publishing Co., LaSalle, Illinois.

have practical bearing you conceive the object of your conception to have. Then your conception of those effects is the WHOLE of your conception of the object."

Questioner: Well, what reason have you for asserting that this is so?

Pragmatist: That is what I specially desire to tell you. But the question had better be postponed until you clearly understand what those reasons profess to prove.

Questioner: What, then, is the *raison d'être* of the doctrine? What advantage is expected from it?

Pragmatist: It will serve to show that almost every proposition of ontological metaphysics is either meaningless gibberish—one word being defined by other words, and they by still others, without any real conception ever being reached—or else is downright absurd; so that all such rubbish being swept away, what will remain of philosophy will be a series of problems capable of investigation by the observational methods of the true sciences—the truth about which can be reached without those interminable misunderstandings and disputes which have made the highest of the positive sciences a mere amusement for idle intellects, a sort of chess—idle pleasure its purpose, and reading out of a book its method. In this regard, pragmaticism is a species of prope-positivism. But what distinguishes it from other species is, first, its retention of a purified philosophy; secondly, it full acceptance of the main body of our instinctive beliefs; and thirdly, its strenuous insistence upon the truth of scholastic realism (or a close approximation to that, well stated by the late Dr. Francis Ellingwood Abbot in the Introduction to his *Scientific Theism*). So, instead of merely jeering at metaphysics, like other prope-positivists, whether by long-drawn-out parodies or otherwise, the pragmaticist extracts from it a precious essence, which will serve to give life and light to cosmology and physics. At the same time, the moral applications of the doctrine are positive and potent; and there are many other uses of it not easily classed. On another occasion, instances may be given to show that it really has these effects.

Questioner: I hardly need to be convinced that your doctrine would wipe out metaphysics. Is it not as obvious that it must wipe out every proposition of science and everything that bears on the conduct of life? For you say that the only meaning that, for you, any assertion bears is that a certain experiment has resulted in a certain way:

nothing else but an experiment enters into the meaning. Tell, me, then, how can an experiment, in itself, reveal anything more than that something once happened to an individual object and that subsequently some other individual event occurred?

Pragmatist: That question is, indeed, to the purpose—the purpose being to correct any misapprehensions of pragmaticism. You speak of an experiment in itself, emphasizing *in itself.* You evidently think of each experiment as isolated from every other. It has not, for example, occurred to you, one might venture to surmise, that every connected series of experiments constitutes a single collective experiment. What are the essential ingredients of an experiment? First, of course, an experimenter of flesh and blood. Secondly, a verifiable hypothesis. This is a proposition [1] relating to the universe environing the experimenter, or to some well-known part of it and affirming or denying of this only some experimental possibility or impossibility. The third indispensable ingredient is a sincere doubt in the experimenter's mind as to the truth of that hypothesis.

Passing over several ingredients on which we need not dwell, the purpose, the plan, and the resolve, we come to the act of choice by which the experimenter singles out certain identifiable objects to be operated upon. The next is the external (or quasi-external) ACT by which he modifies those objects. Next, comes the subsequent *reaction* of the world upon the experimenter in a perception; and finally, his recognition of the teaching of the experiment. While the two chief parts of the event itself are the action and the reaction, yet the unity of essence of the experiment lies in its purpose and plan, the ingredients passed over in the enumeration.

Another thing: in representing the pragmaticist as making rational meaning to consist in an experiment (which you speak of as an event

1. The writer, like most English logicians, invariably uses the word *proposition* not as the Germans define their equivalent, *Satz,* as the language-expression of a judgment *(Urtheil),* but as that which is related to any assertion, whether mental and self-addressed or outwardly expressed, just as any possibility is related to its actualization. The difficulty of the, at best, difficult problem of the essential nature of a Proposition has been increased for the Germans, by their *Urtheil,* counfounding, under one designation, the mental *assertion* with the *assertable.*

in the past), you strikingly fail to catch his attitude of mind. Indeed, it is not in an experiment, but in *experimental phenomena*, that rational meaning is said to consist. When an experimentalist speaks of a *phenomenon*, such as "Hall's phenomenon," "Zeemann's phenomenon," and its modification, "Michelson's phenomenon," or "the chessboard phenomenon," he does not mean any particular event that did happen to somebody in the dead past, but what *surely will* happen to everybody in the living future who shall fulfill certain conditions. The phenomenon consists in the fact that when an experimentalist shall come *to act* according to a certain scheme that he has in mind, then will something else happen, and shatter the doubts of skeptics, like the celestial fire upon the altar of Elijah.

And do not overlook the fact that the pragmaticist maxim says nothing of single experiments or of single experimental phenomena (for what is conditionally true *in futuro* can hardly be singular), but only speaks of *general kinds* of experimental phenomena. Its adherent does not shrink from speaking of general objects as real, since whatever is true represents a real. Now the laws of nature are true.

The rational meaning of every proposition lies in the future. How so? The meaning of a proposition is itself a proposition. Indeed, it is no other than the very proposition of which it is the meaning: it is a translation of it. But of the myriads of forms into which a proposition may be translated, what is that one which is to be called its very meaning? It is, according to the pragmaticist, that form in which the proposition becomes applicable to human conduct, not in these or those special circumstances, nor when one entertains this or that special design, but that form which is most directly applicable to human conduct, not in these or those special circumstances, nor when one entertains this or that special design, but that form which is most directly applicable to self-control under every situation, and to every purpose. This is why he locates the meaning in future time; for future conduct is the only conduct that is subject to self-control. But in order that that form of the proposition which is to be taken as its meaning should be applicable to every situation and to every purpose upon which the proposition has any bearing, it must be simply the general description of all the experimental phenomena which the assertion of the proposition virtually predicts. For an experimental phenomenon is the fact asserted by the proposition that action of a certain description will have a certain kind of experimental result; and experimental results are the only results that can affect human conduct. No doubt, some unchanging idea may come to influence a man more than it had done; but only because some experience equivalent to an experiment has brought its truth home to him more intimately than before. Whenever a man acts purposively, he acts under a belief in some experimental phenomenon. Consequently, the sum of the experimental phenomena that a proposition implies makes up its entire bearing upon human conduct. Your question, then, of how a pragmaticist can attribute any meaning to any assertion other than that of a single occurrence is substantially answered.

Questioner: I see that pragmaticism is a thoroughgoing phenomenalism. Only why should you limit yourself to the phenomena of experimental science rather than embrace all observational science? Experiment, after all, is an uncommunicative informant. It never expiates: it only answers "yes" or "no"; or rather it usually snaps out "No!" or, at best, only utters an inarticulate grunt for the negation of its "no." The typical experimentalist is not much of an observer. It is the student of natural history to whom nature opens the treasury of her confidence, while she treats the cross-examining experimentalist with the reserve he merits. Why should your phenomenalism sound the meagre jew's-harp of experiment rather than the glorious organ of observation?

Pragmaticist: Because pragmaticism is not definable as "thoroughgoing phenomenalism," although the latter doctrine may be a kind of pragmatism. The *richness* of phenomena lies in their sensuous quality. Pragmaticism does not intend to define the phenomenal equivalents of words and general ideas, but, on the contrary, eliminates their sential element, and endeavors to define the rational purport, and this it finds in the purposive bearing of the word or proposition in question.

Questioner: Well, if you choose so to make Doing the Be-all and the End-all of human life, why do you not make meaning to consist simply in doing? Doing has to be done at a certain time upon a certain object. Individual objects and single events cover all reality, as everybody knows, and as a practicalist ought to be the first to insist. Yet, your meaning, as you have described it, is *general*. Thus, it is of the nature of a mere word and not a reality. You say yourself that your meaning of a proposition is only the same proposition in another dress. But a practical man's meaning is the very thing he means. What do you make to be the meaning of "George Washington"?

Pragmaticist: Forcibly put! A good half dozen of your points must certainly be admitted. It must be admitted, in the first place, that if pragmaticism really made Doing to be the Be-all and the End-all of life, that would be its death. For to say that we live for the mere sake of action, as action, regardless of the thought it carries out, would be to say that there is no such thing as rational purport. Secondly, it must be admitted that every proposition professes to be true of a certain real, individual object, often the environing universe. Thirdly, it must be admitted that pragmaticism fails to furnish any translation or meaning of a proper name, or other designation of an individual object. Fourthly, the pragmaticistic meaning is undoubtedly general; and it is equally indisputable that the general is of the nature of a word or sign. Fifthly, it must be admitted that individuals alone exist; and sixthly, it may be admitted that the very meaning of a word or significant object ought to be the very essence of reality of what it signifies. But when those admissions have been unreservedly made, if you find the pragmaticist still constrained most earnestly to deny the force of your objection, you ought to infer that there is some consideration that has escaped you. Putting the admissions together, you will perceive that the pragmaticist grants that a proper name (although it is not customary to say that it has a *meaning*) has a certain denotative function peculiar, in each case, to that name and its equivalents; and that he grants that every assertion contains such a denotative or pointing-out function. In its peculiar individuality, the pragmaticist excludes this from the rational purport of the assertion, although *the like* of it, being common to all assertions, and so, being general and not individual, may enter into the pragmaticistic purport. Whatever exists, *ex-sists,* that is, really acts upon other existents, so obtains a self-identity, and is definitely individual. As to the general, it will be a help to thought to notice that there are two ways of being general. A statue of a soldier on some village monument, in his overcoat and with his musket, is for each of a hundred families the image of its uncle, its sacrifice to the Union. That statue, then, though it is itself single, represents any one man of whom a certain predicate may be true. It is *objectively* general. The word "soldier," whether spoken or written, is general in the same way; while the name "George Washington" is not so. But each of these two terms remains one and the same noun, whether it be spoken or written, and whenever and wherever it be spoken or written. This noun is not an existent thing: it is a *type*, or *form*, to which objects, both those that are externally existent and those which are imagined, may *conform*, but which none of them can exactly be. This is subjective generality. The pragmaticistic purport is general in both ways.

As to reality, one finds it defined in various ways; but if that principle of terminological ethics that was proposed be accepted, the equivocal language will soon disappear. For *realis* and *realitas* are not ancient words. They were invented to be terms of philosophy in the thirteenth century,[2] and the meaning they were intended to express is perfectly clear. That is *real* which has such and such characters, whether anybody thinks it to have those characters or not. At any rate, that is the sense in which the pragmaticist uses the word. Now, just as conduct controlled by ethical reason tends toward fixing certain habits of conduct, the nature of which (as, to illustrate the meaning, peaceable habits and not quarrelsome habits) does not depend upon any accidental circumstances, and *in that sense* may be said to be *destined;* so, thought, controlled by a rational experimental logic, tends to the fixation of certain opinions, equally destined, the nature of which will be the same in the end, however the perversity of thought of whole generations may cause the postponement of the ultimate fixation. If this be so, as every man of us virtually assumes that it is, in regard to each matter the truth of which he seriously discusses, then, according to the adopted definition of "real," the state of things which will be believed in that ultimate opinion is real. But, for the most part, such opinions will be general. Consequently, *some* general objects are real. (Of course, nobody ever thought that *all* generals were real; but the scholastics used to assume that generals were real when they had hardly any, or quite no, experiential evidence to support their assumption; and their fault lay just there, and not in holding that generals could be real.) One is struck with the inexactitide of thought even of analysts of power, when they touch upon modes of being. One will meet, for example, the virtual assumption that what is relative to thought cannot be real. But why not, exactly?

2. See Prantl, *Geschichte der Logik*, III, 91, Anm. 362.

Red is relative to sight, but the fact that this or that is in that relation to vision that we call being red-is not *itself* relative to sight; it is a real fact.

Not only may generals be real, but they may also be *physically efficient,* not in every metaphysical sense, but in the common-sense acception in which human purposes are physically efficient. Aside from metaphysical nonsense, no sane man doubts that if I feel the air in my study to be stuffy, that thought may cause the window to be opened. My thought, be it granted, was an individual event. But what determined it to take the particular determination it did, was in part the general fact that stuffy air is unwholesome, and in part other *Forms,* concerning which Dr. Carus [3] has caused so many men to reflect to advantage—or rather, *by* which, and the general truth concerning which Dr. Carus' mind was determined to the forcible enunciation of so much truth. For truths, on the average, have a greater tendency to get believed than falsities have. Were it otherwise, considering that there are myriads of false hypotheses to account for any given phenomenon, against one sole true one (or if you will have it so, against every true one), the first step toward genuine knowledge must have been next door to a miracle. So, then, when my window was opened, because of the truth that stuffy air is *malsain,* a physical effort was brought into existence by the efficiency of a general and non-existent truth. This has a droll sound because it is unfamiliar; but exact analysis is with it and not against it; and it has besides, the immense advantage of not blinding us to great facts—such as that the ideas "justice" and "truth" are notwithstanding the iniquity of the world, the mightiest of the forces that move it. Generality is, indeed, an indispensable ingredient of reality; for mere individual existence or actuality without any regularity whatever is a nullity. Chaos is pure nothing.

That which any true proposition asserts is *real,* in the sense of being as it is regardless of what you or I may think about it. Let this proposition be a general conditional proposition as to the future, and it is a real general such as is calculated really to influence human conduct; and such the pragmaticist holds to be the rational purport of every concept.

3. "The Foundations of Geometry," by Paul Carus, *The Monist,* XIII, p. 370.

Accordingly, the pragmaticist does not make the *summun bonum* to consist in action, but makes it to consist in that process of evolution whereby the existent comes more and more to embody those generals which were just now said to be *destined,* which is what we strive to express in calling them *reasonable.* In its higher stages, evolution takes place more and more largely through self-control, and this gives the pragmaticist a sort of justification for making the rational purport to be general.

There is much more in elucidation of pragmaticism that might be said to advantage were it not for the dread of fatiguing the reader. It might, for example, have been well to show clearly that the pragmaticist does not attribute any different mode of being to an event in the future from that which he would attribute to a similar event in the past, but only that the practical attitude of the thinker toward the two is different. It would also have been well to show that the pragmaticist does not make Forms to be the *only* realities in the world, any more than he makes the reasonable purport of a word to be the only kind of meaning there is. These things are, however, implicitly involved in what has been said. There is only one remark concerning the pragmaticist's conception of the relation of his formula to the first principles of logic which need detain the reader.

Aristotle's definition of universal predication,[4] which is usually designated (like a papal bull or writ of court, from its opening words), as the *Dictum de omni,* may be translated as follows: "We call a predication (be it affirmative or negative), *universal,* when, and only when, there is nothing among the existent individuals to which the subject affirmatively belongs, but to which the predicate will not likewise be referred (affirmatively or negatively, according as the universal predication is affirmative or negative)." . . .[5] The important words "existent individuals" have been introduced into the translation (which English idiom would not here permit to be literal); but it is plain that "existent individuals" were what Aristotle meant. The other departures from literalness only serve to give modern English forms of expression. Now, it is well known that propositions in formal logic go in pairs, the two of one pair being convertible into another by the

---

4. *Prior Analytics,* 24b, 28–30.
5. [Peirce's citation of the Greek text is omitted.—C. M. S.]

interchange of the ideas of antecedent and consequent, subject and predicate, etc. The parallelism extends so far that it is often assumed to be perfect; but it is not quite so. The proper mate of this sort to the *Dictum de omni* is the following definition of affirmative predication: We call a predication *affirmative* (be it universal or particular) when, and only when, there is nothing among the sensational effects that belong universally to the predicate which will not be (universally or particularly, according as the affirmative predication is universal or particular) said to belong to the subject. Now, this is substantially the essential proposition of pragmaticism. Of course, its parallelism to the *Dictum de omni* will be admitted by a person who admits the truth of pragmaticism.

Suffer me to add one word more on this point. For if one cares at all to know what the pragmaticist theory consists in, one must understand that there is no other part of it to which the pragmaticist attaches quite as much importance as he does to the recognition in his doctrine of the utter inadequacy of action or volition or even of resolve or actual purpose, as materials out of which to construct a conditional purpose or the concept of conditional purpose. Had a purposed article concerning the principle of continuity and synthetizing the ideas of the other articles of a series in the early volumes of *The Monist* ever been written, it would have appeared how, with thorough consistency, that theory involved the recognition that continuity is an indispensable element of reality, and that continuity is simply what generality becomes in the logic of relatives, and thus, like generality, and more than generality, is an affair of thought, and is the essence of thought. Yet even in its truncated condition, an extra-intelligent reader might discern that the theory of those cosmological articles made reality to consist in something more than feeling and action could supply, inasmuch as the primeval chaos, where those two elements were present, was explicitly shown to be pure nothing. Now, the motive for alluding to that theory just here is that in this way one can put in a strong light a position which the pragmaticist holds and must hold, whether that cosmological theory be ultimately sustained or exploded, namely, that the third category—the category of thought, representation, triadic relation, mediation, genuine thirdness, thirdness as such—is an essential ingredient of reality, yet does not by itself constitute reality, since this category (which in that cosmology appears as the element of habit) can have no concrete being without action, as a separate object on which to work its government, just as action cannot exist without the immediate being of feeling on which to act. The truth is that pragmaticism is closely allied to the Hegelian absolute idealism, from which, however, it is sundered by its vigorous denial that the third category (which Hegel degrades to a mere stage of thinking) suffices to make the world, or is even so much as self-sufficient. Had Hegel, instead of regarding the first two stages with his smile of contempt, held on to them as independent or distinct elements of the triune Reality, pragmaticists might have looked up to him as the great vindicator of their truth. (Of course, the external trappings of his doctrine are only here and there of much significance.) For pragmaticism belongs essentially to the triadic class of philosophical doctrines, and is much more essentially so than Hegelianism is. (Indeed, in one passage, at least, Hegel alludes to the triadic form of his exposition as to a mere fashion of dress.)

Milford, Pa.,
September, 1904

22

Royce:

TIME: CONCEPT AND WILL

Josiah Royce (1855–1916) was born in California, studied under Lotze in Germany and under Peirce at Johns Hopkins, where he received his doctorate. He joined the Harvard faculty in 1892 as the protégé of James. His pragmatic idealism was concerned to wed the pragmatic outlook to the tradition of speculative metaphysics which he remolded to emphasize the importance of the individual in the cosmic order of dynamic reality. The Gifford Lectures, which he delivered in 1900–1901, were published in 1904 as *The World and the Individual;* the selection is taken from the first three parts of the chapter entitled "The Temporal and the Eternal." * One chief point of this discussion is to show that "For the finite world in general, then, as for us human beings, the distinction of past and future appears to be coextensive with life and meaning."

I

Time is known to us, both perceptually, as the psychologists would say, and conceptually. That is, we have a relatively direct experience of time at any moment, and we acknowledge the truth of a relatively indirect conception that we possess of the temporal order of the world. But our conception of time far outstrips in its development and in its organization anything that we are able directly to find in the time that is known to our perceptions. Much of the difficulty that appears in our metaphysical views about time is, however, due to lack of naïveté and directness in viewing the temporal aspects of reality. We first emphasize highly artificial aspects of our conception of time. Then we

---

* Josiah Royce, *The World and the Individual,* Second Series (New York: The Macmillan Company, 1904).

wonder how these various aspects can be brought into relation with the rest of the real world. Our efforts to solve our problem lead very easily to contradictions. We fail to observe how, in case of our more direct experience of time and of its meaning, various elements are woven into a certain wholeness,—the very elements which, when our artificial conception of time has sundered them, we are prone to view as irreconcilable with one another and with reality.

Our more direct perceptions of time form a complex sort of consciousness, wherein it is not difficult to distinguish several aspects. For the first, some Change is always occurring in our experience. This change may belong to the facts of any sense, or to our emotions, or to our ideas; but for us to be conscious is to be aware of change. Now this changing character of our experience is never the whole story of any of our clearer and more definite kinds of consciousness. The next aspect of the matter lies in the fact that our consciousness of change, wherever it is definite and wherever it accompanies definite successive acts of attention, goes along with the consciousness that for us something comes first, and something next, or that there is what we call a Succession of events. Of such successions, melodies, rhythms, and series of words or of other simple acts form familiar and typical examples. An elementary consciousness of change without such definite successions we can indeed have; but where we observe clearly what a particular change is, it is a change wherein one fact succeeds another.

A succession, as thus more directly experienced by us, involves a certain well-known relation amongst the events that make up the succession. Together these events form a temporal sequence or order. Each one of them is over and past when the next one comes. And this order of the experienced time-series has a determinate direction. The succession passes *from* each event *to* its successor, and not in the reverse direction; so that herein the observed time relations notoriously differ from what we view as space relations. For if in space b is next to a, we can read the relation equally well as a coexistence of a with b, and as a coexistence of b with a. But in case b succeeds a, as one word succeeds another in a spoken sentence, then the relation is experienced as a passing from a to b, or as a passing over of a into b, in such wise that a is past, as an event, before b comes. This direction of the stream of time forms one of its most notable empirical characters.

It is obviously related to that direction of the acts of the will whose logical aspect interested us in connection with the consideration of our discriminating consciousness.

But side by side with this aspect of the temporal order, as we experience this order, stands still another aspect, whose relation to the former has been persistently pointed out by many psychological writers, and as persistently ignored by many of the metaphysical interpreters of the temporal aspect of the universe. When we more directly experience succession,—as, for instance, when we listen to a musical phrase or to a rhythmic series of drum-beats,—we not only observe that any antecedent member of the series is over and past before the next number comes, but also, and without the least contradiction between these two aspects of our total experience, we observe that this whole succession, with both its former and later members, so far as with relative directness we apprehend the series of drum-beats or of other simple events, is present *at once* to our consciousness, in precisely the sense in which the unity of our knowing mental life always finds present at once many facts. It is, as I must insist, true that for my consciousness *b* is experienced as following *a*, and also that both *a* and *b* are *together* experienced as in this relation of sequence. To say this is no more contradictory than to say that while I experience two parts of a surface as, by virtue of their spatial position, mutually exclusive of each other; I also may experience the fact that both these mutually exclusive parts go together to form one whole surface. The sense in which they form one surface is, of course, not the sense in which, as parts, they exclude each other, and form different surfaces. Well, just so, the sense in which *b*, as successor of *a*, is such, in the series of events in question, that *a* is over and gone when *b* comes, is not the sense in which *a* and *b* are together elements in the whole experienced succession. But that, in *both* of these senses, the relation of *b* to its predecessor *a* is an experienced fact, is a truth that any one can observe for himself.

If I utter a line of verse, such as

"The curfew tolls the knell of parting day,"

the sound of the word *day* succeeds the sound of the word *parting*, and I unquestionably experience the fact that, for me, every earlier word of the line is over and past before the succeeding word or the last word, *day*, comes to be uttered or to be heard. Yet this is unquestionably not my whole consciousness about the succession. For I am certainly *also* aware that the *whole* line of poetry, as a succession of uttered sounds (or, at all events, a considerable portion of the line), is present to me at once, and at this one succession, when I speak the line. For only by virtue of experiencing this wholeness do I observe the rhythm, the music, and the meaning of the line. The sense in which the word *parting* is over before the word *day* comes, is like the sense in which one object in space is *where* any other object is *not*, so that the spatial *presence* of one object excludes the presence of another at that same part of space. Precisely so the presence of the word *day* excludes the presence of the word *parting* from its own place in the temporal succession. And, in our experience of succession, each element is *present* in a particular point of the series, in so far as, with reference to that point, other events of the series are either *past*, that is, over and done with, or are *future*, that is, are later in the series, or are *not yet* *when* this one point of the series is in *this* sense present. Every word of the uttered line of poetry, viewed in its reference to the other words, or to previous and later experiences, is *present* in its own place in the series, is *over and done with* before later events can come, or when they are present, and is *not yet* when the former events of the succession are present. And that all this is true, certainly is a matter of our experience of succession.

But the sense in which, nevertheless, the whole series of the uttered words of the line, or of some considerable portion of the line, is presented to our consciousness *at once,* is precisely the sense in which we apprehend this line as one line, and this succession as one succession. The whole series of words has for us its rhythmic unity, and forms an instance of conscious experience, whose unity we overlook at one glance. And unless we could thus overlook a succession and view at once its serially related and mutually exclusive events, we should never know anything whatever about the existence of succession, and should have no problem about time upon our hands. . . .

I have now characterized the more directly given features in our consciousness of succession. You see, as a result, that we men experience what Professor James, and others, have called our "specious present," as a serial whole, *within* which there are observed temporal differences of former and latter. And this our "specious present" has, when measured by a reference to time-keepers, a length which varies with circumstances, but which appears to be never any very small fraction of a second, and never more than a very few seconds in length. I have earlier referred to this length of our present moments as our characteristic "time-span" of consciousness, and have pointed out how arbitrary a feature and limitation of our consciousness it is. We shall return soon to the question regarding the possible metaphysical significance of this time-span of our own special kind of consciousness.

But it remains here to call closer attention to certain other equally important features of our more direct experience of time-succession. So far, we have spoken, in the main, as if succession were to us a mere matter of given facts, as colors and sounds are given. But all our experience also has relation to the interests whose play and whose success or defeat constitute the life of our will. Every serial succession of which we are conscious therefore has for us some sort of meaning. In it we find our success or our failure. In it our internal meanings [1] are expressed, or hindered, thwarted, or furthered. We are interested in life, even if it be, in idle moments, only the dreary interest of wondering what will happen next, or, in distressed moments, the interest in flying from our present future, or, in despairing moments, of wishing for the end; still more then if, in strenuous moments, our interest is in pursuing our ideal. And our interest in life means our conscious concern in passing on from any temporal present towards its richer fulfilment, or away from its relative insignificance. Now that Direction of temporal succession of which I before made mention, has the most intimate relations to this our interest in our experience. What is earlier in a given succession is related to what is later as being that *from* which we pass *towards* a desired fulfilment, or in search of a more complete expression of our purpose. We are never content in the temporal present in so far as we view it as temporal, that is, as an event in a series. For such a present has its meaning as a transition from its predecessors towards its successors.

Our temporal form of experience is thus peculiarly the form of the

1. [See p. 357 for explanation of this term; also *The World and the Individual*, vol. I, pp. 24–26.—C. M. S.]

Will as such. Space often seems to spread out before us what we take to be the mere contents of our world; but time gives the form for the expression of all our meanings. Facts, in so far as, with an abstractly false Realism, we sunder them from their meanings, therefore tend to be viewed as merely in relations of coexistence; and the space-world is the favorite region of Realism. But ideas, when conscious, assume the consciously temporal form of inner existence, and appear to us as constructive processes. The visible world, when viewed as at rest, therefore interests us little in comparision with the same world when we take note of its movements, changes, successions. As the kitten ignores the dead leaves until the wind stirs them, but then chases them—so facts in general tend to appear to us all dead and indifferent when we disregard their processes. But in the movement of things lies for us, just as truly as in her small way for the kitten, all the glory and the tragedy, all the life and the meaning of our observed universe. This concern, this interest in the changing, binds us then to the lower animals, as it doubtless also binds us to beings of far higher than human grade. We watch the moving and tend to neglect the apparently changeless objects about us. And that is why narrative is so much more easily effective than description in the poetic arts; and why, if you want to win the attention of the child or of the general public, you must tell the story rather than portray coexistent truths, and must fill time with series of events, rather than merely crowd the space of experience or of imagination with manifold but undramatic details. For space furnishes indeed the stage and the scenery of the universe, but the world's play occurs in time.

Now all these familiar considerations remind us of certain of the most essential characters of our experience of time. Time, whatever else it is, is given to us as that within whose successions, in so far as for us they have a direct interest and meaning, every event, springing from, yet forsaking, its predecessors, aims on, towards its own fulfilment and extinction in the coming of its successors. Our experience of time is thus for us essentially an experience of longing, of pursuit, of restlessness. . . . But as to the facts, every part of a succession is present in so far as when it is, that which is *no longer* and that which is *not yet* both of them stand in essentially significant, or, if you will, in essentially practical relations to this present. It is true, of course, that when we view relatively indifferent time-series, such as the ticking of a watch or the dropping of rain upon the roof, we can disregard this more significant aspect of succession; and speak of the endless flight of time as an incomprehensible brute fact of experience, and as in so far seemingly meaningless. But no series of experiences upon which attention is fixed is wholly indifferent to us; and the temporal aspect of such series always involves some element of expectancy and some sense of something that no longer is; and both these conscious attitudes color our interest in the presented succession, and give the whole the meaning of life. Time is thus indeed the form of practical activity; and its whole character, and especially that direction of its succession of which we have spoken, are determined accordingly.

II

I have dwelt long upon the time consciousness of our relatively direct experience, because here lies the basis for every deeper comprehension of the metaphysics both of time and of eternity. Our ordinary conception of time as an universal form of existence in the external world, is altogether founded upon a generalization, whose origin is in us men largely and obviously social, but whose materials are derived from our inner experience of the succession of significant events. The conceived relations of Past, Present, and Future in the real world of common-sense metaphysics, appear indeed, at first sight, vastly to transcend anything that we ourselves have ever observed in our inner experience. The infinite and irrevocable past that no longer is, the expected infinite future that has as yet no existence, how remote these ideal constructions, supposed to be valid for all gods and men and things, seem at first sight from the brief and significant series of successive events that occur within the brief span of our actual human consciousness. Yet, as we saw in the ninth lecture of our former Series,[2] common sense, as soon as questioned about special cases, actually conceives the Being of both the past and the future as so intimately related to the Being of the present that every definite conception of the real processes of the world, whether these processes are viewed as physical or as historical or explicitly as ethical, depends

2. [I.e., "Universality and Unity," vol. I, *The World and the Individual.*— C. M. S.]

upon taking the past, the present, and the future as constituting a single whole, whose parts have no true Being except in their linkage. As a fact, moreover, the term *present,* when applied to characterize a moment or an event in the time-stream of the real world, never means, in any significant application, the indivisible present of an ideal mathematical time. The present time, in case of the world at large, has an unity altogether similar to that of the present moment of our inner consciousness. We may speak of the present minute, hour, day, year, century. If we use the term *present* regarding any one of these divisions of time, but regard this time not as the experienced form of the inner succession of our own mental events, but as the time of the real world in which we ourselves form a part, then we indeed conceive that this present is world-embracing, and that suns move, light radiates between stars, the deeds of all men occur, and the minds of all men are conscious, in this same present time of which we thus make mention. Moreover, we usually view the world-time in question in terms of the conceptions of the World of Description, and so we conceive it as infinitely divisible, as measurable by various mathematical and physical devices, and as a continuous stream of occurrence. Yet in whatever sense we speak of the real present time of the world, this present, whether it is the present second, or the present century, or the present geological period, it is, for our conception, as truly a divisible and connected whole region of time, within which a succession of events takes place, as it is a world-embracing and connected time, within whose span the whole universe of present events is comprised. A mathematically indivisible present time, possessing no length, is simply no time at all. Whoever says, "In the universe at large only the present state of things is real, only the present movement of the stars, the present streamings of radiant light, the present deeds and thoughts of men are real; the whole past is dead; the whole future is not yet,"—any such reporter of the temporal existence of the universe may be invited to state how long his real present of the time-world is. If he replies, "The present moment is the absolutely indivisible and ideal boundary between present and future,"—then one may rejoin at once that in a mathematically indivisible instant, having no length, no event happens, nothing endures, no thought or deed takes place,—in brief, nothing whatever temporally exists,—and that, too, whatever conception you may have of Being. But if the real present is a divisible portion of time, then it contains within itself succession, precisely as the "specious present" of psychological time contains such internal succession. But in the case, within the real present of the time-world, there are already contained the distinctions that, in case of the time of experience, we have heretofore observed. If, in what you choose to call the present moment of the world's history, deeds are accomplished, suns actually move from place to place, light waves traverse the ether, and men's lives pass from stage to stage, then *within* what you thus call the present there are distinguishable and more elementary events, arranged in series, such that when any conceived element, or mere elementary portion of any series is taken in relation to its predecessors and successors, it is *not yet* when its antecedents are taken as temporally present, and is *past and gone* when its successors are viewed as present. The world's time is thus in all respects a generalized and extended image and correspondent of the observed time of our inner experience. In the time of our more direct experience, we find a twofold way in which we can significantly call a portion of time a present moment. The present, in our inner experience, means a whole series of events grasped by somebody as having some unity for his consciousness, and as having its own single internal meaning. This was what we meant by the present experience of this musical phrase, this spoken line of verse, this series of rhythmic beats. But, in the other sense of the word, an element within any such whole is present in so far as this element has antecedents and successors, so that, they are *no longer* or *not yet* when it is temporally viewed as present, while in turn, in so far as any one of them is viewed as the present element, this element itself is either *not yet* or *no longer*. But precisely so, in the conceptual time of our real world, the Present means any section of the time-stream in so far as, with reference to anybody's consciousness, it is viewed as having relation to this unity of consciousness, and as in a single whole of meaning with this unity. Usually by "our time," or "the real time in which we now live," we mean no very long period of the conceived time-stream of the real world. But we never mean the indivisible *now* of an ideal mathematical time, because, in such an indivisible time-instant, nothing could happen, or endure, or genuinely exist. But within the present, if conceived as a section of the time-stream, there are internal differences of present, past, and future.

For, in a similar fashion, as the actual or supposed length of the "specious present" of our perceptual time is something arbitrary, determined by our peculiar human type of consciousness, so the length of the portion of conceptual time which we call the *present,* in the first sense of that term, namely, in the sense in which we speak of the "present age," is an arbitrary length determined in this case, however, by our more freely chosen interest in some unity which gives relative wholeness and meaning to this present. If usually the "present age" is no very long time, still, at our pleasure, or in the service of some such unity of meaning as the history of civilization, or the study of geology, may suggest, we may conceive the present as extending over many centuries, or over a hundred thousand years. On the other hand, within the unity of this first present, any distinguishable event or element of an event is *present,* in the second, and more strictly temporal sense in so far as it has predecessors and successors, whereof the first are *no longer,* and the latter *not yet,* when this more elementary event is viewed as happening.

Nor does the parallelism between the perceptual and the conceptual time cease here. The perceptual time was the form in which meaning, and the practically significant aspects of consciousness, get their expression. The same is true of the conceptual time, when viewed in its relations to the real world. Not only is the time of human history, or of any explicitly teleological series of events, obviously the form in which the facts win their particular type of conceived meaning; but even the time of physical science gets its essential characters, as a conception, through considerations that can only be interpreted in terms of the Will, or of our interest in the meaning of the world's happenings.

For the conceived time-series, even when viewed in relation to the World of Description, still differs in constitution from the constitution of a line in space, or from the characters belonging to a mathematically describable physical movement of a body, in ways which can only be expressed in terms of significance. Notoriously, conceptual time has often been described as correspondent in structure to the structure of a line, or as correspondent again, in character, to the character of an uniformly flowing stream, or of some other uniform movement. But a line can be traversed in either direction, while conceptual time is supposed to permit but one way of passing from one instant to another in its course. An uniform flow, or other motion, has, like time, a fixed direction, but might be conceived as returning into itself without detriment to its uniformity. Thus an ideally regular watch "keeps time," as we say, by virtue of the uniformity of its motion; but its hands return ever again to the same places on the face; while the years of conceptual time return not again. And finally, if one supposed an ideally uniform physical flow or streaming in one rectilinear direction only, and in an infinite Euclidean space, the character of this movement might so far be supposed to correspond to that of an ideally conceived mathematical time; except for one thing. The uniformity and unchangeableness of the conceived physical flow would be a merely given character, dependent, perhaps, upon the fact that the physical movement in question was conceived as meeting with no obstacle or external hindrance; but the direction of the flow of time is a character essential to the very conception of time. And this direction of the flow of time can only be expressed in its true necessity by saying that in case of the world's time, as in the case of the time of our inner experience, we conceive the past as leading towards, as aiming in the direction of the future, in such wise that the future depends for its meaning upon the past, and the past in its turn has its meaning as a process expectant of the future. In brief, only in terms of Will, and only by virtue of the significant relations of the stages of a teleological process, has time, whether in our inner experience, or in the conceived world order as a whole, any meaning. Time is the form of the Will; and the real world is a temporal world in so far as, in various regions of that world, seeking differs from attainment, pursuit is external to its own goal, the imperfect tends towards its own perfection, or in brief, the internal meanings of finite life gradually win, in successive stages, their union with their own External Meaning.[3] The general justification for this whole view of the time of the real world is furnished by our idealistic interpretation of Being. The special grounds for regarding the particular Being of time itself as in this special way teleological, are furnished by the foregoing analysis of our own experience of time, and by the fact that the conceptual time in terms of which we interpret the order of the world at large, is

3. [See p. 357 for explanation of this term; also *The World and the Individual*, vol. I, p. 26.—C. M. S.]

fashioned, so to speak, after the model of the time of our own experience.

III

... And so, first, the real world of our Idealism has to be viewed by us men as a temporal order. For it is a world where purposes are fulfilled, or where finite internal meanings reach their final expression, and attain unity with external meanings. Now in so far as any idea, as a finite Internal Meaning, still seeks its own Other, and consciously pursues the Other, in the way in which, as we have all along seen, every finite idea does pursue its Other, this Other is in part viewed as something beyond, *towards* which the striving is directed. But our human experience of temporal succession is, as we have seen, just such an experience of a pursuit directed towards a goal. And such pursuit demands, as an essential part or aspect of the striving in question, a consciousness that agrees in its most essential respect with our own experience of time. Hence, our only way of expressing the general structure of our idealistic realm of Being is to say that wherever an idea exists as a finite idea, still in pursuit of its goal, there appears to be some essentially temporal aspect belonging to the consciousness in question. To my mind, therefore, time, as the form of the will, is (in so far as we can undertake to define at all the detailed structure of finite reality) to be viewed as the most pervasive form of all finite experience, whether human or extra-human. In pursuing its goals, the Self lives in time. And, to our view, every real being in the universe, in so far as it has not won union with the ideal, is pursuing that ideal; and, accordingly, so far as we can see, is living in time. . . .

23

Santayana:

"SENTIMENTAL TIME"

George Santayana (1863–1952) was born in Madrid, raised in Boston, and studied at Harvard. After writing his dissertation on Lotze under Royce, he taught at Harvard for five years before returning to Europe, where he spent most of his adult years, retiring toward the end to a convent in Rome, where he died. A leading figure in the development of 'critical realism', he is perhaps best known as the proponent of a new naturalism. The selection is the second half of the chapter entitled "Pictorial Space and Sentimental Time," in *The Realm of Matter.** Defending the primordiality of physical space and time, his examination of 'sentimental' or subjective time was concerned with the ways in which it echoes the transitivity of the natural world.

I speak here impulsively of the future; but if the return to chaos were complete, if substance, which is the principle of continuity, were itself destroyed, then there would be no sense in saying that time in the second cosmos was future from the point of view of the first. This firstness and secondness would be reversible, and imputed only because I, from the outside, happened to think of one world before the other. It would be substance in me that alone would actually have passed from the thought of one to the thought of the other, and only my intuition would have connected their essences in a specious time. As the space of neither world would lie in the space of the other, so the histories of the two worlds would run irrelevant courses, and be neither before nor after each other, nor simultaneous. For by physical time I understand an order of derivation integral to the flux of matter; so that if two worlds had no material connection, and neither was in

* Reprinted by permission of Charles Scribner's Sons from *The Realm of Matter* by George Santayana (1930).

any of its parts derived from the other, they could not possibly have positions in the same physical time. The same essence of succession might be exhibited in both; the same kind of temporal vistas might perplex the sentimental inhabitants of each of them; but no date in one would coincide with a date in the other, nor would their respective temporal scales and rates of precipitation have any common measure.

The notion that there is and can be but one time, and that half of it is always intrinsically past and the other half always intrinsically future, belongs to the normal pathology of an animal mind: it marks the egotistical outlook of an active being endowed with imagination. Such a being will project the moral contrast produced by his momentary absorption in action upon the conditions and history of that action, and upon the universe at large. A perspective of hope and one of reminiscence continually divide for him a specious eternity; and for him the dramatic centre of existence, though always at a different point in physical time, will always be precisely in himself.

Presentness is the coming, lasting, or passing away of an essence, either in matter or in intuition. This presentness is a character intrinsic to all existence, since an essence would not be exemplified in any particular instance unless it came into, or went out of, a medium alien to it. Such coming and going, with the interval (if any) between, constitute the exemplification of that essence, either in the realm of matter or in that of spirit. Thus presentness, taken absolutely, is another name for the actuality which every event possesses in its own day, and which gives it its place for ever in the realm of truth. But taken relatively, as it is more natural to take it, presentness is rather a name for the middle position which every moment of existence oc-cupies between its source and its results. This presentness is pervasive; a moment does not fail to be eternally present because it never was and never can be present at any other moment. This is the ubiquity of actuality, or if you will, of selfishness: because that very leap and cry which brings each moment to birth cuts it off from everything else; the rest becomes, and must for ever remain, external to it. And this division is as requisite as this actuality: the moment does not, as if by some sin or delusion, cut itself off from some placid unitary contin-uum in which it ought to have remained embedded; it is not a frag-ment torn from eternity. On the contrary, its place in eternity is established by that severance. It is one wave in a sea that is nothing but waves. That from which it cuts itself off cuts itself off from it in turn, by the same necessity of its being; and every part of that remainder, with the very same leap and cry, cuts itself off from every other part.

Now life is a rich and complex instance of this precipitation and self-assertion, and of this pervasive presentness. Intuition, when it arises, arises within a physical moment, and expresses a passing condition of the psyche. Feelings and thoughts are parts of natural events; they belong to the self which generates them in one of its physical phases. Intuition creates a synthesis in present sensibility; it is an act of attention occurring here and now. It has not, intrinsically, any miraculous transcendence, as if it spontaneously revealed distant things as they are or were or shall be. Yet in expressing the moment intuition evokes essences; and these essences, coming as they come in the heat of action, and attributed as they are to the objects of physical pursuit or physical attention, may bring tidings of facts at any distance. Within the life of the organism this distance is primarily a distance in physical time—a distance which intuition synthesises in the feeling of duration. For the animal psyche is retentive and wound up to go on; she is full of survivals and preparations. This gathered experience and this potentiality work within her automatically: but sometimes she becomes aware of them in part, in so far as she learns to project given essences and to develop spatial and temporal perspectives within the specious field of the moment.

This feat is made easy by the frequent complexity of the specious field, in which one feature may be seen to vanish or to appear whilst others persist. The essence of change includes a direction in the felt substitution of term for term; it therefore includes the notion of the earlier absence of something now given, and of the earlier presence of something now absent. This is the very sense of existence and of time, and the key to all intelligence and dominion over reality. It contains *the principle* of transitive knowledge, since the present is aware of the past, and I in this condition think of myself in another condition; and it supplies *an instance* of transitive cognition which is knowledge in the sense of being true; because the essence of transition given in intuition is elicited directly by actual partial transitions in the state of the psyche, and this psyche is a congruent part of nature; so that when the essence of transition given now is projected over all physical time, it is plausibly projected and has every chance of being true.

Thus sentimental time is a genuine, if poetical, version of the march of existence, even as pictorial space is a genuine, if poetical, version of its distribution. The views taken are short, especially towards the future, but being extensible they suggest well enough the unfathomable depths of physical time in both directions; and if the views, being views, must be taken from some arbitrary point, they may be exchanged for one another, thus annulling the bias of each, in so far as the others contradict it. I am far from wishing to assert that the remainder or resultant will be the essence of physical time; but for human purposes a just view enough is obtained if we remember that each *now* and *here* is called so only by one voice, and that all other voices call it a *then* and a *there*. It is inevitable, and yet outrageous, that every day and year, every opinion and interest, should think itself alone truly alive and alone basking in the noon of reality. The present is indeed a true focus, an actual consummation, in so far as spirit is awake in it; and it could not have a universal scope if it had not a particular foundation. To uproot intuition from its soil in animal life would be to kill it, or if somehow it were alleged to survive, to render it homeless in the realm of matter. The body is not a prison only, but a watchtower; and the spirit would live in darkness or die of inanition if it could not open the trap-door of action, downwards towards its foundations and food in matter; or if it could not open the two windows of time, eastward and westward, towards rising or sinking constellations. It is no cruel fate, but its own nature, that imprisons life in the moment, and prevents it from taking just cognisance of any other time save by a great effort of intelligence.

If the present could see the past and the future truly, it would not feel its own pre-eminence; it would substitute their intrinsic essences for the essences of their effects within its own compass, or for its anticipations of them. It would thereby cease to be this waking moment in this animal life, and would rise into a preternatural impartiality and ubiquity. For the double aerial perspective in which a living moment sees the future and the past is a false perspective; it paints in evanescent colours what is in fact a steady procession of realities, all equally vivid and complete; it renders the past faded and dead, and the future uncertain and non-existent. But this is egregious egotism and animal blindness. The past is not faded, except in the eye of the present, and the future is not ambiguous, except to ignorant conjecture. Yet is is only by conjecture or confused reminiscence that the present is able to conceive them. The present had to choose between misrepresentation and blindness: for if it had not seen the past and future in a selfish perspective, it could not have been aware of them at all. In the romantic guise of what is not yet or what is no longer, the fleeting moment is able to recognise outlying existences, and to indicate to its own spirit the direction in which they lie. Flux by its very essence, cannot be synthesised; it must be undergone. The animal egotism which gives it a centre now and not then, contrary to fact, yet enables the passage from then to now, and from now onwards to another now, to be given in a dramatic synthesis—a synthesis which actual succession excludes. But without synthesis there is no intuition, no feeling: so that the sentimental perspectives of time are the only available forms in which a physical flux could be reported to the spirit.

The least sentimental term in sentimental time is the term *now,* because it marks the junction of fancy with action. *Now* is often a word of command; it leans towards the future, and seems to be the voice of the present summoning the next moment to arise, and pouncing upon it when it does so. For *now* has in it emotionally all the cheeriness of material change: it comes out of the past as if impatient at not having come sooner, and it passes into the future with alacrity, as if confident of losing nothing by moving on. For it is evident that actual succession can contain nothing but *nows,* so that *now* in a certain way is immortal. But this immortality is only a continual reiteration, a series of moments each without self-possession and without assurance of any other moment; so that if ever the *now* loses its indicative practical force and becomes introspective, it becomes acutely sentimental, a perpetual hope unrealised and a perpetual dying.

Various other modes of felt time cluster round this travelling *now. Just now,* for instance, is already retrospective, and marks a condition of things no longer quite in hand, already an object of intent and of questionable knowledge, yet still so near, that any mistake about it would be more like an illusion of sense than like a false assertion. *Just now* is like a spoken word lingering in the ear, so distinctly that what it was may still be observed, and described by inspection; and yet the description arising is distinguishable from it, and the two may be compared within the field of intuition. For though the voice is silent, I am still able to hear it inwardly; and its increasing uncertainty, its

evanescence, is a notable part of what I perceive. At last the rever-
beration becomes inaudible, and I have only an image formed at
intervals by silently repeating the word. If I am able to repeat it, it is
not because I can mentally remember it, but I remember it mentally
because I am able to repeat it physically. This muted repetition is
thereafter my physical memory of what that sound was at its loudest;
it suggests the glorious *then*, contrasted with this silent and empty *now*.

The psyche is full of potential experience—pictures which she is
ready to paint and would wish to hang—but the question is where, in
those vast vague galleries of fancied time, she shall hang them. When
the picture is composed and, so to speak, in her hand, but she can find
no hook for it in the direction of action, she says *not yet;* and unless she
drops and forgets it, she may go on to say *soon,* or *some day;* or if,
diabolically, the place it was about to occupy appears otherwise filled,
she may cry, *too late.* There are corresponding removes and vacilla-
tions in the direction of the past, where active impulses are not engaged. A risen image looks for points of attachment in the realm of
truth; the sentimental mind would hardly entertain it, unless it could
pass for somebody's ghost. A place can be found for almost anything
in the fog of distance, and the most miraculous tale begins with *once
upon a time.* There is hardly more definiteness, with added melan-
choly, in saying *long ago.* The event has sunk, for sentiment, below the
horizon, into the night where everything sleeps which, though theoretically credited, is not relevant to present action.

All, however, is not flaccidity and weakness in sentimental time; it
has its noble notes in the sublime *always* and the tragic *never.* These
terms, when true, describe facts in physical time, and their force
comes from that backing, but not their sentimental colour; for there is
nothing tragic in the mere absence of anything from the universe, and
nothing sublime in the presence there of such elements as may happen
to be pervasive, like matter and change. The sublimity or tragedy
comes from projecting sentimental time, with its human centre, upon
the canvas of nature. *Never* then proclaims an ultimate despair or
relief. It professes to banish from the whole past or future some
essence now vividly present to the mind, tempting or horrible; and the
living roots of that possibility here in the soul render the denial of it in
nature violent and marvellous. That something arresting to thought
should be absent from the field of action causes a cruel division in the

animal spirit; for this spirit can live only by grace of material circumstances, yet, being a spirit, it can live only by distinguishing
eternal essences. Its joy lies in being perpetually promoted to intuition
and confirmed in it; but it is distracted if its intuitions are thwarted,
and expel or contradict one another. Therefore, whenever the spirit
finds some steadfast feature in the world, it breathes, even in the midst
of action and discovery, its native air. That which exists always, even if
only a type of a law, seems almost rasied above existence; *always*
assimilates time to eternity. Imagination then flatters itself that it
dominates the universe, as it was in the beginning, is now, and ever
shall be. In fact, imagination can never synthesise anything but its
own vistas, in a specious, not a physical time; but, if the vista had
happened to be true, and if physical time had repeated endlessly the
feature now singly conceived, thought would have marvellously
domesticated that crawling monster, prescribing, or seeming to prescribe, how and how far it should uncoil itself. In respect to the whole
flux of existence such a boast is gratuitous; but it may be justified in
respect to that field of action on which the spirit immediately depends.

Spirit, however scattered its occasions and instances may be, is
always synthetic in its intellectual energy or actuality; it gives the form
of totality to its world. The frontiers may be vague and the features
confused, but they could not be confused or vague if they were not
features and frontiers of a single scene. Synthesis is a prerequisite to
the sensation of change. An image of motion or a feeling of lapse is a
single feeling and a single image. That exciting experience in which
now, now, keeps ringing like a bell would not be a strained experience
of the mind, nor more than so many fresh experiences, if the iteration
were not synthesised, each stroke being received by a sensorium
already alive, and as it were elastic, so that the dying strokes and
others imminent formed with the fresh stroke a single temporal
landscape, a wide indefinite *now* open to intuition, which psychologists call the specious present. This intuition no doubt has its
basis in the datable physical moment, in a definite phase of the
organism; but the temporal landscape given in it is ideal, and has no
date and no clear limits; indeed, it is a miniature of all time, as
imagination might survey and picture it. A few more details on the
same canvas, a closer attention to the remoter vistas, and this specious
present might contain all knowledge. All that the most learned his-

torian or the deepest theologian has ever actually conceived must have come to him in the specious present. Unless action concentrates the attention on the forward edge of this prospect, on the budding *now,* the whole leans rather towards the past, and grows more and more extended. It becomes full of things vain for our present purpose. Even when action or scientific observation rescue the specious present from this sentimentality, they leave it ideal, and its field a purely sensuous unit.

Action and scientific observation, though framed within the specious present in perception, ignore it in practice; but arts like music and eloquence are directed (without knowing it) to enriching this specious present and rendering it, in some climax, so overwhelmingly pregnant and brilliant, that scientific observation and action become impossible; as if the flux of things had culminated, collected all its treasures, and wished, at last for a moment, to stop and possess them. Under such tension, when the spirit takes flight towards its proper objects, essence and truth, the body, which cannot follow, bursts into irrelevant action, such as tears or applause or laughter or contortions. Synthesis, which must be material before it can be spiritual, has then brought to a head more currents than can flow into any useful channel, and some random outlet must be found for that physical fullness and that physical impotence. Meantime intuition shines in all its glory. On lesser occasions, when attention lights up some particular path in the field of action, spirit, being all eagerness and appetition, almost seems to be a form of energy or source of motion in matter. But the fact that it is always wholly contemplative becomes evident in its richer moments; for then the specious present extends far beyond the urgent occasion for action, and may even drop its conscious relevance to the person or the hour; and at the same time its vistas will cease to be sentimental, because the survey of them will have become intellectual and impartially receptive.

The specious present is dateless, but it is temporal;[1] it is the vaguely

1. For the exact force of these terms, see *Scepticism and Animal Faith,*p. 270: [A given essence containing no specious temporal progression or perspective between its parts would be *timeless.* Colour, for instance, or number, is timeless. . . .

    A being that should have no external temporal relations and no locus limited foreground of sentimental time. In it the precipitation proper to existence, the ticking of the great clock is clearly audible: the bodily feeling of the moment is complicated in it by after-images of instants receding, and strained by anticipation of instants to come. As the psyche can synthesise in memory the fading impressions that reverberate within her, so she can prefigure and rehearse the acts for which she is making ready. A pleasant rumble and a large assurance thus pervade animal consciousness in its calmer moments. There is no knife-edge here of present time, nor any exact frontiers, but a virtual possession of the past and even of the future; because the psyche is magnificently confident not only that a future will be forthcoming, but that its paces will be familiar, and easy to deal with; in fact she is quietly putting irons in the fire, and means to have a hand in shaping it. The general character of the future is felt to be as sure as the ebb and flow of the tide, or as the rest of a sentence half spoken; and this assurance, which might seem groundless, is not so, because the psyche is not a detached spirit, examining the world by the light of some absolute logic. She is herself a material engine, a part of that substantial flux which exists only by bridging its intervals, and is defined as much by what it becomes as by what it may have been. That future which is in part her product is, in that measure, her substance transformed, or by its action transforming the surrounding substance.

At bottom, then , the future is no new event, but the rest of this transformation. In an organism like the psyche, it is the present pregnancy of matter that determines the further development of its forms. Of course this development may be cut short at any point by contacts with other organisms, or with the inorganic; and then, if the psyche survives, she will be diverted in some new direction, at first violent and unpredictable. But if she could understand fully all the substance of which she is a portion, she would dominate in the specious present the remotest future of her universe, as well as its

---

in physical time would be *dateless*. Thus every given essence and every specious present is dateless, internally considered, and taken transcendentally, that is, as a station for viewing other things or a unit framing them in. Though dateless, the specious present is not timeless, and an instant, though timeless, is not dateless.]

deepest past; for physical time is nothing but the deployment of substance, and the essence of this substance is the form which, if free, it would realise in its deployment. The psyche, then, is not radically deceived in her somnolent confidence in the future, nor in her conviction that she helps to determine it. The future may be largely foreseen, or brought about as intended, even as the past may be largely remembered; the psyche has materials within the moment for both those temporal vistas. It is her egotism here that is in error rather than her faith; for the vivacity of her wishes or momentary fancies may obscure her sense for things at a distance; and beyond the chance limits of the specious present, she may think the future indeterminate and unreal, and the past dead.

The dead past? Certainly all years prior to A.D. 1920 are past at the time when I write these words, and all later years are future, and to me uncertain; yet the reader knows that one of the years perhaps long subsequent to 1920 is as living and sure as 1920 is to me now. My problematic future will be his living present, and both presently a third man's irrevocable past. There is no moment in the whole course of nature (unless there be a first and a last moment) which is not both future and past for an infinity of other moments. In itself, by virtue of its emergence in a world of change, each moment is unstably present, or in the act of elapsing; and by virtue of its position in the order of generation, both pastness and futurity pervade it eternally. Such double abysmal absence is the price which existence pays for its momentary emphasis. Futurity and pastness, the reproach of being not yet and being no longer, fall upon it from different quarters, like lights of different colours; and these same colours, like the red light and the green light of a ship, it itself carries and spreads in opposite directions. And these shafts cross one another with a sort of correspondence in contradiction: the moments which any moment calls past call it future, and those which it calls future call it past.

Contradictory epithets of this sort are compatible when they are seen to be relative; but it must be understood that they are the relative aspects of something which has an absolute nature of its own, to be the foundation of those relations. And the absolute nature of moments is to be present: a moment which was not present in itself could not be truly past or future in relation to other moments. What I call the past and what I call the future are truly past or future *from here;* but if they were *only* past or *only* future, it would be an egregious error on my part to believe that they were past or future at all, for they would exist only in my present memory or expectation. In their pastness and futurity they would be merely specious, and they would be nothing but parts of a present image. If I pretended that they recalled or forecast anything, I should be deceived; for nothing of that kind, either in the past or in the future, would ever rejoice in presentness and exist on its own account. Thus only false memory and false expectation end in events intrinsically past or intrinsically future—that is to say, intrinsically sentimental. False legends and false hopes indeed have their being only in perspective; their only substance is the thought of them now, and it is only as absent that they are ever present.

Romantic idealism, which saturates modern speculation even when not avowed, if it were not halting and ambiguous, would reduce all time to the picture of time, and establish the solipsism of the present moment. And the present moment indeed includes and absorbs all sentimental time: apart from its vistas, neither the past nor the future would be at all romantic or unreal. Common sense admits this in respect to the past. The equal reality of the past with the present, and its fixity of essence, so that ideas of it may be false or true, are not seriously questioned. But in respect to the future, imagination is less respectful of fact, and flounders in an abyss to which it attributes its own vagueness. The prevision which the psyche often has of its future actions, as in planning or calculation, ought to give steadiness to thoughts of the future, in as much as here the causes of future events are already found at work; and the ancients, who were not sentimental, cultivated policy and prophecy, endeavouring to define the future through its signs and causes, as we define the past through its memorials. But modern anticipations are based rather on supposed "laws", or on intuition of divine purposes; that is, on forms found in the perspective of events, as synthesised in imagination: whereby physical time is asked to march to the music which sentimental time may pipe for it.

In reality, nature moves in a time of her own, everywhere equally present, of which sentimental time is a momentary echo; for sometimes a single pulse of substance may become conscious of its motion, and may fantastically endeavour to embrace the true past and the true future, necessarily external to it, in a single view. This sentimental agony fancy then transfers from its own flutterings to the brisk precipitation and the large somnolence of the general flux, which is neither regretful nor perturbed, and not intent on prolonging one of its phases rather than another. Animal life itself shares this conformity with the flux of things, whenever animal life is perfect; and even perfect intuition shares it, by entirely transmuting motion into light: so that both in rude health and in pure contemplation it either forgets the past and the future altogether, or ceases to be sentimental about them, and learns to feel their direct and native reality, as they feel it themselves.

24.

Dewey:

"TIME AND INDIVIDUALITY"

The long career of John Dewey (1859-1952) began in Vermont, reached maturity in Chicago and in New York where it ended. It has touched virtually every aspect of philosophic inquiry to which he continually brought a sense of temporal context. Impressed with the central import of the scientific experience, Dewey urged that its lesson be taken up in the deliberate use of intelligence to mold a future instead of merely acquiescing before one. A firm proponent of a naturalism in which man is seen as part of the natural order, he has brought the humanistic strains of pragmatic thinking together in this essay * which casts a different light on his work than commentaries generally suggest. Focusing on the significance of time in and for human experience, he has effectively argued here that it is the foundational concept undergirding the pragmatic themes of contextual individualism, control, and the use of intelligence in the facing of an open future.

The Greeks had a saying "Count no man happy till after his death." The adage was a way of calling attention to the uncertainties of life. No one knows what a year or even a day may bring forth. The healthy become ill; the rich poor; the mighty are cast down; fame changes to obloquy. Men live at the mercy of forces they cannot control. Belief in fortune and luck, good or evil, is one of the most widespread and persistent of human beliefs. Chance has been deified by many peoples. Fate has been set up as an overlord to whom even the Gods must bow. Belief in a Goddess of Luck is in ill repute among pious folks but

* John Dewey. "Time and Individuality." *Time and Its Mysteries*, Series II (New York: New York Universtiy Press, 1940). By permission.

their belief in providence is a tribute to the fact no individual controls his own destiny.

The uncertainty of life and one's final lot has always been associated with mutability, while unforeseen and uncontrollable change has been linked with time. Time is the tooth that gnaws; it is the destroyer; we are born only to die and every day brings us one day nearer death. This attitude is not confined to the ignorant and vulgar. It is the root of what is sometimes called the instinctive belief in immortality. Everything perishes in time but men are unable to believe that perishing is the last word. For centuries poets made the uncertainty which time brings with it the theme of their discourse—read Shakespeare's sonnets. Nothing stays; life is fleeting and all earthly things are transitory.

It was not then for metaphysical reasons that classic philosophy maintained that change, and consequently time, are marks of inferior reality, holding that true and ultimate reality is immutable and eternal. Human reasons, all too human, have given birth to the idea that over and beyond the lower realm of things that shift like the sands on the seashore there is the kingdom of the unchanging, of the complete, the perfect. The grounds for the belief are couched in the technical language of philosophy, but the cause for the grounds is the heart's desire for surcease from change, struggle, and uncertainty. The eternal and immutable is the consummation of mortal man's quest for certainty.

It is not strange then that philosophies which have been at odds on every other point have been one in the conviction that the ultimately real is fixed and unchanging, even though they have been as far apart as the poles in their ideas of its constitution. The idealist has found it in a realm of rational ideas; the materialist in the laws of matter. The mechanist pins his faith to eternal atoms and to unmoved and unmoving space. The teleologist finds that all change is subservient to fixed ends and final goals, which are the one steadfast thing in the universe, conferring upon changing things whatever meaning and value they possess. The typical realist attributes to unchanging essences a greater degree of reality than belongs to existences; the modern mathematical realist finds the stability his heart desires in the immunity of the realm of possibilities from vicissitude. Although classic rationalism looked askance at experience and empirical things because of their continual subjection to alteration, yet strangely enough traditional sensational empiricism relegated time to a secondary role. Sensations appeared and disappeared but in their own nature they were as fixed as were Newtonian atoms—of which indeed they were mental copies. Ideas were but weakened copies of sensory impressions and had no inherent forward power and application. The passage of time dimmed their vividness and caused their decay. Because of their subjection to the tooth of time, they were denied productive force.

In the late eighteenth and the greater part of the nineteenth centuries appeared the first marked cultural shift in the attitude taken toward change. Under the names of indefinite perfectability, progress, and evolution, the movement of things in the universe itself and of the universe as a whole began to take on a beneficent instead of a hateful aspect. Not every change was regarded as a sign of advance but the general trend of change, cosmic and social, was thought to be toward the better. Aside from the Christian idea of a millennium of good and bliss to be finally wrought by supernatural means, the Golden Age for the first time in history was placed in the future instead of at the beginning, and change and time were assigned a benevolent role.

Even if the new optimism was not adequately grounded, there were sufficient causes for its occurrence as there are for all great changes in intellectual climate. The rise of new science in the seventeenth century laid hold upon general culture in the next century. Its popular effect was not great, but its influence upon the intellectual elite, even upon those who were not themselves engaged in scientific inquiry, was prodigious. The enlightment, the *éclaircissement*, the *Aufklärung*—names which in the three most advanced countries of Europe testified to the widespread belief that at last light had dawned, that dissipation of the darkness of ignorance, superstition, and bigotry was at hand, and the triumph of reason was assured—for reason was the counterpart in man of the laws of nature which science was disclosing. The reign of law in the natural world was to be followed by the reign of law in human affairs. A vista of the indefinite perfectability of man was opened. It played a large part in that optimistic theory of automatic evolution which later found its classic formulation in the philosophy of Herbert Spencer. The faith may have been pathetic but it has its own nobility.

At last, time was thought to be working on the side of the good instead of as a destructive agent. Things were moving to an event which was divine, even if far off.

This new philosophy, however, was far from giving the temporal an inherent position and function in the constitution of things. Change was working on the side of man but only because of *fixed* laws which governed the changes that take place. There was hope in change just because the laws that govern it do not change. The locus of the immutable was shifted to scientific natural law, but the faith and hope of philosophers and intellectuals were still tied to the unchanging. The belief that "evolution" is identical with progress was based upon trust in laws which, being fixed, worked automatically toward the final end of freedom, justice, and brotherhood, the natural consequences of the reign of reason.

Not till the late nineteenth century was the doctrine of the subordination of time and change seriously challenged. Bergson and William James, animated by different motives and proceeding by different methods, then installed change at the very heart of things. Bergson took his stand on the primacy of life and consciousness, which are notoriously in a state of flux. He assimilated that which is completely real in the natural world to them, conceiving the static as that which life leaves behind as a deposit as it moves on. From this point of view he criticized mechanistic and teleological theories on the ground that both are guilty of the same error, although from opposite points. Fixed laws which govern change and fixed ends toward which changes tend are both the products of a backward look, one that ignores the forward movement of life. They apply only to that which life has produced and has then left behind in its ongoing vital creative course, a course whose behavior and outcome are unpredictable both mechanically and from the standpoint of ends. The intellect is at home in that which is fixed only because it is done and over with, for intellect is itself just as much a deposit of *past* life as is the matter to which it is congenial. Intuition alone articulates in the forward thrust of life and alone lays hold of reality.

The animating purpose of James was, on the other hand, primarily moral and artistic. It is expressed, in his phrase, "block universe," employed as a term of adverse criticism. Mechanism and idealism were abhorrent to him because they both hold to a closed universe in which there is no room for novelty and adventure. Both sacrifice individuality and all the values, moral and aesthetic, which hang upon individuality, for according to absolute idealism, as to mechanistic materialism, the individual is simply a part determined by the whole of which he is a part. Only a philosophy of pluralism, of genuine indetermination, and of change which is real and intrinsic gives significance to individuality. It alone justifies struggle in creative activity and gives opportunity for the emergency of the genuinely new.

It was reserved, however, for the present century to give birth to the out-and-out assertion in systematic form that reality *is* process, and that laws as well as things develop in the processes of unceasing change. The modern Heraclitean is Alfred North Whitehead, but he is Heraclitus with a change. The doctrine of the latter, while it held that all things flow like a river and that change is so continuous that a man cannot step into the same river even once (since it changes as he steps), nevertheless also held that there is a fixed order which controls the ebb and flow of the universal tide.

My theme, however, is not historical, nor is it to argue in behalf of any one of the various doctrines regarding time that have been advanced. The purpose of the history just roughly sketched is to indicate that the nature of time and change has now become in its own right a philosophical problem of the first importance. It is of time as a *problem* that I wish to speak. The aspect of the problem that will be considered is the connection of time with individuality, as the later is exemplified in the living organism and especially in human beings.

Take the account of the life of any person, whether the account is a biography or an autobiography. The story begins with birth, a temporal incident; it extends to include the temporal existence of parents and ancestry. It does not end with death, for it takes in the influence upon subsequent events of the words and deeds of the one whose life is told. Everything recorded is an historical event; it is something temporal. The individual whose life history is told, be it Socrates or Nero, St. Francis or Abraham Lincoln, is an extensive event; or, if you prefer, it is a course of events each of which takes up into itself something of what went before and leads on to that which comes after. The skill, the art, of the biographer is displayed in his ability to discover and portray the subtle ways, hidden often from the individual himself, in which one event grows out of those which preceded and enters into those which follow. The human individual is himself a history, a career, and for this reason his biography can be related only as a temporal event. That which comes later explains the earlier quite as truly as the earlier explains the later. Take the individual Abraham Lincoln at one year, at five years, at ten years, at thirty years of age, and imagine everything wiped out, no matter how minutely his life is recorded up to the date set. It is plain beyond the need of words that we then have not his biography but only a fragment of it, while the significance of that fragment is undisclosed. For he did not just exist in a time which eternally surrounded him, but time was the heart of his existence.

Temporal seriality is the very essence, then, of the human individual. It is impossible for a biographer in writing, say the story of the first thirty years of the life of Lincoln, not to bear in mind his later career. Lincoln as an individual *is* a history; any particular event cut off from that history ceases to be a part of his life as an individual. As Lincoln is a particular development in time, so is every other human individual. Individuality is the uniqueness of the history, of the career, not something given once for all at the beginning which then proceeds to unroll as a ball of yarn may be unwound. Lincoln made history. But it is just as true that he made himself as an individual in the history he made.

I have been speaking about human individuality. Now an important part of the problem of time is that what is true of the human individual does not seem to be true of physical individuals. The common saying "as like as two peas" is a virtual denial to one kind of vegetable life of the kind of individuality that marks human beings. It is hard to conceive of the individuality of a given pea in terms of a unique history or career; such individuality as it appears to possess seems to be due in part to spatial separateness and in part to peculiarities that are externally caused. The same thing holds true of lower forms of animal life. Most persons would resent denial of some sort of unique individuality to their own dogs, but would be slow to attribute it to worms, clams, and bees. Indeed, it seems to be an exclusive prerogative of the romantic novelist to find anything in the way of a unique career in animal lives in general.

When we come to inanimate elements, the prevailing view has been that time and sequential change are entirely foreign to their nature.

According to this view they do not have careers; they simply change their relations in space. We have only to think of the classic conception of atoms. The Newtonian atom, for example, moved and was moved, thus changing its position in space, but it was interchangeable in its own being. What it was at the beginning or without any beginning it is always and forever. Owing to the impact of other things it changes its direction and velocity of motion so that it comes closer and further away from other things. But all this was believed to be external to its own substantial being. It had no development, no history, because it had no potentialities. In itself it was like a God, the same yesterday, today, and forever. Time did not enter into its being either to corrode or to develop it. Nevertheless, as an ultimate element it was supposed to have some sort of individuality, to be itself and not something else. Time, in physical science, has been simply a measure of motion in space.

Now, this apparently complete unlikeness in kind between the human and the physical individual is a part of the problem of time. Some philosophers have been content to note the difference and to make it the ground for affirming a sheer dualism between man and other things, a ground for assigning to man a spiritual being in contrast with material things. Others, fewer in numbers, have sought to explain away the seeming disparity, holding that the apparent uniqueness of human individuality is specious, being in fact the effect of the vast number of physical molecules, themselves complex, which make up his being, so that what looks like genuine temporal change or development is really but a function of the number and complexity of changes of constituent fixed elements. Of late, there have been a few daring souls who have held that temporal quality and historical career are a mark of everything, including atomic elements, to which individuality may be attributed.

I shall mention some of the reasons from the side of physical science that have led to this third idea. The first reason is the growing recognition that scientific objects are purely relational and have nothing to do with the intrinsic qualities of individual things and nothing to say about them. The meaning of this statement appears most clearly in the case of scientific laws. It is now a commonplace that a physical law states a correlation of changes or of ways and manners of change. The law of gravitation, for example, states a relation which holds between bodies with respect to distance and mass. It needs no argument to show that distance is a relation. Mass was long regarded as an inherent property of ultimate and individual elements. But even the Newtonian conception was obliged to recognize that mass could be defined only in terms of inertia and that inertia could be defined only in terms, on the one hand, of the resistance it offered to the impact of other bodies, and, on the other hand, of its capacity to exercise impact upon them, impact being measured in terms of motion with respect to acceleration. The idea that mass is an inherent property which caused inertia and momentum was simply a holdover from an old metaphysical idea of force. As far as the findings of science are concerned, independent of the intrusion of metaphysical ideas, mass is inertia-momentum and these are strictly measures and relations. The discovery that mass changes with velocity, a discovery made when minute bodies came under consideration, finally forced surrender of the notion that mass is a fixed and inalienable possession of ultimate elements or individuals, so that time is now considered to be their fourth dimension.

It may be remarked incidentally that the recognition of the relational character of scientific objects completely eliminates an old metaphysical issue. One of the outstanding problems created by the rise of modern science was due to the fact that scientific definitions and descriptions are framed in terms in which qualities play no part. Qualities were wholly superfluous. As long as the idea persisted (an inheritance from Greek metaphysical science) that the business of knowledge is to penetrate into the inner being of objects, the existence of qualities like colors, sounds, etc., was embarrassing. The usual way of dealing with them is to declare that they are merely subjective, existing only in the consciousness of individual knowers. Given the old idea that the purpose of knowledge (represented at its best in science) is to penetrate into the heart of reality and reveal its "true" nature, the conclusion was a logical one. The discovery that the objects of scientific knowledge are purely relational shows that the problem is an artificial one. It was "solved" by the discovery that it needed no solution, since fulfillment of the function and business of science compels disregard of qualities. Using the older language, it was seen that so-called primary qualities are no more inherent properties of ultimate objects than are so-called secondary qualities of odors, sounds, and colors, since the former are also strictly relational; or, as Locke stated in his moments of clear insight, are "retainers" of objects in their connections with other things. The discovery of the nonscientific because of the empirically unverifiable and unnecessary character of absolute space, absolute motion, and absolute time gave the final *coup de grâce* to the traditional idea that solidity, mass, size, etc., are inherent possessions of ultimate individuals.

The revolution in scientific ideas just mentioned is primarily logical. It is due to recognition that the very method of physical science, with its primary standard units of mass, space, and time, is concerned with measurement of relations of change, not with individuals as such. This acknowledgement brought with it a further idea which, in spite of the resistance made to it by adherents of older metaphysical views, is making constant headway. This idea is that laws which purport to be statements of what actually occurs are statistical in character as distinct from so-called dynamic laws that are abstract and mathematical, and disguised definitions. Recognition of the statistical nature of physical laws was first effected in the case of gases when it became evident that generalizations regarding the behavior of swarms of molecules were not descriptions or predictions of the behavior of any individual particle. A single molecule is not and cannot be a gas. It is consequently absurd to suppose that the scientific law is about the elementary constituents of a gas. It is a statement of what happens when a very large number of such constituents interact with one another under certain conditions.

Statistical statements are of the nature of probability formulations. No insurance company makes any prediction as to what will happen to any given person in respect to death, or to any building with respect to destruction by fire. Insurance is conducted upon the basis of observation that out of a large number of persons of a given age such and such a proportionate number will probably live one year more, another proportionate number two years, and so on, while premiums are adjusted on the basis of these probability estimates. The validity of the estimates depends, as in the case of a swarm of molecules, upon the existence of a sufficiently large number of individuals, a knowledge which is a matter of the relative frequency of events of a certain kind to the total number of events which occur. No statement is made about what will take place in the case of an *individual*. The application of scientific formulations of the principle of probability statistically determined is thus a logical corollary of the principle already stated, that the subject matter of scientific findings is relational, not individual. It is for this reason that it is safe to predict the ultimate triumph of the statistical doctrine.

The third scientific consideration is found in Heisenberg's principle of uncertainty or indeterminacy, which may be regarded as a generalization of the ideas already stated. In form, this principle seems to be limited in its application. Classical science was based upon the belief that it is possible to formulate both the position and the velocity at one time of any given particle. It followed that knowledge of the position and velocity of a given number of particles would enable the future behavior of the whole collection to be accurately predicted. The principle of Heisenberg is that given the determination of position, its velocity can be stated only as of a certain order of probability, while if its velocity is determined the correlative factor of position can be stated only as of a certain order of probability. Both cannot be determined at once, from which it follows necessarily that the future of the whole collection cannot possibly be foretold except in terms of some order of probability.

Because of the fundamental place of the conceptions of position and velocity in physical science the principle is not limited in scope but is of the broadest possible significance.

Given the classic conception, Laplace stated its logical outcome when he said "we may conceive the present state of the universe as the effect of its past and the cause of its future. An intellect who at any given instant knew all the forces of animate nature and the mutual positions of the beings who compose it . . . could condense into a single formula the movement both of the greatest body in universe and of its lightest atom. Nothing would be uncertain to such an intellect, for the future, even as the past would be ever present before his eyes." No more sweeping statement of the complete irrelevancy of time to the physical world and of the complete unreality for individuals of time could well be uttered. But the principle of indeterminacy annihilates the premises from which the conclusion follows. The principle is thus a way of acknowledging the pertinency of real time to physical beings. The utmost possible regarding an individual is a statement as to some order of probability about the future. Heisenberg's principle has been seized upon as a basis for wild statements to the effect that the doctrine of arbitrary free will and totally uncaused activity are now scientifically substantiated. Its actual force and significance is generalization of the idea that the individual is a temporal career whose future canot be *logically* deduced from its past.

As long as scientific knowledge was supposed to be concerned with individuals in their own intrinsic nature, there was no way to bridge the gap between the career of human individuals and that of physical individuals, save by holding that the seeming fundamental place of development and hence of time in the life histories of the former is only seeming or specious. The unescapable conclusion is that as human individuality can be understood only in terms of time as fundamental reality, so for physical individuals time is not simply a measure of predetermined changes in mutual positions, but is something that enters into their being. Laws do not "govern" the activity of individuals. They are a formulation of the frequency-distributions of the behavior of large number of individuals engaged in interactions with one another.

This statement does not mean that physical and human individuality are identical, nor that the things which appear to us to be nonliving have the distinguishing characteristic of organisms. The difference between the inanimate and the animate is not so easily wiped out. But it does show that there is no fixed gap between them. The conclusion which most naturally follows, without indulging in premature speculation, is that the principle of a developing career applies to all things in nature, as well as to human beings—that they are born, undergo qualitative changes, and finally die, giving place to other individuals. The idea of development applied to nature involves differences of forms and qualities as surely as it rules out absolute breaches of continuity. The differences between the amoeba and the human organism are genuinely there even if we accept the idea of organic evolution of species. Indeed, to deny the reality of the differences and their immense significance would be to deny the very idea of development. To wipe out differences because of denial of complete breaks and the need for intervention of some outside power is just as surely a way to deny development as is assertion of gaps which can be bridged only by the intervention of some supernatural force. It is then in terms of development, or if one prefers the more grandiose term, evolution, that I shall further discuss the problem of time.

The issue involved is perhaps the most fundamental one in philosophy at the present time. Are the changes which go on in the world simply external redistributions, rearrangements in space of what previously existed, or are they genuine qualitative changes such as apparently take place in the physiological development of an organism, from the union of ovum and sperm to maturity, and as apparently take place in the personal life career of individuals? When the question is raised, certain misapprehensions must be first guarded against. Development and evolution have historically been eulogistically interpreted. They have been thought of as necessarily proceeding from the lower to the higher, from the relatively worse to the relatively better. But this property was read in from outside moral and theological preoccupations. The real issue is that stated above: Is what happens simply a spatial rearrangement of what existed previously or does it involve something qualitatively new? From this point of view, cancer is as genuinely a physiological development as is growth in vigor; criminals as well as heroes are a social development; the emergence of totalitarian states is a social evolution out of constitutional states independently of whether we like or approve them.

If we accept the intrinsic connection of time with individuality, they are not mere redistributions of what existed before.

Since it is a *problem* I am presenting, I shall assume that genuine transformations occur, and consider its implications. First and negatively, the idea (which is often identified with the essential meaning of evolution) is excluded that development is a process of unfolding what was previously implicit or latent. Positively it is implied that potentiality is a category of existence, for development cannot occur unless an individual has powers or capacities that are not actualized at a given time. But is also means that these powers are not unfolded from within, but are called out through interaction with other things. While it is necessary to revive the category of potentiality as a characteristic of individuality, it has to be revived in a different form from that of its classic Aristotelian formulation. According to that view, potentialities are connected with a fixed end which the individual endeavors by its own nature or essence to actualize, although its success in actualization depended upon the cooperation of external things and hence might be thwarted by the "accidents" of its surroundings—as not every acorn becomes a tree and few if any acorns become the typical oak.

When the idea that development is due to some indwelling end which tends to control the series of changes passed through is abandoned, potentialities must be thought of in terms of consequences of interactions with other things. Hence potentialities cannot be *known* till *after* the interactions have occurred. There are at a given time unactualized potentialities in an individual because and in as far as there are in existence other things with which it has not as yet interacted. Potentialities of milk are known today, for example, that were not known a generation ago, because milk has been brought into interaction with things other than organisms, and hence now has other than furnishing nutriment consequence. It is now predicted that in the future human beings will be wearing clothes made of glass and that the clothes will be cleaned by throwing them into a hot furnace. Whether this particular prediction is fulfilled or not makes no difference to its value as an illustration. Every new scientific discovery leads to some mode of technology that did not previously exist. As things are brought by new procedures into new contacts and new interactions, new consequences are produced and the power to produce these new consequences is a recognized potentiality of the thing in question. The idea that potentialities are inherent and fixed by relation to a predetermined end was a product of a highly restricted state of technology. Because of this restriction, the only potentialities recognized were those consequences which were customary in the given state of culture and were accordingly taken to be "natural." When the only possible use of milk was as an article of food, it was "natural" to suppose that it had an inherent tendency to serve that particular end. With the use of milk as a plastic, and with no one able to tell what future consequences may be produced by new techniques which bring it into new interactions, the only reasonable conclusion is that potentialities are not fixed and intrinsic, but are a matter of an indefinite range of interactions in which an individual may engage.

Return for a moment to the human individual. It is impossible to think of the historical career, which is the special individuality constituting Abraham Lincoln, apart from the particular conditions in which he lived. He did not create, for example, the conditions that formed the issues of States' rights and of slavery, the issues that influenced his development. What his being as an individual would have been without these interacting conditions it is idle to speculate upon. The conditions did not form him from without as wax is supposed to be shaped by external pressure. There is no such thing as interaction that is merely a one-way movement. There were many other persons living under much the same conditions whose careers were very different, because conditions acted upon them and were acted upon by them in different ways. Hence there is no account possible of Lincoln's life that does not portray him interacting day by day with special conditions, with his parents, his wife and children, his neighbors, his economic conditions, his school facilities, the incidents of his profession as a lawyer, and so on. The career which is his unique individuality is the series of interactions in which he was created to be what he was by the ways in which he responded to the occasions with which he was presented. One cannot leave out either conditions as *opportunities* nor yet unique ways of responding to them. An occasion is an opportunity only when it is an evocation of a specific event, while a response is not a necessary effect of a casue but is a way of using an occasion to render it a constituent of an ongoing unique history.

Individuality conceived as a temporal development involves uncertainty, indeterminacy, or contingency. Individuality is the source of whatever is unpredictable in the world. The indeterminate is not change in the sense of violation of law, for laws state probable correlations of change and these probabilities exist no matter what the source of change may be. When a change occurs *after* it has occured, it belongs to the observable world and is connected with other changes. The nomination of Lincoln for the presidency, his election, his Emancipation Proclamation, his assassination, after they took place can be shown to be related to other events; they can also be shown to have a certain connection with Lincoln's own past. But there was nothing in Lincoln's own life to cause by itself the conjunction of circumstances which brought about any one of these events. As far as he as an individual was concerned, the events were contingent, and as far as the conjunction of circumstances was concerned, his behavior at any given time in response to them was also contingent, or if you please fortuitous.

At critical junctures, his response could not be predicted either from his own past or from the nature of the circumstances, except as a probability. To say this is not arbitrarily to introduce mere chance into the world. It is to say that genuine individuality exists; that individuality is pregnant with new developments; that time is real. If we knew enough about Shakespeare's life we could doubtless show *after* Hamlet was produced how it is connected with other things. We could link it with sources; we could connect its mood with specific experiences of its author, and so on. But no one with the fullest knowledge of Shakespeare's past could have predicted the drama as it stands. If they could have done so, they would have been able to write it. Not even Shakespeare himself could have told in advance just what he was going to say—not if he was an individual, not a nodal point in the spatial redistribution of what already existed.

The mystery of time is thus the mystery of the existence of real individuals. It is a mystery because it is a mystery that anything which exists is just what it is. We are given to forgetting, with our insistence upon causation and upon the necessity of things happening as they do happen, that things exist as just what they qualitatively are. We can account for a change by relating it to other changes, but existences we have to accept for just what they are. Given a butterfly or an earthquake as an event, as a change, we can at least in theory find out and state its connection with other changes. But the individual butterfly or earthquake remains just the unique existence which it is. We forget in explaining its occurrence that it is only the *occurrence* that is explained, not the thing itself. We forget that in explaining the occurrence we are compelled to fall back on other individual things that have just the unique qualities they do have. Go as far back as we please in accounting for present conditions and we still come upon the mystery of things being just what they are.

Their occurrence, their manifestation, may be accounted for in terms of other occurrences, but their own quality of existence is final and opaque. The mystery is that the world is as it is—a mystery that is the source of all joy and all sorrow, of all hope and fear, and the source of development both creative and degenerative. The contingency of all into which time enters is the source of pathos, comedy, and tragedy. Genuine time, if it exists as anything else except the measure of motions in space, is all one with the existence of individuals as individuals, with the creative, with the occurrence of unpredictable novelties. Everything that can be said contrary to this conclusion is but a reminder that an individual may lose his individuality, for individuals become imprisoned in routine and fall to the level of mechanisms. Genuine time then ceases to be an integral element in their being. Our behavior becomes predictable because it is but an external rearrangement of what went before.

In conclusion, I would like to point out two considerations that seem to me to follow, two morals, if you wish to give them that name. I said earlier that the traditional idea of progress and evolution was based upon belief that the fixed structure of the universe is such as automatically brings it about. This optimistic and fatalistic idea is now at a discount. It is easy in the present state of the world to deny all validity whatever to the idea of progress, since so much of the human world seems bent on demonstrating the truth of the old theological doctrine of the fall of man. But the real conclusion is that, while progress is not inevitable, it is up to men as individuals to bring it about. Change is going to occur anyway, and the problem is the control of change in a given direction. The direction, the quality of change, is a matter of individuality. Surrender of individuality by the many of some one who is taken to be a superindividual explains the retrograde movement of society. Dictatorships and totalitarian states, and belief in the inevitability of this or that result coming to pass are, strange as it may sound, ways of denying the reality of time and the creativeness of the individual. Freedom of thought and of expression are not mere rights to be claimed. They have their roots deep in the existence of individuals as developing careers in time. Their denial and abrogation is an application of individuality and a virtual rejection of time as opportunity.

The ground of democratic ideas and practices is faith in the potentialities of individuals, faith in the capacity for positive developments if proper conditions are provided. The weakness of the philosophy originally advanced to justify the democratic movement was that it took individuality to be something given ready-made; that is, in abstraction from time, instead of as a power to develop.

The other conclusion is that art is the complement of science.

Science as I have said is concerned wholly with relations, not with individuals. Art, on the other hand, is not only the disclosure of the individuality of the artist but is also a manifestation of individuality as creative of the future, in an unprecedented response to conditions as they were in the past. Some artists in their vision of might be but is not have been conscious rebels. But conscious protest and revolt is not the form which the labor of the artist in creation of the future must necessarily take. Discontent with things as they are is normally the expression of vision of what may be and is not, art in being the manifestation of individuality is this prophetic vision. To regiment artists, to make them servants of some particular cause does violence to the very springs of artistic creation. But it does more than that. It betrays the very cause of a better future it would serve, for in its subjection of the individuality of the artist it annihilates the source of that which is genuinely new. Were the regimentation successful, it would cause the future to be but a rearrangement of the past.

The artist in realizing his own individuality reveals potentialities hitherto unrealized. This revelation is the inspiration of other individuals to make the potentialities real, for it is not sheer revolt against things as they are which stirs human endeavor to its depths, but vision of what might be and is not. Subordination of the artists to any special cause no matter how worthy does violence not only to the artist but to the living source of a new and better future. Art is not the possession of the few who are recognized writers, painters, musicians; it is the authentic expression of any and all individuality. Those who have the gift of creative expression in unusually large measure disclose the meaning of the individuality of others to those others. In participating in the work of art, they become artists in their activity. They learn to know and honor individuality in whatever form it appears. The fountains of creative activity are discovered and released. The free individuality which is the source of art is also the final source of creative development in time.

VII.

The Structure of Experiential Time

We can only experience the world, or ourselves, in the ways in which we are able to do so; just as the structure of one's spectacles is crucial to the perceptual reports he offers, so the temporal structure of his experiential outlook is a fundamental condition of the nature of his consciousness, of awareness, of his capacities and their limitations; the experiences we enjoy are conditioned in advance by the capacities we have for them. That statements such as those in the preceding sentence have often been dismissed as 'trivial' or unimportantly obvious is significant just because philosophic thought has too often neglected to face such pervasive commonplaces with seriousness, to think through their implications for the guidance of our further thinking, or to make sure of its own grounding first principles before seeking their applications in diverse enterprises.

If we grant that the nature of our experience says something about the kind of experiences we may have and the kind of experiencers we are, it would seem that questions about how our experiences are structured become foundational. When we acknowledge the pervasiveness of the temporal in our experiencing, we are impelled to examine the structuring of experiential time as the grounding of all the experiences we may have. For it is out of temporally structured human experiencing that men have developed the languages they use for communicating, the logical systems they use for thinking, and the arts, sciences, and other activities to which they devote the times of their lives.

If we are to talk about the world from the human point of view—a point of view within which we are necessarily bound—we must first, in any rigorously systematic thinking, examine the ways in which that human point of view is formed and functions. By taking the question of time as our lead question and guide, we see that the demand for an examination of the structure of experiential time is not new in philosophic history. It seems to have been called for by Augustine's interior examination of his experience of time, by the modern tradition stemming from Descartes which resurrected his principle of interiority, from Leibniz's conception of the self as a monadic center of dynamic experiencing, from Kant's temporalization of cognitive consciousness, Bergson's and Whitehead's descriptions of two kinds of time experience, McTaggart's of two kinds of temporal discourse, and James's observation that the experiential present is a spread, a "saddle-back," not a point, and our direct link with the world in which we are.

In turning to the way in which experiential time is structured, we find two distinct methods of study developing almost side-by-side. On first glance they seem opposed in many ways but, sharing something of a common Kantian heritage, it is not surprising that their results tend to a large degree of convergence.[1] Whether or not there will be, in the end, a fundamental philosophic difference between them still remains to be seen.

The genetic epistemology developed by Jean Piaget appears to honor man's essential temporality while attempting to understand it; working from the thesis that the notion of time is not a full-born intuition but a slowly developed concept, it seeks to chronicle the progressive development of the concept of time in the time of the child's development. Proceeding by carefully designed experiments, Piaget, observing behavior patterns of children at different developmental stages, has delineated definitive stages in the development of the ways in which time consciousness manifests itself. The hope is to understand the genesis and mode of development of the concept itself. By understanding how this, or any other, fundamental concept is actually brought to maturity in normal developmental patterns, it is felt that insight can be attained into what that concept really means as it comes to function in our usage of it.

The phenomenological examination is concerned, not so much with the stages of development but, with the internal structuring of the temporal outlook of the developed adult who must, in the end, ask the questions about conceptual development in terms of his own developed outlook. In contrast to the genetic approach of external observation of behavior patterns under experimental conditions, the phenomenologist has sought to systematize introspection itself, so that we may, indeed, face the question of how we experience our own experience. He seeks to unveil the way in which the temporal modes—past, present, and future—are indeed unified into the unity of one's experiential outlook.

Working from the fact that whatever is given to us is originally given in the experiential present, the question is then just how this present is constituted. The question is how the experiential present, knowing nothing of moments and instants, unites past and future into itself, claims for its content the objects appearing within it, and abides as continually changing and experientially authentic. For experiential time, functioning as it does in dynamic durational spreads, inherently involves the capacity to transcend any notion of momentary time.

The phenomenological method—which is not to be confused with any particular philosophic doctrine—was developed by Edmund Husserl in mid-career. Trained as a mathematician, he sought to bring to philosophy the rigor of a deductive science. This rigor was to be applied to the examination of consciousness itself and, in a set of early lectures, particularly to the way in which consciousness constitutes its moving experiential present.

Stimulated by Franz Brentano, Husserl drew heavily on work that had been done by Descartes, Hume, Kant, and James. Insisting on the primacy, not of what is experienced but, of the experiencing self, he declared:

> Whether we like it or not, whether (for whatever prejudices) it may sound monstrous or not, this (the "I am") is the fundamental fact to which I have to stand up, which, as a philosopher, I must never blink for a moment. For philosophical children this may be the dark corner haunted by the specters of solipsism or even of psychologism and relativism. The true philosopher, instead of running away from them, will prefer to illuminate the dark corner.[2]

Husserl called upon philosophy, as Spiegelberg pointed out, to honor Augustine's methodological admonition: "Don't go abroad. Truth dwells inside man."[3] Those who took up his questions have honored this injunction as they have, in somewhat divergent ways, tried to further our insight into the constitution of temporal consciousness.

Prominent among those who carried this time analysis forward are

Eugène Minkowski and Martin Heidegger. Contemporaries who were born only four years apart, they apparently have had little if any contact or reciprocal influence. They came to philosophy from different educational backgrounds (medicine and theology), pursued different philosophical interests (psychology and ontology), and claimed different mentors (Bergson and Husserl). They yet succeeded in existentializing phenomenology and thus making it directly relevant to lived experience. The general outcome of their analyses of temporal consciousness thus tend to converge. They were both concerned to examine the 'lived present'. But Minkowski has provided an interesting and often insightful discussion of the 'lived past' as well; and, Heidegger, directly facing the question of the structuring of the temporal outlook, has argued forcefully that it is organized and directed under the aegis of futurity.

Jean Piaget is an eminent Swiss psychologist whose philosophic education and familiarity with philosophic texts and issues seem to have influenced the focus of his concerns. His developmental studies have been focused on concepts—such as time, causality, intelligence, and reality—which are usually regarded as essentially philosophic in character. The impetus for his study of the genesis of the concept of time seems to have come out of some discussions he had had with Albert Einstein; but it is clear, from his repeated criticisms of specific aspects of the work of Kant and Bergson, that much of the conceptual structure which he brings to his own work comes from them.

He has sharply criticized many contemporary philosophers for ignoring psychological and sociological factors that bear on philosophic issues; he has urged the importance of subjecting such topics of discussion to experimental examination instead of merely speculative consideration. The "first principle of genetic epistemology," he has explained in an exposition of his methodological outlook, is "to take psychology seriously" and to consult research rather than inspiration when questions of psychological fact arise.[4] Indeed, his methodological 'manifesto', delivered as a series of lectures at Columbia University in 1968, might well be read in many philosophical circles with profit. With eloquence, he has condemned the tendency of some contemporary thinkers to try to resolve important issues by falling back on logical formalism instead of attending to experience itself.

Citing both Chomsky's rationalism and logical positivism as two contemporary cases in point, he has urged that meaningful discussion concerning questions of human knowledge cannot be legitimately founded on any kind of formalization.[5]

Turning his perspective to actually functioning human knowledge, Piaget sees it as "essentially active. . . . Knowing an object does not mean copying it—it means acting upon it. It means constructing systems of transformations that can be carried out on or with this object." [6] His approach, then, to the question of the development of the concept of time is to examine the ways in which time awareness is manifested in active commerce with aspects of the world at different stages of one's development.

If time were *completely* intuitive, as Kant has often been interpreted to have claimed, Piaget observes that there would be no real problem of development. But this is not the case: young children have no demonstrable grip on the idea of time. Therefore, the study of this conceptual development must start at an age "when the child does not yet suspect that time is common to all phenomena." [7] What is found to have been developed is a concept of time that depends on "the operational co-ordination of the [child's observed] motions themselves;" what is demonstrated is that "the relations between simultaneity, succession and duration must first be constructed, one by one." [8]

In developing the concept of time, he has shown that we must make a crucial distinction between the *temporal schema* and the *temporal content*. Temporal content is, indeed, irreversible; it is usually referred to when we say that 'the course of time is irreversible'. But when we do so, we are really referring, not to the concept of time itself, but to the sequence of events which we see in temporal form. Empirically, then, time is unidirectional—as the moving frame of the sequence of events we observe. But the rational time of thinking must be *imaginatively* reversible. The concept, itself, must be reversible in thought just because "succession involves two directions. . . . It is thus, that though a past event cannot be resurrected, we can nevertheless reconstruct the past in thought. . . ." This "failure to distinguish between irreversible events and reversible mechanisms of operational time," by which we bring the sequence of events into the unity of experience, is not only a stage in the child's development, as Piaget has pointed out;[9] it is also a failure that goes deep into a number of otherwise erudite discussions.

As we develop a rational conception of time, we find that we can imaginatively reverse it so as to recover a notion of the past as well as the present. Without this reversibility we could not really have any rational understanding of the notions of succession or of continuity. Our operational notion of time, then, is one that is constructed as a unique schema which is somewhat akin to the construction of logical and arithmetic ones; it does not come from an observation of the physical world as such but from *our* operational relations with it. The construction of time is then comparable to the construction of number; psychological and physical time conceptions arise out of the same kinds of operations or interactions. Bergson's thesis, then, that "lived durations ... are the 'very stuff of reality'" (p. 480) is not incorrect after all—but it applies to physical as well as psychological time.

But this is to indicate that time and space are seen together. "Space is a still of time, while time is space in motion. . . ." Thus Piaget has severely criticized the complete divorce between the two which Bergson had advanced.[10] For, basic to the conceptual distinction of temporal predicates, Piaget's claim is the demonstration of the notion of velocity. Taking velocity as experientially primal, time then emerges as one of its conceptually distinguished elements. But this means that time is discovered as such in actual operational structures, which we learn to employ. It is not an 'object' of contemplation but a rationally distinguished abstraction from the experience of velocity. But, if this is the case, then the sense of time does not develop as the measure of spatial motion in proportion to temporal measure;[11] for the notion of time is itself an intellectual construction based upon our reflective abstractions from the coordination of actions in dealing with varying velocities.

Awareness of time, then, in conceptual form, emerges in the seven- or eight-year-old child by abstraction from the concrete unity of his lived experience—in which space, time, motion, and his operational unity with the world in which he functions, appear to be inextricably involved with each other, as in a synthetic whole. The process of rational abstraction is a developed capacity. It enables one to transcend his spatial position, "freeing oneself from the present" (p. 259) by rising from a merely 'empirical' time, which is the passive observation of events, to the use of 'rational' time. By developing intellectual insight into the concept of time itself, we are enabled to use it as the prototype of all reversible operations of thought, to trace back a sequence from effect to cause, to operate in correlation with other people and with the processes of the physical world.

One may, of course, grandly dismiss Piaget's work as *merely* the description of a learning procedure which overlooks an honored distinction between the order in which knowledge is acquired and the nature of the knowledge that is acquired. Such a distinction has its own utility; but, if pressed too far, exposes its traditional if debatable ontological presumption, the irrelevance of time to being.

If my development is part of me, if I am in any sense my history, then the genetic development of the concepts that presently constitute my outlook is an essential part of the being that I am, and certainly colors the way in which those concepts function and have thereby been shaped in my use of them. The meaning, interpretation, value, and utility of any concept is shaped by the way in which the individual has incorporated it into the being that he is.

In a more generalized way, we can also note that if human children must develop a rational concept of time, and do so in certain describable stages—which Piaget's work would argue—this genetic study tells us something about the kinds of beings we are and the way in which we interrelate with others. To examine concepts without regard to their modes of becoming known, being shaped by the uses to which they are put and the ways in which they influence those who use them, is to seek to deal with the concepts of gods, not of time-bound men. If all of our concepts are, in any sense, temporal, they must not only be able to deal with the world appearing in the form of time, as Kant's schematism taught; they must also themselves be subject to temporal development: for, they are part of the intellectual equipment of human beings whose way of being is to develop in temporal form as they respond to and take part in the re-formation of the temporal situations in which they find themselves.

One additional point Piaget has made carries special significance. His distinction between 'empirical' and 'rational' time appears, at first blush, as merely a new 'twist' to the developing list of temporal dualisms—from Descartes, Bergson, McTaggart, Royce, and Whitehead. Speaking entirely on the side of the *experiencer* (not the experienc*ed)*, Piaget distinguishes the 'empirical' time, which is the passive observation of sequences of events, from the 'rational' time, which is the intellectual grasping of time conceptually. But this is no longer merely the stance of an external passive observer of sequences on a screen. To grasp time conceptually, to be able to reverse sequences in thought, is to be able to *use* the notions of past and future and not merely to note a temporal landscape. This is to say that when time is used conceptually, we no longer regard it as inherently serial —and the traditional schema of dots or dashes on a line no longer seems fitting.[12]

Piaget's notion of 'rational' time inherently seems to involve the conceptual use of past and future—but this means that experiential time cannot be reduced to simple serial order. Somehow, past and future conceptually enter into the living present as I, now, reconstruct a past event in thought for purposes of future employment. In some manner that is certainly far from clear, this notion of 'rational' time, and its essential attribute of reversibility, suggests that experiential time cannot be thought of as a simple stream placidly flowing without interruption from its past to its future. The direction to the future may, indeed, be crucial—but at least it seems that the reaching back into the past demands that we picture it as having turbulence and eddies along the way.

However far Piaget may have taken us in this direction, it is nevertheless clear that his experimental program has not been 'value-neutral'; it is subject to the serious criticism that he did not subject his own particular concept of time, the development of which he was studying, to any explicit examination. Piaget has been concerned to trace the development of the schematic structure of the adult who is able to use an 'objective' concept of quantifiable sequential time. His conclusion is that this capacity to schematize, and thus experientially assimilate, the given in these sequential terms emerges during the seventh or eighth year of life. This is, however, to have started with a refined notion of chronological time sequence as a mathematical construction and to trace *its* emergence. If explicitly stated, this does not in any way detract from Piaget's accomplishment; on the contrary, it enables us to understand how finely developed and sophisticated is the notion of measurable clock-time which we take for granted every day. But to equate *this* particular conception with all possible temporal conceptualization is, however, a serious philosophic blunder. Piaget has argued that the young child has no concept of time; what he seems to have established is that the child has no concept of quantifiable time, that this slowly develops as the child slowly moves towards adolescence.

But this criticism suggests that quantifiable time is not experientially primary, for even in infancy the individual begins to distinguish the modes of experiential time, as is clearly manifest in the very early emergence of anticipatory behavior. The ability to anticipate satisfaction or disappointment and then act on the basis of that anticipation, is to distinguish the present from the future and to discern in the living present signs of possible future states which are taken into the present as grounds for anticipating future presents. Such anticipatory behavior is clearly temporally structured, as it brings into one focus the present that is given, the similarities to past analogous states which are recalled or remembered. Even the young child cannot quantify such temporal modes—just because they are unquantifiable.

Such anticipatory behavior, emerging much earlier than the ability to rationally handle quantifiable time, effectively argues that genetically it is a more fundamental or primary form of temporal experience—from which the capacity to calculate later emerges. It is to an analysis of the primacy, the nature and the structuring of this fundamental experience of temporality that phenomenologists have turned their attention and the force of their philosophic method. This perspectival primordiality of the synthesis of past-and-present-and-future is the central focus of the phenomenological investigations of experiential time. We are led to such consideration not only by the early emergence of non-quantifiable, but nevertheless temporal, anticipatory behavior. We are also led to such consideration by Piaget's own genetically developed concept of 'rational' reversible time. For both lead us to the notion of a present as something of a field, structured by the events experienced as transpiring and not as a point on a line. Just what this notion of an extended present might mean has so far been left in obscurity. It is to this kind of clarification that the phenomenological analyses of temporal experience are largely directed.

We need, then, to question whether this explicit and self-conscious conceptualization—Piaget's 'rational' time—is necessary for us to face the interweaving of past and future in experience. At least in the developed adult, are conceptualization and simple perception really that distinct? Isn't thought, in some form, present in even the simplest perception as the movement from past to future of the perceived event is the object of awareness? How, indeed, is this possible? How is it that our consciousness of what we claim to perceive contains more time-spread than what is literally before us? How, for example, can we have any experience of the simplest duration, such as Bergson had regarded as the fundament of experience, unless we can 'at any moment' be aware of more than we literally see? Kant had already argued that we bring memory into the consciousness of the present perception. James, ostensibly eschewing Kant's 'transcendental machinery', yet insisted that perceptual time, itself, has breadth and takes time. In one glance, of even a discernible second, there is somehow present a literal 'past' and a literal 'not-yet' of expectation; the momentary split second of the literal 'now' has no real experiential meaning. One reason why James had called the experiential present a 'specious present' was just because its content, that to which the awareness of the perception claims to refer, is not literally present.

It is to this kind of consideration that Husserl's early analysis of time consciousness was directed. "Indifferent to the question of empirical genesis," he declared, what is of interest and concern is the understanding of "lived experiences as regards their objective sense and their descriptive content. . . . *lived experiences* of time." As he was quick to note, the analysis of the actually lived time experience, no matter how reduced in extent, is no simple matter. At the very least, he noted that any perception of an object "itself has temporality, that perception of duration itself presupposes duration of perception, and that perception of any temporal configuration whatsoever itself has its temporal form." [13] The problem that is posed by a phrase such as 'specious present' is, therefore, quite complex. For spread out in temporal form are both the object that is perceived as enduring while it is being seen, and the perceptual attention itself. Proceeding to a careful step-by-step analysis of the perceptual act itself, Husserl asked how our experiences of temporal passage are themselves experienced.

In order to understand how the phenomenon of temporality arises within consciousness, how the pervasively temporal relations with perceptual objects are maintained, Husserl undertook a rigorous internal examination of this experiential phenomenon, a phenomenological investigation. Starting with the awareness of the perceptual object, the investigation will ultimately lead to its grounding in an investigation of the awareness I have of my own self in its activity of being. When pursued long enough, the investigation will lead in a rigorous and step-by-step fashion to the ground presupposition of the living self in the lived world.

This program suggests the picture of the Leibnizian monad concerning itself, not with the 'reflections' of the 'external' world constituting the content of consciousness, but with the internal structure by which that content is constituted; by 'bracketing' such content, it then undertakes to examine the 'reflecting' structure itself. Foreshadowed in what Kant had once referred to as the "dissection of the faculty of the understanding itself," [14] Husserl's method was fashioned to reveal the essential experiential structuring of the ways in which we do, indeed, help to form our own experiences.

So, with the question of time, he set out to examine the way in which the so-called 'specious present' actually arises in human experience, how we are able to maintain a continuing perception that continually refers to more than the instantaneity of what is before it in any 'click'; literally broader in awareness than it literally is in any discernible now, these discriminated nows themselves are continually in the state of transformation into other nows. One way to approach his analysis, then, is to view it as an attempt to explain a puzzle—how an object which is intrinsically dynamic can retain its continuing identity in a consciousness which is inherently temporal and continually in the process of changing.

One way to begin is to note that in listening to a melody, I do not literally hear the melody; what is actually being heard are the tones literally present to my ear. The first explanation of the melody is to regard it as an interpretive mental act in which there is an imaginative holding together of the past tones as memory with the present audition and the expectation of what is next to come. This, Husserl, pointed out, will not do:

Every tone itself has a temporal extension: with the actual sounding I hear it as now. With its continued sounding, however, it has an ever new now, and the tone actually preceding is changing into something past. Therefore, I hear at any instant only the actual phase of the tone, and the Objectivity of the whole enduring tone is constituted in an act-continuum which in part is memory, in the smallest punctual part is perception, and in a more extensive part expectation.[15]

Each heard tone does *not* retain its 'integrity' as, unchanged, it merely recedes further and further into the past; it *does* continually change as its temporal 'distance' from the immediate now changes, in its continuity of 'running off'.

Merleau–Ponty has summarized Husserl's study of our consciousness of time as well as anyone; indeed, one could do no better than to refer back to it for an overall view in order to maintain one's bearings while working through the, perhaps excessive, rigor of Husserl's study:

> . . . the field of presence [is] the primary experience in which time and its dimensions make their appearance unalloyed, with no intervening distance and with absolute self-evidence. It is here that we see a future sliding into the present and on into the past. Nor are these three dimensions given to us through discrete acts. . . . The present itself, in the narrow sense, is not posited. The paper, my fountain-pen, are indeed there for me, but I do not explicitly perceive them. I do not so much perceive objects as reckon with an environment; I seek support in my tools, and am my task rather than confronting it. Husserl uses the terms protentions and retentions for the intentionalities which anchor me to an environment. They do not run from a central *I*, but from my perceptual field itself, so to speak, which draws along in its wake its own horizon of retentions, and bites into the future with its protentions. I do not pass through a series of instances of now, the images of which I preserve and which, placed end to end, make a line. With the arrival of every moment, its predecessor undergoes a change: I shall have it in hand and it is still there, but already it is sinking away below the level of presents; in order to retain it, I need to reach through a thin layer of time. It is still the preceding moment, and I have the power to recapture it as it was just now; I am not cut off from it, but it would not belong to the past unless something had altered it. It is beginning to be outlined against, or

projected upon, my present, whereas it *was* my present a moment ago. When a third moment arrives, the second undergoes a new modification; from being a retention it becomes the retention of a retention, and the layer of time between it and me thickens. One can, as Husserl does, represent this phenomenon diagrammatically. In order to make it complete, the symmetrical perspective of protensions would have to be added. *Time is not a line, but a network of intentionalities.*[16]

It is, then, apparent that perception (in its structured richness and complexity) is taken as the key to what transpires in the flux of appearances. For it is perception, actual or possible, that bears the referential intentionalities of our cognitive processes. In our examination of time consciousness, we note two streams of intentionality. The first is that of the flow of time itself which is constituted, as a flow, of unity and multiplicity by means of retention. The second is, of course, the reference to the object that appears, the temporal object, and is concerned more with its phases than with the continuum of temporal experience in which it appears.

It would, however, be an error to regard these two levels of intentionality as though they were independent. They are, rather, discernible only as related and involved in the unity of experience, which is indeed constituted as a temporal synthesis. The perceptual object, which is taken as 'permanent' across many perceptions, appears to consciousness in the dynamic flux portrayed by the double intentionality of continuum and flux. Thus the primacy of perception which is inherently temporal yields, at the outset, an experiential synthesis whose perceptual object is also in temporal form. By thus examining our mode of experiencing, we thus discover time as the time of consciousness; the ideas which form the content of consciousness refer to or 'intend' perceptual objects; consciousness, therefore, is essentially intentional insofar as it is always consciousness of, or awareness of, something other than itself. Thus, in pointing out the double nature of this essential temporal intentionality, Husserl develops the notion of the essential intentionality of consciousness, its qualification by temporal references, and the way in which human temporality is itself formed in terms of 'retention' and 'protention', two temporal categories which undergo development in his own later work and become fundamental in the time analyses that followed him.

This is to say that, in marked contrast to almost all of his predecessors, Husserl's examination was concerned with analysis of the experiencing itself—and not with a speculative metaphysic concerning what is derived from it:

> What we accept, however, is not the existence of a world-time, the existence of a concrete duration, and the like, but time and duration appearing as such. These, however, are absolute data which it would be senseless to call into question. To be sure, we also assume an existing time; this, however, is not the time of the world of experience but the *immanent time* of the flow of consciousness.[17]

Seeking to penetrate to the foundations of our experience, his highly rigorous and often tedious analysis of what is involved in time-consciousness makes some very important distinctions, which have generally been overlooked. In a manner somewhat reminiscent of James's notion of the 'fringe', Husserl tries to make sense of the phenomenon of 'running-off' as a crucial aspect of the ambiguity of the boundaries of the present; the distinction between 'retention' and 'recollection' or 'secondary remembrance' is, in a way, a key to his analysis of the nature of the spread-present which we experience. In his insistence that there is 'protension' in 'recollection', that perception, itself, has a forward-looking character, he is trying to give grounding to the kind of intimation of some kind of priority of futurity (which had already been suggested by Bergson [18] and made quite explicit by Peirce and by pragmatism generally).

What comes out of his very rigorous and conscientious and difficult analysis is the demonstration of the very real complexity of the so-called 'specious present'. We are no longer able to consider viewing as a merely passive affair or refer to the perceptual present in a glib and simple-minded manner. For in the most passive and uninvolved viewing itself, there is an active continuity of synthesizing of a continuing dynamic chain of perspectives which all focus on the present object to which perceptual attention is being directed and to which the description claims to refer. It is in terms of this complex nexus that meaning and awareness are intertwined as the constituting of the moving present is always bringing past and future into its ken and intentionality of focus. Husserl's analysis has largely focused attention on the phenomenon of 'running off' into the past and the way in which 'pastness' is constructed *out of the present;* the protentional propulsion toward the future has been but noted and has not been accorded any real examination.

In a very real way, this pioneering investigation has shown how our experience of time is itself intrinsically structured in temporal terms, how fundamental to human consciousness in its perceptual behavior is its essentially temporal and temporalizing constitution. But his model is still the perceptor whose perception is irrelevant to the perceived object and his span is still the 'specious present' which has now presumably been explained. If we are to go beyond this we must, as Royce had suggested of James, seek out wider horizons. If we are to understand how the immediate perceptual activity is itself undertaken, how it focuses a wider perspective into the 'moment' at hand, we need an existential extension of the phenomenological method so that we can finally begin to comprehend the nature of actual experience as it manifests itself in the wider and yet coherent spreads of lived time.

Although Husserl's lectures on the 'phenomenology of internal time consciousness' were not published until 1928 (under Heidegger's editorship), they had been originally delivered as lectures in 1904–1905, with additional material added during the five years that followed. Originally trained as a mathematician, Husserl had brought to his work the kind of rigor one expects from a deductive science. Quite different in personality and outlook, then, from Henri Bergson, he recognized the close affinity between the work he was already doing and what, without his knowledge, Bergson had already initiated; when Husserl first heard of Bergson's work on time, intuition, and the given of consciousness, he was reported to have exclaimed, "We are the true Bergsonians." [19] But this suggests that the phenomenological movement in France, which has since 'merged' with that deriving from Husserl, has had its own Bergsonian origins.[20]

The prime thinker who brought French phenomenology to the fore was Eugène Minkowski. Born in St. Petersburg, he studied medicine in Munich and, during the first World War, volunteered for service in the French army, settling in Paris after the war. Primarily influenced in his philosophic development by Bergson, and somewhat by Max Scheler, he brought an explicit phenomenology to France. His continuing import comes from his pioneering efforts to bring phenomenological and psychiatric studies into contact with each other.

His *Lived Time (Le Temps Vécu)* was first published in 1933, but it was written in the late 1920's and utilized material developed over the preceding decade. That this work had been composed before the appearance of Heidegger's *Being and Time* is quite significant. For the many parallels in the work of these two very different persons, who might be thought of as 'second generation' phenomenologists in their respective traditions, does not reflect a common background, a shared source, a reciprocity of influence or, in view of Heidegger's ontological rather than Minkowski's psychological interests, any mutuality of developmental direction. The parallels are many: most notable are their reinstitution of the experiential import of space (but under temporal control), their existentializing of the phenomenological method, their analyses of the experiential present as a most complex affair which is no simple 'specious present', the concern with the philosophic import of the phenomenon of death, and the decisive shift to the priority of futurity in the formation of human temporality.

Minkowski worked largely from Bergson's *Time and Free Will* and perhaps from *Matter and Memory* as well. He developed Bergson's implicit phenomenology as he also developed his focus on lived time as duration. Finding that important aspects of human experience are inherently unquantifiable, Minkowski reasoned that the understanding of man depends upon intuition and lived experience. This suggests that he insisted on remaining in what Husserl would have called the "natural standpoint." [21] His prime departures from Bergson seem to have been his professional reliance on psychological observation (which suggests contrasting comparisons with Piaget), and the sharp Bergsonian distinction between time and space; in life, Minkowski noted, they are rarely sharply separated.

Although he followed the Bergsonian lead which pointed to the prefiguring of the future in the present,[22] he pursued it in a systematic manner and argued that "The future is the most important modality of time. The past is closer to knowledge than to life." [23] In effect, agreeing with the thrust of Husserl's analysis of the present as being a most complex act, he pointed out that the present itself must contain duration, and agreed with the French psychologist, Pierre Janet, that duration is more fundamental than even memory—which develops from it as its narration.[24] But he continued the Bergsonian distrust of overintellectualizing the interpretation of experience, of yielding to the constant temptations to spatialize the non-spatial and to quantify that which cannot be authentically quantified.

The reason for the import of the future Minkowski found to be in the phenomenon of activity itself; for activity is the sign of the temporal, and for a living being activity is "duration oriented toward the future." [25] In contrast, the present almost seems fleeting and tenuous while it is unfolding, and the past continually stretches further and further away. But the future is continually before me; it is this 'pushing forward' towards it that engages my energies and my concerns and lets me find myself in unity with my environment and my world. Indeed, there are two different ways in which the future enters into my horizon of life: the first is that activity which, indeed, 'pushes' toward the future and thereby helps to create it; its opposite is not passivity but expectation, which marks one's attitude as he merely lets the future come to him.

This priority of the future in the forming of the living present points us to the significance of death for life. As it always hovers before us, it finitizes our living and enables us to live as active beings. Although it may appear to some as illogical, death, which is the negation of life, is necessary for meaning, significance, and direction; its impending denial of continuity enables us to live the living present which we essentially structure in terms of a thrust to a future which may be indefinite but which is definitely finite.

As we live towards the future, we live the living present, but we also must live the past. From the outlook of the lived perspective, it is not the past that yields the present, but the present that yields the past: memory arises out of duration as it spreads behind us. Plato had urged that 'knowledge is recollection', and the tradition associating knowledge and memory is a long one. Minkowski carries this forward and urges that "all memory contains *knowledge*" (p. 506), and this is one of the ways in which we may differentiate the past from the future. Only the future may fulfill a present expectation but only memory can now confirm a cognitive claim.

But the past is not necessarily dead. It is not merely a repository of past wisdom but, as we are able to live the past, we make it part of our present experiencing. This is to say that the way in which one lives his past is not reducible to his memory. Memory is always surrounded by a wider area into which it fades off—into a 'before' that is forgotten; and from this awareness of the forgotten we first intuit the past. In contrast to the lived present in which diverse experiences are integrated into experiential unity, the past differentiates those experiences which we have recovered from the forgotten. The past is seen as 'organized', broken down into 'various facets'; the unity of experience is differentiated into the 'things that it is about', into topics, subjects, tendencies, facts. Compared to the living present, the past always seems as something less, "as amputated from something, because it is characterized by being cut up and carved out" (p. 515). What has been, has been classified. When the three modes of time are unified into one, then, it is often being done under the aegis of the past; the present, as an agglomeration of facts, is only such because it is "destined to be inscribed into the past" (p. 516). To treat the present or the future in terms of the past, as in forecasting a specific set of unintegrated facts, is to "build the future on the model of the past" as science and religion both often do (p. 517).

The past, then, is not merely that part of time which came before the present; experientially, it has come *after* the present. And, similarly, the present is not that part of a linear series somehow hovering between a past that once was and a future that is not yet. The living present is a different *way* of living than the past. It integrates; it does not disintegrate. It opens up the possibilities of the future; it does not close them into the notion of a dead necessity. It invites us forth into what yet can be; it does not confine us into the narrowness of the merely factual. But this is to say that the usual way of speaking of the past is to misunderstand it.

The lived past is a reconstruction out of the forgotten; it is the dynamic bringing of the forgotten back into light and life. The lived past is "not nothingness or some kind of surrogate of a spatial order" (p. 514). It is dynamic, moving, with time *in* it. The past enters into the present, not as the dead past, a past present which has gone on to oblivion. The living past enters the present as remembrance, as the rescuing of the forgotten out of "the past [that] not only has been but *is no longer"* (p. 517); it is then that "Negation has penetrated time" (p. 517). Urging us to listen to life rather than logic, to appreciate the nuances of affirmation in negation, it is, Minkowski urges, when we understand that the past is the negation of the present, in a sense which the future cannot begin to hold, that memory, and the many problems of memory, arise. As it rescues from the forgotten, as it brings knowledge into the present, it is selective, discriminating and, indeed, prospective; it is to be found in the phenomenon of surpassing; for the lived present does not fear the voids it skips over to bring back a selected 'has been and *is no longer';* as one retrieves one's past, one brings into the present what one needs and ought to know, so that the fullness of the future, which radiates the living present with meaning, may, indeed, yet be.

A most careful and painstaking analysis of experiential time—its formation, structure, and significance—is to be found in Martin Heidegger's major work, *Being and Time.* His detailed discussion of time and temporality constitutes the second half of that book. His understanding of time builds on, and yet undergirds (as in a transcendental argument), his analysis of the nature of human existence generally. Bringing together several of the themes we have already encountered into synoptic unity, his prime argument is that human existence, the nature of the individual self, our conception of history, and our ordinary modes of reckoning with time as with space are all grounded and grounded together in the human mode of organizing its temporal perspectives.

He has pursued a generally phenomenological mode of inquiry; his basic search has been for the structures in terms of which one organizes his experience (although his reason has been, as we shall see, for the construction of a new 'fundamental ontology'). He has insisted on dealing with living experience in its integral unity—in which there are no really sharp lines between intellect and emotion, between my self and 'my' world, between one moment and the next. In looking at my experience, the first discovery is that I do not live in momentary bits and pieces gluing together separate specious presents. Nor do I find that my experiential time is really akin to a line with past, present, and future clearly delineated from each other in the living. My experience is found, by me, to be in stretches of lived time, of durational spreads, of temporally defined situations, defined as I define them for myself.

If Heidegger's time analysis is to disclose the structure of my experiential temporality, we can see him as having taken Husserl's analysis of the observational present forward into the analysis of lived duration. But there is a crucial difference, aside from the time-spread involved: rather than take the focus of the 'observer' as paradigmatic, Heidegger works on the thesis of an active involvement *in* the situations which we define for ourselves. If we remember that Kant's description of cognitive experience always involved the active questioning of what appeared, "constraining nature to give answer to questions," one can perhaps better see Heidegger as taking forth the tri-partite structure of Kant's notion of cognitive experience and expanding it into that of experience generally.[26]

However this derivation may be, either one—from Kant, from Husserl, or from both—suggests at the outset that Heidegger's analysis of the living present must involve the integral unity of past and present and future in one temporal synthesis. His essential 'picture' is that of a moving field of focus, directed by what is of interest or concern, which at any 'point' lights up what is yet-ahead of us—like the light on a miner's cap which *now* lights up the path on which he *will* walk, bringing his habits from his *past* into the walking. Similarly, as I drive along at night, my headlights leave me in darkness but illuminate the road ahead; what I am doing as I am driving—turning the wheel to steer, speeding up, slowing down, braking—all take place in terms of my judgment of what I see ahead of me, in terms of my interpretation of the 'where I will be next'. The driving itself is explained by me to myself in terms of where I am going, why I am going there, and what I intend to do after I will have arrived.

Whether I define my present in the momentary terms of the highway ahead of me, or the larger duration of the trip itself, my whole living situation is intelligible only in terms of the future into which I am going, which I have 'brought back' into my immediate present so that I know how to act. Into this situation, I bring my *relevant* past, my knowledge of how to drive or to choose the route; other aspects of my past are left in 'inventory', unnoticed and almost deliberately forgotten. What I have brought out of my past into present consciousness —aside from random daydreaming—is precisely that part of my past which is pertinent to the future which I am, in a very real sense, bringing toward me, as I am creating my 'future' present.

This temporal situation has been brought together by what Heidegger calls 'anticipatory resoluteness'; having anticipated a specific future (out of all the possibilities I could have selected), of arriving safely at my destination for whatever reason I decide to do so, I have effectively committed part of my future to ensuring the anticipated outcome; other possible uses of *this* time have been brushed out of my life; with resolution I have committed my being and my time (and I can only commit them together) to this particular anticipated result. My future, in a real sense, 'comes toward' me as I drive to and arrive at my destination (see p. 520). My future is, then, not merely a remote point on a time-line; it is not a 'now that is not-yet' for it *is now functioning in my present* as I drive along; otherwise, why would I be in this car on this road instead of *being involved in* some other activity?

As I drive along, certain directional signs seem to 'stand out' from the general scene and almost demand my attention; other signs—of movies, hotels, or other towns—that are of no pertinent interest to me are hardly noticed or remembered, even if graphically more obvious to the neutral observer. But this is to say that the landmarks and signposts I note as marking the territory through which I am driving are primarily those, out of countless other possible items for attention, responding to my own interests. What I *notice* as I am driving are those landscape features that somehow have bearing on this future of 'getting to my destination and what I anticipate doing after I get there', or that remind me of some earlier experience. They are selected by me from the myriad range of possible things I could notice, in terms of their temporal relevance to my present outlook. My 'noticings' are my response to the way they 'stand-out' from the general scene and seem to demand my attention. But this is to say that the perceptual content of my journey is structured, somehow, by the 'ecstatical' (or, 'standing-out') nature that this particular synthesis of future-and-thereby-selected-past somehow picks out for me to see. This temporal synthesis of mine, that constitutes my present driving, is somehow structured for me in terms of *this* selected future and not some other one (see p. 524).

That the content of my experiencing as I am driving along is structured for me in this way tells me much about the kind of being I am. I am able to focus my attention and shut out of perceptual attention the thousands of visual and auditory sightings that are about me. I am able to focus my attention and direct my activity in terms of why I want to arrive at that destination and what I want to do after I get there. I am able to lock away and forget all those myriad past experiences which might have bearing on other contexts but no useful bearing on this present intention. Having resolved on this course of action, I have in that same decision eliminated alternative courses of action from entering my being. In choosing, I have taken another step in the ongoing (but not endless) creating of my own self. I am precisely *this* individual who is characterized by *these* interests, aversions, predilections, loves, and hates. But this is a way of saying that in a very real sense, I *am* my world of interests and concerns. I am not a disembodied mind, somehow worrying about how I am connected to my body and looking out on an 'external' world to which I am hopelessly irrelevant. I am the living center of *my* world; in a very real and meaningful way, I am not a spatial point, but involved in my being is the ambiguously drawn 'area' or nexus of involvements that are me; I am the 'me' my friends know with their expectations of how I look and what I might do; I am the living biography that I bring along with me; I am my hopes, fears, aspirations, ideals, and cynicisms. I am 'here' or 'there'—temporally and spatially—in the world of my concerns.

This is to say that I am, to use Heidegger's term, my *Dasein.* Using an old German word which traditionally has meant 'existence'—*Das Dasein Gottes* = 'The existence of God'—he has chosen to take it in its most literal sense: the *da* meaning 'here' or 'there' in the sense of my environment, context, neighborhood, surroundings, and the *sein,* the infinitive verb, 'to be'. I am my 'to-be-there' in the world of my concerns. I am not merely this biological entity. I am a living center of experien*cing.* And experiencing, as a time-consuming activity, means that my concerns, interests, and activities are all structured by me in my continuing forming of situational presents which bring selected possible future states and pertinent pastness into the continuity of my living present.

Heidegger uses the word *Dasein* instead of 'man', 'human', or other possible nouns. He does this, not only because he does not want us to bring along some other connotation when using the word 'man', but also because he feels the continual need we have to remind ourselves that each of us is not merely a biological entity with a precise geographic location. He uses the word 'Dasein' also as a continuing reminder that I am not a 'thing' but a special kind of entity whose mode of being is to be activity and whose activity is always problematical—for 'Dasein' is essentially an infinitive term and, in literal translation, would be replaced with '(the) to-be-there'. My areas of concerns are not merely the spatial circle within which I find myselves; they reach out to my friend half way around the world, and also to the destination to which I am now driving while ignoring whatever may be before me except the links between my destination and myself.

How, then, do I structure this world of my concerns? I do so in terms of the lacks I see about me, the 'not-yet' that I want actualized, the hope that what I enjoyed yesterday will be available tomorrow; I do it in terms of those 'things' I am able to care about, and this capacity is expressed in the specific concerns I enumerate. But, then, my capacity for caring, for building up concerns and molding myself in terms of them is essentially temporal, is essentially built in terms of my hopes and fears and prayers about the future. In the first half of his book, Heidegger has seen this fundamental capacity 'to care' as the organizing center of the many specific activities that make us the individuals we are. In turning to the nature of temporality, then, he has turned to the question of 'how is it possible for me to structure my individual being in terms of this capacity to care?'; his reply is the analysis of temporality which begins for us under the heading, "Temporality as the Ontological Meaning of Care."

In view of the fact that our pasts cannot be altered—although our present *uses* of elements in the past can change their present meanings for us—the immediately present situation is one in which we find ourselves, as though we had been 'thrown' (see p. 521) into it. For the only temporal perspective which provides for deliberate decision, for self-creation, for action is the future. If we face the possible futures open to us with deliberate decision—annointing one with acceptance as anticipation (thus relegating alternates into no-longer-viable-possibilities)—we always start from where we are, from the situation into which we find ourselves as having been 'thrown' or 'cast'. Our only constructive, deliberative decision can relate to the query of 'to where from here?'. No other options are ever open.

All deliberate decision, then, is decision about what to do, from here on in—into the future. We may, indeed, decide to reassess the past, to rescue it from the 'dead' past by retrieving its lessons and bringing them into the present; but we can do so only under the aegis of futurity, just because our decision must be stated in the future tense. Such deliberate decision and commitment which recognizes the futural nature of all intention is the mark of Heidegger's use of the word 'authenticity'; it is not meant to have a moral ring to it, although it is admittedly hard to refrain from the suspicion that some value judgments are involved.

For Heidegger's claim is that he is concerned, at least in these discussions, not with what we ought to do or want to do—but with the way in which we do, indeed, construct the continuity of being, which is in each case *my* being. We do so because in every question, judgment, decision each of us is applying the question of 'to-be-or-not-to-be'—'*am* I happy?', '*is* he truly a friend?', '*will* he *be* as loyal in the future as he *was* in the past?', 'she *is* attractive', 'I *am now* hurrying because I *will be* late', 'because *it was* raining, it *is now* wet', '*is it* good or bad?', '*will* I regret and feel guilty about the action I am about to undertake now?'.

Because the 'to-be-or-not-to-be' outlook of an individual is crucial to the way he works in and maintains his world, the ways in which he treats himself and others, Heidegger points out that the being-structure of an individual's outlook is the ontology fundamental to his structuring of *his* world (in which he necessarily includes himself); this structure is his *fundamental ontology*. This is to contrast the traditional use of 'ontology' to describe the being-structure of the world as it presumably really is in itself. Heidegger's use of the term is, then, deliberately post-Kantian; as 'internalized', it thus means the being-structure of *my* interpretive outlook in terms of which *I* construct *my* 'picture' of *my* world.

If there are, indeed, certain ways of facing the question of to-be-or-not-to-be (about the cigarette being smoked, the book I am discarding, my loved ones, my pet cat, the steak I am about to consume, and all the other entities in my world), then those ways of facing the question of the to-be, of what is and what is not to be, have a kind of categorial status. If we all construct our temporal perspectives in much the same way—even as the content differs; if temporal synthesis is man's way of organizing his experiential being under the aegis of futurity; then the elements universally applicable as forms of human experience are 'existential categories', the categories by which the individuated existences of human beings are formed. Just as a modern skyscraper, no matter what its height, shape, color, or function may be, is supported by an internal framework, so each of us uses the same internal 'beaming' or 'structure' in building the existing self that he is. As structural components of human existence, as a self-conscious, questioning, temporalizing way of being, these existential categories or *existentials* function as the basic ontological components which each individual fashions in his own peculiar way.

But why coin a new word such as 'existentiale' instead of using a traditional word like 'category'? Indeed, why this seemingly perverse persistence in creating new terms, constructions, and circumlocutions that almost seem designed to drive the reader away? Whatever we may think of the result, however, it is important to see that it was at least a philosophic experiment motivated by some important philosophic considerations; if Heidegger's language will not do, it behooves us to come up with another way of meeting at least four issues of concern.

First, he has taken the Kantian distinction between 'persons' and 'things' with utmost seriousness. To treat persons as though they were things without first explicitly establishing that this is the case is certainly unwarrantable. But much thinking has succumbed to this fundamental 'category mistake' because of our language which has too freely used words derived from the description of inanimate things to describe persons. Hence, Heidegger leaves the word 'category' to describe the logical clusters of predicates justifiably used to describe the world of things (which are quantifiable and hopefully wholly explicable by physicists) and creates a parallel word, 'existentiale', to describe the logical clusters of predicates justifiably used to describe the existential modes of persons.

Second, he has developed Kant's distinction between the 'transcendental' and the 'empirical' into the similar distinction between the 'ontological' and the 'ontic'. The reasons for this development need not concern us here, but the advantage of the Heideggerian distinction, and the consequent development of separate words for the ontological and ontic versions of similar states (given in the appended table) [27] can be very helpful once mastered—because at a glance, we know whether what is being discussed is a facet of existence as a universal characteristic that is necessarily presupposed as an a priori of all particular existents—or a particular individual manifestation of one.

Third, if we are always looking out from the perspective of a person, our conceptual language, if it is to be experientially authentic, should conform to the nature of our experiencing. But our conceptual language has been built around the naming of *kinds* of things, and these genus-and-species names have been presumed, like a Platonic Idea, to be beyond time and change. The content of consciousness, in which our experiencing transpires, however, is always in flux. Our language has thus reflected the general inauthenticity of most philosophic reflection which has notably failed to square itself with the structuring of the experienc*ing* out of which it arises. As Heidegger has, himself, noted, "not only most of the words, but, above all the 'grammar' " are lacking for the new perspective he seeks to explicate. If we are to take the principle of Kant's Copernican Revolution with full seriousness and construct our ontology in terms of the human experiencing which produces it, then "we have to struggle against the same difficulty which keeps all ontological terminology in its grip." [28] For if our conceptualizing of a dynamic experience is essentially in terms of verbs (which, as Aristotle had already noted, connote not only an action but also its time),[29] then an experientially authentic language would focus on verbs as grammatical subject. Heidegger's language, neologisms, and circumlocutions are, in most cases, attempts to do precisely this, to convey in language the time-ness of what he talks about. This can be done, perhaps, with somewhat less linguistic butchery in German than in English—which has a tendency to translate deliberately used German infinitives such as *sein* ('to be') into substantive nouns such as 'Being' instead of '(the) to be'—(but even German treats such verb-subjects as nouns by giving them a grammatical gender!). For example, the German title of Heidegger's book is *Sein und Zeit,* the *literal* English translation of which would be *To Be and Time;* in a very real way that more literal rendering conveys something of the problematic notion of 'Being' he tries to express more than the English substantive, which immediately makes us think that it is some kind of peculiar eleatic thing.

When we approach things from the way in which we experience them, we find (and, this is a point developed at length in the early part of his book) that we generally notice things in terms of their utilities to us (such as the signposts on the road while driving) and only later depersonalize their relationship and treat them as neutral things which can be 'objectively' regarded, as science, for example, does. The two words which the English translators have chosen for the German terms are 'ready-to-hand' for the notion of utility and 'present-at-hand' for depersonalized 'objective' just being-there.

One of Heidegger's essential arguments is that our temporal syntheses which organize our experiences, always arise from our encounter with things as 'ready-to-hand', as useful, as future-pointing, as always pointing us to our necessary involvement with other people and with things in the world; I, therefore, always find myself, not as a disembodied consciousness, but as one whose essential structure is to-be-in-the-world as part of my own nature. This being-in-the-world is essentially constituted by my temporality.

Having presented his thesis concerning the priority of futurity in the formation of the presentness of consciousness, Heidegger proceeds to demonstrate its truth by exhibiting it in the ways in which we handle our everyday lives. For if I, as Dasein, exist as suggested, then my essential temporality must disclose to me the world of other people and of things in which I find myself. Of most importance in this regard is the analysis of the human understanding—which, he suggests, is *the* locus of possibility-retrieval and selection *from the future,* as it brings futurity, in terms of conceived possibilities, into the ken of the present. In a similar way, what is called in English 'state-of-mind', but which really refers to one's emotive set, is where the 'thrown' situation is revealed to one. Because of the importance which some contemporary thinkers attach to the phenomenon of language, we should also note Heidegger's rooting of language *in* one's mode of being.

After a careful examination of the way in which each of the 'ec-stases' or 'modes' of temporality are manifested in the synthesis of everyday living, he moves to two somewhat more theoretical the-ses—the derivations of the entire notions of historicality and of quantifiable time from this subjective temporality in which individual selfhood is to be found. Although it is usually good practice to read a book in the order in which the author has written it, I think that here, just because of the ground already covered, it might be wiser to go directly to his derivation of clock- and calendar-time from the time of personal experience.

His essential argument is given by the title of the chapter in which this discussion takes place: "Temporality and Within-Time-Ness as the Source of the Ordinary Conception of Time." In the course of his complete discussion, he refers specifically to both Aristotle and Augustine (see pp. 539, 546) as having intimated what he is now developing.

Clocks and calendars arose, he points out, because they filled a need. "The reckoning is prior to such equipment, and is what makes anything like the use of clocks possible at all. . . . Reckoning with time is an elemental kind of behaviour . . ." [30] which has an essentially *"public character* on the basis of Dasein's ecstatical being-in-the-world."* [31] This public time is thus a public construction, in terms of events common to a generalized experience 'in' which the ready-to-hand and the present-at-hand may be encountered together; it is expressed in a dating system which always arises in terms of what is or has been encountered within the world of human experience. Because this public time, which arises out of the need our futurity imposes for a common rubric of coordination of efforts, the notion of 'now' is developed as a mode of meeting. But this is to say that our mechanisms for measuring the time 'in which' we find ourselves are not merely so that we can 'locate' the 'now' in which we do the measuring; they are, so that they may be used in a common utilization of our common futurities, in the transformation of selected possibilities into the actuality we usually call the real. [32]

If Heidegger's analyses carry a basic validity with them, it would seem that he has brought an age-old discussion to a new level from which it henceforth must proceed. As one of his most severe critics has with the greatest reluctance said, Heidegger's consideration of the nature of experiential time is a "very acute piece of work which will remain a lasting contribution to philosophy." [33] He has brought the phenomenological analysis of something like the 'specious present' into an examination of just how time is formed as meaningful durational temporality in actual human existing. His discussion of time is not the perfunctory checking off of just one more item on an agenda of metaphysical 'topics' that one somehow has an obligation to discuss. He has seen that it is foundational to all human experience—to the experience itself as well as to what can appear as the referred-to content of that experience. In effect, he has provided a foundation which justifies the pragmatists' insight into the priority of futurity, for what McTaggart had seen as the experiential priority of what he called the A series, for Hegel's insistence that nature always appears in temporal form. If his analysis rings true, he has given us an understanding of how future and past merge into the 'extended' present that is a field and not a point, how we construct our experiences in forward-reaching temporal terms. We begin to understand just why the ancients always thought of time as derivative from the idea of eternity which they had, in fact, derived from their own essential temporality,[34] why even "in the beginning" man conceived God to have created that which-is in temporal form.

In a later essay, he stated well the underlying perception of the import of time which emerges from his time analysis. At least within the perspective of the human (and what other perspective do we have?), " 'Time' is called the first name of the truth of Being, and this truth is the presence of Being and thus Being itself." [35] In short, if Heidegger's understanding of the nature of temporality rings true, we have the beginning of an explanation of the integrity of our temporalizing experience *and* of our present attempt to understand it as well.

It has often been said of Kant that whether one agrees with him or not, one must face what he has had to say; one can no longer reason without him, without taking him into account. Heidegger's analysis of human temporality suggests that, in terms of time, we must come to terms with what he has had to say. We can no longer detour around him. If we are to comprehend the nature of time in human experiencing, we must take his account into account. Whether it is to be accepted, corrected, or repudiated, it would seem that, because of its extensive reach and foundational nature, it provides the ground from which our discussions must henceforth proceed.

25.

Piaget:

DEVELOPING THE CONCEPT OF TIME

Jean Piaget (1896–   ) is a philosophically astute Swiss psychologist, who has undertaken a number of studies of the genetic
development of prime concepts (e.g., causality, number, speed,
reality, motion, and space). His philosophic outlook is perhaps
most clearly stated in his 1968 Columbia University lectures,
Genetic Epistemology and in his *Insights and Illusions of Philosophy,* the English translation of which was published in 1971. The
selection is taken from the final chapter, "Conclusions," in his
landmark study, *The Child's Conception of Time,** which was
concerned to establish the definitive stages in the development of
the concept of quantifiable time.

From all the preceding discussions, we have learned that time is the
co-ordination of motions at different velocities—motions of external
objects in the case of physical time, and of the subject in the case of
psychological time. When we say motions, we are thinking of real
motions, and not of the displacements or ideal movements of
geometry. The latter are simply changes of position or 'placements', in
which the velocity can be neglected: that is why displacement is a
spatial concept and why time only appears with real motions, i.e. with
velocities. While the conception of time is not yet grasped operationally, i.e. as the ratio of the distance covered (or the work done, etc.) to
the velocity, the temporal order is confused with the spatial order and
duration with the path traversed. Conversely, before the temporal
order has been constructed, the idea of velocity is often bound up with

* From "Conclusions" of *The Child's Conception of Time,* by Jean Piaget, translated by A. J. Pomerans, (©) 1969 Routledge & Kegan Paul, Ltd., Basic Books, Inc.,
Publishers, New York. (Translated from the French, *Le Développement de la Notion
de Temps chez l'Enfant* [Presses Universitaires de France, 1927]). By permission.

that of overtaking, i.e. with a purely spatial intuition involving a change in the respective positions of two moving bodies. The construction of time proper therefore begins with the correlation of velocities, be it in the case of human activity or of external motions.

I

The most elementary form of time is found at the sensory-motor level, and has been described elsewhere.[1] When it cries with hunger, the baby has its first experience of duration (waiting time) and when, trying to reach a distant object, it first gets hold of an appropriate tool (support or stick) it establishes a primitive order of succession between the means and the end. However, this does not mean that time, at the sensory-motor level, is constructed into a homogeneous scheme even on the purely unconscious and practical plane: all the baby does is to correlate the succession or duration of particular actions with spatial displacements. At the primitive level, where objects are not yet endowed with permanence, the succession of events simply gives rise to motor responses, and later to those egocentric schemas that we have called 'subjective series': that is why, when it sees a person go out of the door, the baby will look for him by the side of its cot (where the person has just been). In spatial terms, therefore, displacements of objects are not yet 'grouped' into trajectories independent of the ego; and in temporal terms, there is a reversal of the order of succession, as if the 'watch' represented by the object suddenly ran backwards and its motion lacked all continuity. It is only once spatial groups of displacements have been constructed, that time itself can become objectivized or rather decentred on the practical plane and, it goes without saying, in one direction only. Thus, one-year-olds will look at the door or wave good-bye as soon as the name of a recent visitor is mentioned: the temporal construction of a course of events therefore goes hand in hand with its spatial construction. However, it cannot be stressed enough that, on the practical or sensory-motor plane, each action still has a time of its own, so that there are as many temporal series as there are schemes of action.

When, with the acquisition of language and verbo-motor concepts,

---

1. *La Construction du Réel chez l'Enfant,* Chapter IV.

the child's intelligence leaves the sensory-motor for the plane of thought, temporal concepts assume the form we have described as stage I. This includes a preliminary period (from 1½ to 4 years) during which the child's reactions must be observed directly rather than elicited by questions, and it was in this way that Decroly and Mlle Degand were able to correlate progress in temporal conceptions with the child's use of language.[2] In particular, they observed a gradual extension of temporal notions to embrace both the future and the past, and also the gradual emergence (which C. and W. Stern had described previously) of relative conceptions by which the future becomes transformed into the present (tomorrow is changed into today) and the present into the past (today is changed into yesterday).

We see, therefore, that, in accordance with a general law governing the transition from the sensory-motor level to that of nascent thought (see *La Construction du Réel chez l'Enfant*, 'Conclusions'), the child begins to re-interpret what it had previously learned in action. Thus, having learned to anticipate successions on the practical plane (by localizing them in the immediate present) or to take certain durations into account, he must now reconstruct these concepts by substituting virtual for real actions, and appropriate signs and mental representations for their purely perceptive characteristics.

That is why, at the age of 4–5, children still have difficulty in arranging a series of symbols (and even drawings) into a simple time sequence, though they are perfectly capable of constructing such a sequence on the purely practical plane. Thus these children will be able to predict that, if water is run from one vessel into another, the level will drop in the first and rise in the second, but they remain incapable of seriating what drawings they themselves have made of this process (Chapter One, §2 and §3).

In short, at stage I, the child must reconstruct into concepts his elementary ideas of succession and duration. Now, at that stage his constructions remain exceedingly primitive: true, they are abstracted from their particular context and generalized by the very fact of their conceptualization, but they do not yet lead to the differentiation between temporal and spatial structures. In effect, time at stage I is simply the order of succession and the colligation of durations of a

---

2. O. Decroly, *Etudes de Psychogenèse*, 1932, Chapter IV.

single series of linear events, irrespective of its own velocity or its intersections with other series with different velocities. In other words, the child watching a man walk along will say that the man reached B after A, and C after B; he might also say that the man took longer to go from A to C than from A to B. But in this particular case, it is clear that the temporal succession coincides with the spatial, and the durations with the displacements, so that the child can give the correct answer by relying on purely spatial considerations. Now, if we ask the same child to compare a motion from A to C with another motion, along the same path AC, but at a different velocity, he will be at a complete loss, i.e. he will be quite unable to tell us which of the two bodies will reach its destination first.

Time at stage I is therefore a localized time in the double sense that it varies from one motion to the next, and that it is confused with the spatial order. It is, one might say, a time without velocities, or a time that is homogeneous only so long as all the velocities are uniform. As soon, however, as actions at different velocities are introduced, the terms 'before' and 'after' lose all meaning or else preserve their purely spatial sense (Chapter Three); simultaneity is denied (Chapter Four); the equality of two synchronous durations ceases to make sense (Chapter Five); the colligation of durations can no longer be performed (Chapters Six and Seven)—nor, *a fortiori,* can the measurement of time (Chapter Eight). Even the concept of age, which would seem to be based on what the child has heard from adults, is interpreted spatially, inasmuch as differences in growth-rate lead to failure in grasping the order of succession of births and the conservation of age differences (Chapter Nine). Finally, psychological time, too, may be assessed by such spatial criteria as the results of a particular action (Chapter Ten). All these findings lead us to the same conclusions, namely that the construction of homogeneous time involves the co-ordination of velocities, and that the temporal ideas prevailing before this co-ordination is achieved, must necessarily be bound up with spatial intuitions.

It is evident that while temporal concepts remain intuitive, the concept of velocity remains intuitive as well. This is a point to which we shall be returning elsewhere.[3] When one moving body overtakes

---

3. See *Les Notions de mouvement et de vitesse chez l'Enfant.*

another, or continues further along the same path, all small children are agreed that it goes 'faster': to elementary intuitive thought, velocity is therefore equivalent to the process of overtaking. As soon as this process becomes invisible (e.g. when the two bodies move in two tunnels of different length) or when there is no overtaking (i.e. when the two bodies move in opposite, or even in the same direction but on two concentric courses, the outer one of which is visibly greater) estimates of velocity become vague. It follows that velocity is not yet recognized as a relationship between time and the space traversed, and this cannot, in fact, happen before time as such has been constructed. Indeed, we are merely being tautological since, if time is, in fact, the co-ordination of velocities (or of real motions), velocity must remain a fragmentary intuition before this co-ordination is achieved. In brief, it is the simultaneous construction of the operational conception of velocity (the ratio $v = s/t$) and of the operational conception of time ($t = s/v$) which alone enables the child to compare velocities when there is no visible overtaking, and also durations when the velocities differ.

But why precisely does the intuitive grasp of time remain spatialized for so long? To answer that question we need merely generalize our remarks about psychological time (Chapter Ten).

Grasping time is tantamount to freeing oneself from the present, to transcending space by a mobile effort, i.e. by reversible operations. To follow time along the simple and irreversible course of events is simply to live it without taking cognizance of it. To know it, on the other hand, is to retrace it in either direction. Rational time is therefore reversible, whereas empirical time is irreversible, and the former cannot embrace the latter unless this fundamental contrast is fully taken into account. Hence it is easy to understand why young children should have such difficulty in handling temporal concepts.

It is a characteristic of primitive thought that it treats as absolutes the particular perspectives it happens to be dwelling upon, and that it consequently fails to 'group'. This initial 'realism' is both a form of egocentrism, since it places current states of consciousness at the centre of everything, and also a form of irreversibility, because, in it, moment succeeds moment without leading to the construction of a general flux. More precisely, egocentrism and irreversibility are one and the same thing, and characterize the state of 'innocence' which

precedes the phase of critical construction. In the field of psychological time, they mean living purely in the present and assessing
the past exclusively by its results: whence the many problems of
'reflection' (in the literal sense of the term) we have been discussing,
and also the inability to arrive at the correct order of succession and
colligation of durations. In other words, the operational construction
of inner time calls for the correlation of one's own time not only with
that of others, but also with physical time, within a reversible system
that has ceased to be egocentric and is no longer bound up with
current events. In the field of physical time, too, reversibility takes the
child beyond egocentric and local time, i.e. beyond the irreversible
time characteristic of the motion of a single body, in which differences
in velocity can be ignored. We repeat that egocentrism and irreversibility are but two complementary aspects of one and the same lack
of co-ordination, and this fact alone explains the characteristic
property of primitive time, i.e. the lack of differentiation between
temporal and spatial successions.

This lack is gradually made good as the child passes from stage I to
stage II, i.e. to the stage of 'articulated intuitions'. Here progress in
intuitive regulation helps to reduce the excessive deformations that
spring from the irreversible centrations we have just been describing.
Articulated intuition thus marks the beginning of decentration and so
prepares the way for operations. In the case of durations, it leads to
the appreciation that time is inversely proportional to velocity and
thus opens the way for the correct colligations. In the case of successions it leads to the anticipation and reconstruction of the motions
themselves and deflects the attention from their end points. In short,
intuitive decentrations introduce corrections and these, in turn, lead
to certain correlations. However, as we saw, the correlation of velocity
and duration does not automatically introduce the correct order of
succession, or *vice versa:* these rudimentary correlations, far from
being operations, are simply articulated intuitions, i.e. intuitions subject to relatively constant regulations.

II

Much as primitive intuitions of time are examples of irreversible
thought, so operational time is the prototype of reversible thought,

and perhaps the clearest example of the way in which rational operations tend, by their very construction, to take the form of 'groupings'.
Thus while the operational fusion of the colligation of classes and the
seriation of asymmetrical relations involved in the construction of
number leads directly to a 'group' in the proper sense of the word, i.e.
to a mathematical system, the colligation of durations and the seriation of the asymmetrical relations of succession do not become fused
straightaway; they constitute two distinct logical 'groupings', in one-
to-one correspondence on the qualitative plane, but capable of being
combined on the quantitative plane.

From the psychological point of view, the construction of temporal
'groupings' which marks the transition from stage II to stage III, is
remarkable for two paradoxical facts on which we have dwelled at
some length. The first is that the child succeeds in constructing one
and the same system of temporal groupings in two distinct ways:
sometimes he will discover successions before he is able to colligate
durations, at other times he takes the opposite path; in both cases,
however, he arrives at the same operational result, i.e. he learns to
base successions on durations and *vice versa.* Now, here we have a
most curious fact. In other spheres, too, we encounter distinct 'types',
for example the abstract and visual types of mathematics, but they
never express the same truths in distinct languages. The second fact
worthy of notice is the relatively short period of transition between
stages II and III, i.e. the relatively quick operational construction of
time. True, there is also a sub-stage II B, but we gain the impression
that the moment when the child first succeeds in organizing a complete temporal system is so sudden that we can never actually put our
finger on it: often he will correct an error, and in so doing trigger off a
total process the speed of which is far greater than that of any conscious processes.

Now this total solution of the problem of time can be summed up in
a single formula: operational time is constructed as soon as the order
of successions is deduced from the colligation of durations and *vice
versa.* . . .

II. Let us now look at the qualitative operations involved in the
grouping of *durations.* . . . the interval is independent of the order of
events or of its mental reconstruction. . . .

The reader will recall how children tackle the problem of duration. At first, they assess durations by the path traversed, irrespective of the velocities: whence the negation of the equality of synchronous durations, the impossibility of colligating durations as such, and the absence of homogeneous time. Then, at stage II, they either discover the temporal (as distinct from the spatial) order of succession but fail to apply it directly to the durations, or else they discover that time is inversely proportional to velocity but fail to relate this relation to the order of succession. Stage III, and with it the grouping of durations, begins as soon as the child grasps that, of two bodies starting simultaneously, the one that moves for a longer time is also the one that stops last, or that age depends on the succession of births. Psychologically as well as logically, therefore, durations are partly treated as a system of intervals based on the grouping of successions. They nevertheless constitute an independent grouping from successions, in that their elementary relations are symmetrical and their formative operation is commutative.

In brief, the grouping of durations is engendered by the grouping of successions. However, successions can also be reconstructed from durations, and this fact fully accords with the psychological process. Now, as soon as the two systems can be correlated, both of them become operational at once, whence the impossibility of telling which determines the organization of the other at the beginning of stage III. The operational system involved in qualitative time thus constitutes an inseparable whole—psychologically as well as logically.

IIb. But duration is not simply an interval between two successive events—it can also be defined in terms of velocity, and this is a fact that quite a large number of our subjects take into account.

Now, it is true that the size of the intervals defined under II reduces to the set of instantaneous positions which a moving body occupies between the privileged states chosen to describe its motion. But these instantaneous positions are points in time without any duration, which comes back to defining intervals by distance alone, i.e. independently of the dynamic factors involved. In reality, the length of the interval between two successive points in time is nothing other than the motion itself related to its velocity.

This aspect of duration is, as we have just said, taken into consideration by quite a few of our subjects. To begin with, they treat duration as a simple function of the path traversed or of the work done, but as soon as articulated intuition enables them to dissociate actions and motions from the results, they treat duration as being inversely proportional to the velocity. In quantitative terms, this would mean that they put $t = s/v$, but since qualitative evaluations alone are involved, they simply make use of the logical expression $t = s \times (- v)$, meaning that (1) with equal distances any increase in time is equivalent to a decrease of velocity and *vice versa;* (2) with equal velocities, any increase in time is equivalent to an increase in distance, and (3) with equal times, any increase in distance is equivalent to an increase in time and *vice versa.*

From the quantitative point of view, all this is quite obvious, simply because time can only be expressed in terms of motions with a fixed velocity. However, it is important to stress that the same thing happens with qualitative time: we cannot see or perceive time as such since, unlike space or velocity, it does not impinge upon our senses. All we can perceive is events, i.e. motions, actions, speeds, and their results. Thus temporal successions are determined by the order of events, and durations either by the motions, i.e. by distances covered at given velocities, or else by actions, i.e. the work done at a given rate. Qualitative time, at the operational stage is based on velocity no less than quantitative time. It, too, is the relation between velocity and motion, or between activity and work, the only difference being that it is expressed by simple seriations and above all by simple colligations in the absence of a mobile unit by which alone the two can be combined. However, in both cases, time is essentially the co-ordination of velocities.

Now, since velocity is itself a relation between distance and time, how do we, in fact, help matters by asserting that duration is derived from it? Here we must bear in mind the distinction between the qualitative approach which reflects the actual construction of concepts, and the quantitative approach by which these concepts are given their simplest form. From the qualitative point of view, velocity, i.e. all judgements involving such terms as 'more rapidly', does not, in fact, involve the existence of durations but simply that of simultaneities: of two motions α and β which begin and end simultaneously, the more rapid is the one that covers the greater distance; and, if one of them should come to an end before the other, we need only determine the position of the other at that very moment, to obtain the velocity once again in terms of the path traversed. Needless to say, these remarks do not represent an exhaustive analysis of the qualitative conception of velocity; they simply show how the child manages to group velocities independently of durations. . . .

In brief, to define durations in terms of distances and velocities (or work and power) once again comes back to defining them as intervals between successive, instantaneous events, but with this difference, that the intervals are conceived as functions of their content, i.e. of the actions and motions whose co-ordination constitutes time.

III. We have so far ignored quantitative time as such. From the logical point of view, the construction of quantitative time is comparable to that of number: the iteration of the unit of duration results from the operational fusion of the colligation of durations (classes) with the seriation of successions (asymmetrical relations). However, in the case of qualitative time, these two operations are complementary, i.e. can be derived from each other, but do not become fused into a single grouping. . . .

In brief, in the sphere of time as in all other spatial and physical spheres, measurement appears as a synthesis of two fundamental systems of operation: displacement and partition. The reader may recall that number is the synthesis of the colligation of classes and the seriation of asymmetrical relations. Now similarly, when logico-arithmetical operations are replaced by spatio-temporal operations, and when the colligations of classes becomes the partition or colligation of parts into hierarchic wholes, and the seriation of relations becomes a spatio-temporal succession or placement (including changes in placement or displacement), measurement will result from the possible substitution of parts by their own displacement or by the displacement of a standard part chosen as a common unit. In the case of time, this unit is a motion at constant velocity that can be reproduced at will, i.e. that can be displaced in time and synchronized with the partial durations to be measured.

Now this is precisely how quantitative time is elaborated as the child develops its mental powers. To begin with, the child has great difficulties in admitting that a given partial duration, e.g. the time necessary for water to run between two levels (Chapters One and

Two), can be equal to a prior or subsequent duration: before he can do so, he must be able to divorce the duration from its qualitative context and feel free to reproduce it in a context that did not exist during the first duration, and that subsequently abolishes the earlier context. When faced with problems of this kind (Chapter Two, §3) children react as if they had been asked to equate an hour spent at enjoyable play with an hour spent on tedious calculations. And, indeed, the two are not comparable unless they have first been divorced from the actual events and related to, say, the motion of a clock.

This leaves us with the construction of the concept of motion at uniform velocity, which, as we saw, is an indispensable adjunct to the elaboration of quantitative time. Now, this construction seems to introduce a vicious circle: while the measurement of time rests on uniform velocity, the latter rests on the fact that two equal distances are covered in two equal durations. Hence how is it possible to establish that a given motion is uniform if we lack a unit of time? This particular psycho-genetic problem is the more interesting in that the same vicious circle appears in the scientific measurement of time: the adjustment of clocks rests on the regularity of natural motions, i.e. on the isochronism of small oscillations and the majestic periodicity of celestial orbits, but all we can say about this natural chronology is based on our own chronometry.[4] In point of fact, and this is characteristic of the operational organization of thought: the child discovers the conservation of uniform velocity simultaneously with the measurement of time, and by the identical operations (see our forthcoming *Les notions de mouvement et de vitesse chez l'enfant).*

IV. This new correlation between the construction of the concepts of velocity and of time, leads us to the psychological processes involved in the operational grouping of temporal relations and in the elaboration of the three fundamental attributes of rational time: homogeneity, continuity, and uniformity.

Now, just as the intuitive conception of time results from the egocentric and irreversible thought of young children, so the operational construction of time is the direct result of reversible correla-

4. See J. G. Juvet, *La Structure de nouvelles théories physique,* Paris (Alcan), 1935.

tions. The reversibility of thought is, in fact, marked by the correction of two tendencies, or, if you like, by the decentration of two types of centration. On the one hand, whereas the natural tendency of thought is to follow the course of the action itself, reversibility involves the retracing of that course: whence the operational construction of the concept of succession or order. On the other hand, while the personal point of view constitutes a privileged centration, reversibility, in the field of symmetrical relationships, leads to the construction of reciprocal viewpoints: whence the emergence of the concept of synchronous durations. In brief, the two chief results of decentration and the resulting reversibility of temporal concepts are the unfolding of time in two directions, after the discovery that the present is but a single moment in a continuous process, and the co-ordination of all the intersecting trajectories that, at any given moment, form a common medium to a host of simultaneous events.

Even on the qualitative plane, therefore, temporal operations lead to two remarkable results: they render time homogeneous and they also make it continuous. Quantitative operations, for their part, help to render the flow of time uniform (at least in the case of the small velocities characteristic of our everyday world).

Since homogeneous time is common to all phenomena, it is no longer the local time of intuition. But homogeneity does not imply the uniformity of successive durations: time could be common to the entire universe, even if its flow were constantly accelerated or slowed down, and even if it varied from one epoch to the next. The homogeneity of time results from synchronizations and other qualitative operations, and since such operations are limited to the colligation of the partial duration α or α into a total duration in the form $\alpha + \alpha' = \beta; \beta + \beta' = \gamma$; etc., they cannot ensure the uniformity of successive durations—quantitative operations alone can tell us anything at all about the relation between α and α', β', etc.

As for the continuity of time, it is a remarkable fact that it, like homogeneity, should not be taken for granted at all levels of mental development: for young children, in effect, time is discontinuous as well as local, since it stops with any partial motion. That is why adults are thought to have stopped ageing, why a tree is thought to age if it still grows but not otherwise, etc. It is only with the introduction of operational time that duration is treated as a continuous flux, which shows that, far from being an intuitive concept, the continuity of time calls for a special construction. Now this construction is simply the system of qualitative colligations, which leads to the partition of durations and ensures that it can be continued indefinitely and at all times. True, the various topological interpretations of the continuum introduce extensive quantity (e.g. Dedekind's or Cantor's axiom) or even metric quantity (Archimedes' axiom). But since the idea of continuous time is not grasped at the earlier stages of mental development, it follows that the mind must construct a qualitative continuum (intensive quantity) based on colligations (proposition 6) before the latter can give rise to mathematical quantification.

As for the uniform flow of time, it is based on uniform velocity, and its construction therefore calls for quantative rather than qualitative operations. However since, in the temporal as in all other spheres we have been investigating (number, mass, weight and volume), quantitative and extensive operations emerge the moment the grouping of qualitative or intensive operations has been achieved, the uniformity of time is recognized just as soon as its homogeneity and continuity have been constructed.

Reversibility of thought thus helps the child to unfold successions or asymmetries in two directions and, by gradual progress in decentration, to construct a general grouping, both qualitative and quantitative, of temporal relations that ensure the homogeneity, continuity and uniformity of time (on our scale). As Kant put it so profoundly, time is not a concept, i.e. a class of objects, but a unique schema, common to all objects, or, if you like, a formal object or structure. However, on the grounds that time is not a logical class, Kant argued that it is an 'intuition' (see Chapter Two, §2), i.e. an 'a priori form of sensibility' like space, and hence unlike the categories of the understanding, e.g. unlike quantity. Now, genetic analysis has led us to a quite different conclusion, namely that time must be *constructed* into a unique scheme by operations and, moreover, by the same groupings and groups as go into the construction of logical and arithmetical forms. The only difference is that, with time, the operations are not wholly logical (colligation of classes or seriation of relations) or arithmetical (correlation of invariant objects) but infra-logical (partitions and displacements), i.e. identical with the operations used in the very construction of objects, or rather in their colligation into that total object which is the universe of space-time.[5] This is why time, though forming a unique object, or one of its structures, is operational nevertheless. This is equally true of space, with which, however, we are not concerned in this volume.

III

When it comes to psychological time, finally, we saw that it is not simply intuitive, as so many authorities claim, but that it involves the same operations as physical time: the evaluation of 'lived' duration calls for a host of conscious or unconscious comparisons that lead to continuous progress from the level of perceptive or intuitive regulations to that of operational grouping.

The seriation of instants, first of all, is as essential in psychological as it is in physical time. The well-known idea of the 'flow of consciousness' should not be allowed to disguise the fact that every particular moment in this inner flux does not represent a point on a line, but a multiple and complex state resulting from the intermingling of a great many diverse currents. At any particular moment, we can be happy about our work, unhappy about the political situation, confident about the welfare of a near relative, etc., all at once, so that each slice of our inner time continuum appears as a tissue of simultaneous events, or as a snapshot. The reconstruction of a series of inner events thus invariably involves the process of co-seriation.

But it is in respect of durations that the operational character of psychological time is most often overlooked. This is due to the common error of confusing the implicit qualitative operations with intuitions, and of the explicit qualitative operations with measurements: since inner durations generally lack a common measure, we imagine that they do not involve operational colligations. However, it must be clear that, whenever we are able to arrange internal events in their order O, A, B, C, etc., we are introducing the duration α (between O and A), α' (between A and B), β' (between B and C), etc. Now while we may not be able to evaluate these durations in numerical terms, or even tell if they are uniform, or what precisely is the relation between

5. For a further discussion of infra-logical or spatio-temporal operations see *Le Développement des Quantités chez l'Enfant,* 'Conclusions'.

α, α' and β, we do know that $\alpha + \alpha' = \beta$ (β being the duration between O and B); $\beta + \beta' = \gamma$ (the duration between O and C), etc., and hence that $\alpha < \beta < \gamma$ ... etc., i.e. that these durations can be colligated. One might say that this is very little knowledge, indeed, but it is, in fact, all that is needed for the logic of classes in general. And above all, it is this knowledge which, joined to the seriation of successive moments, enables the child to construct physical time before it can tell hours and minutes.

Nor is that all. Lived durations are not simply intervals but, as Bergson so rightly put it, the 'very stuff of reality'. However, this in no way differentiates them from physical durations, since the real content of both is identical, i.e. the work done at a given rate (cf. proposition 5). True, in the case of psychological time, the work does not take the form of a distance traversed, because inner time is not spatialized, nor is it usually measurable since we never count our ideas or perceptions, but it can nevertheless be assessed in terms of plus or minus. 'Time is creation, or it is nothing at all', Bergson said, and this is perfectly true, provided only we remember that mental, unlike physical, creation can only be translated into duration in terms of power (and hence of rapidity). That this translation is subject to systematic errors, as a result of which intense work seems short while it is being done and long in retrospect, no one will deny, but these illusions are partly corrected—thanks precisely to those operational comparisons which the mind performs incessantly and almost automatically.

Operations in psychological time would therefore seem to be mainly of a qualitative kind. Does that mean that there is no such thing as quantitative inner time? Bergson borrowed most of his imagery from music and, whenever this master of introspection wished to show that creative duration involved irreducibly intuitive and anti-rational factors, he did so in terms of melody, rhythm and symphony. But what else is music than an inner type of mathematics? Long before Pythagoras discovered the numerical ratios which determine the principal musical intervals, ancient shepherds, singing their songs or playing an air on their pipes, busily constructed musical scales and realized, without being able to put it into so many words, that a minim equals two crotchets and a crotchet equals two quavers. Musical rhythm is, in fact, the most intuitive of all time measurements and is most certainly not imposed on us from outside.[6] The same is true of stress in common speech and quite particularly of metre in poetry. Here, too, it was not the theorists who invented the metre but the bards, thus showing that there is no contradiction between elementary arithmetic and the expression of rhythms in inner life. The case of metre even provides us with a good example of the continuous links between perceptive rhythms and spontaneous temporal operations.

All this points to the common nature of temporal operations in all spheres, and to the close relationship between psychological and physical time: both are co-ordinations of motions with different velocities, and both involve the same 'groupings'. This is only to be expected since both are derived from practical or sensory-motor time which, in its turn, is based on objective relations and on personal actions. As the external universe is gradually differentiated from the inner universe, so objects and actions become differentiated as well, but remain closely interrelated.

It goes without saying that the development of psychological time involves physical time, since the co-ordination of actions performed at different rates presupposes that some work has been done in the first place, and since all work is sooner or later incorporated into the external world. Hence personal memory is the memory of things and actions in the external world as much, if not more so than, the memory of things and actions in the inner world. What is far less clear is that physical time implies psychological time: the succession of psychological phenomena can only be grasped by an observer who goes beyond them and so resurrects a physical time that is no longer. Stueckelberg, in a recent study, has even tried to show that since mechanical time is reversible, and the time of thermodynamics and of micro-physics is subject to fluctuations, the direction of physical time can only be determined by the correlation of external trajectories with a series of psychological or biological memories, which latter alone have an unequivocal direction in time. It is significant that a physicist

---

6. In his suggestive 'Sur les operations de la composition musicale' *(Archives de Psychologie,* Vol. XXVII, p. 186), A. Mercier has tried to show that tone and rhythm represent two fundamental musical 'groups'.

should have felt the need to base physical on psychological time, whereas psychologists, who know about the active reconstructions of external events that go into every act of memory, tend to look upon physical time as the basis of inner time. In fact, the two are closely interrelated and both alike involve reconstructions of the causal order of events. Time, in both cases, is therefore the co-ordination of motions, and the direction of its flow can only be deduced from the causal chain, because causes necessarily precede their effects. Now, if causality is the general system of operations enabling us to correlate physical events, it is clear that before we can establish the existence of a causal relationship by experiment, we must first be able to correlate our measurements and this involves appealing to our memory or to reconstructions characteristic of psychological time. This is precisely what we mean when we say that physical time implies psychological time, and *vice versa*.

As for the time of relativity theory, far from being an exception to this general rule,[7] it involves the co-ordination of motions and their velocities even more clearly than the rest. Let us recall first of all that relativity theory never reverses the order of events in terms of the observer's viewpoint: if A precedes B when considered from a certain point of view, it can never follow B when considered from a different standpoint, but will at most be simultaneous with it. Einstein's refinements of the concept of time bear solely on non-simultaneity at a distance, and consequently on the dilation of durations at very great velocities. Now both these consequences follow directly from our definition of simultaneity as a limiting case of succession, i.e. as the result of two signalling motions in opposite directions, whose relative successions cancel out (proposition 2a). Simultaneity must therefore be relative to an organic or physical instrument (moving eye or optical signals, etc.). Now, since the relative velocity of light is constant and so constitutes a kind of absolute standard, simultaneity, in the case of great velocities, depends purely on the relative motions of the ob-

---

7. It is significant that Bergson, far from applauding the fact that Einsteinian time presents physics with a much closer model of psychological time (we might say of Bergsonian time) than Newtonian time did, challenged relativity theory with the claim that relative time was a characteristic of life alone.

server and the phenomenon he observes, as well as on their distance apart. And if simultaneities are indeed relative to velocities, it follows that the measurement of durations will itself depend on the co-ordination of these velocities. Relativistic time is therefore simply an extension, to the case of very great velocities and quite particularly to the velocity of light, of a principle that applies at the humblest level in the construction of physical and psychological time, a principle that, as we saw, lies at the very root of the time conceptions of very young children.[8]

8. The reader will forgive us if, in a book devoted to the development of time concepts in children, we ignore the subject of microphysical time, with which we shall be dealing in a separate work.

26

Husserl:

THE CONSTITUTION OF THE PRESENT

Edmund Husserl (1859–1938) is generally regarded as the progenitor of contemporary phenomenology. His *Phenomenology of Internal Time-Consciousness* * is the first systematic phenomenological study of just how our consciousness of time is experienced. The central portion of this short but difficult work is the manuscript, edited by Edith Stein, of Husserl's 1904–1905 lectures. Together with material coming down to 1910, the book was finally published under Martin Heidegger's editorship in 1928. Husserl, seeing that all experience actually transpires in the 'living present', sought to work out its essential constitutional structure, which is crucial to any theory of awareness. The selection, taken from the section entitled "The Analysis of Time-Consciousness," is designed to highlight the major steps of his argument.[1]

§8. Immanent Temporal Objects [*Zeitobjekte*]
and Their Modes of Appearance

... The sound is given; that is, I am conscious of it as now, and I am so conscious of it "as long as" I am conscious of any of its phases as now. But if any temporal phase (corresponding to a temporal point of the duration of the sound) is an actual now (with the exception of the

---

* Edmund Husserl, *The Phenomenology of Internal Time-Consciousness*, ed. Martin Heidegger, trans., James S. Churchill (C) 1964. Used by permission of Indiana University Press, Bloomington, Indiana.

1. I am deeply indebted to Prof. José Huertas-Jourda of Wilfrid Laurier University and its Center for Advanced Research in Phenomenology for his invaluable help in choosing the passages constituting the selection.—C. M. S.

beginning point), then I am conscious of a continuity of phases as "before," and I am conscious of the whole interval of the temporal duration from the beginning-point to the now-point as an expired duration. I am not yet conscious, however, of the remaining interval of the duration. At the end-point, I am conscious of this point itself as a now-point and of the whole duration as expired (in other words, the end-point is the beginning point of a new interval of time which is no longer an interval of sound). "During" this whole flux of consciousness, I am conscious of one and the same sound as enduring, as enduring now. "Beforehand" (supposing it was not expected, for example) I was not conscious of it. "Afterward" I am "still" conscious of it "for a while" in "retention" as having been. It can be arrested and in a fixating regard [*fixierenden Blick*] be fixed and abiding. The whole interval of duration of the sound or "the" sound in its extension is something dead, so to speak, a no longer living production, a structure animated by no productive point of the now. This structure, however, is continually modified and sinks back into emptiness [*Leere*]. The modification of the entire interval then is an analogous one, essentially identical with that modification which, during the period of actuality, the expired portion of the duration undergoes in the passage of consciousness to ever new productions.

What we have described here is the manner in which the immanent-temporal Object "appears" in a continuous flux, i.e., how it is "given." To describe this manner does not mean to describe the temporal duration itself, for it is the same sound with its duration that belongs to it, which, although not described, to be sure, is presupposed in the description. The same duration is present, actual, self-generating duration and then is past, "expired" duration, still known or produced in recollection "as if" it were new. The same sound which is heard now is, from the point of view of the flux of consciousness which follows it, past, its duration expired. To my consciousness, points of temporal duration recede, as points of a stationary object in space recede when I "go away from the object." The object retains its place; even so does the sound retain its time. Its temporal point is unmoved, but the sound vanishes into the remoteness of consciousness; the distance from the generative now becomes ever greater. The sound itself is the same, but "in the way that" it appears, the sound is continually different.

§ 9. The Consciousness of the Appearances of Immanent Objects [*Objekte*]

... We must now examine more closely what we find here and can describe as the phenomena of temporally constitutive consciousness, that consciousness in which temporal objects with their temporal determinations are constituted. We distinguish the enduring, immanent Object in its modal setting [*das Objekt im Wie*], the way in which we are conscious of it as actually present or as past. Every temporal being "appears" in one or another continually changing mode of running-off, and the "Object in the mode of running-off" is in this change always something other, even though we still say that the Object and every point of its time and this time itself are one and the same. The "Object in the mode of running-off" we cannot term a form of consciousness (any more than we can call a spatial phenomenon, a body in its appearance from one side or the other, from far or near, a form of consciousness). "Consciousness," "lived experience," refers to an Object by means of an appearance in which "the Object in its modal setting" subsists. Obviously, we must recognize talk of "intentionality" as ambiguous, depending on whether we have in mind the relation of the appearance to what appears or the relation of consciouness on the one hand to "what appears in its modal setting" and on the other to what merely appears.

§ 10. The Continua of Running-off Phenomena —The Diagram of Time

... To begin with, we emphasize that modes of running-off of an immanent temporal Object have a beginning, that is to say, a source-point. This is the mode of running-off with which the immanent Object begins to be. It is characterized as now. In the continuous line of advance, we find something remarkable, namely, that every subsequent phase of running-off is itself a continuity, and one constantly expanding, a continuity of pasts. The continuity of the modes of running-off of the duration of the Object we contrast to the continuity of the modes of running-off of each point of the duration which obviously is enclosed in the continuity of those first modes of

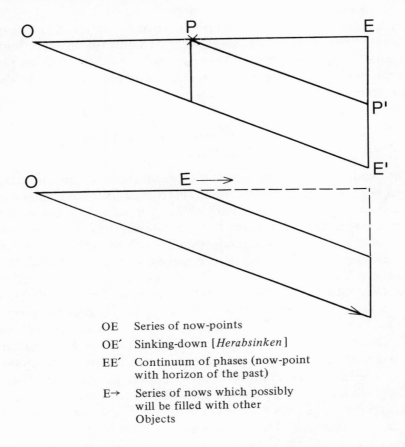

OE   Series of now-points

OE′   Sinking-down [*Herabsinken*]

EE′   Continuum of phases (now-point
        with horizon of the past)

E→   Series of nows which possibly
        will be filled with other
        Objects

running-off; therefore, the continuity of running-off of an enduring Object is a continuum whose phases are the continua of the modes of running-off of the different temporal points of the duration of the Object. If we go along the concrete continuity, we advance in continuous modifications, and in this process the mode of running-off is constantly modified, i.e., along the continuity of running-off of the temporal points concerned. Since a new now is always presenting itself, each now is changed into a past, and thus the entire continuity of the running-off of the pasts of the preceding points moves uniformly "downward" into the depths of the past. In our figure the solid horizontal line illustrates the modes of running-off of the enduring Object. These modes extend from a point O on for a definite interval which has the last now as an end-point. Then the series of modes of running-off begins which no longer contains a now (of this duration). The duration is no longer actual but past and constantly sinks deeper into the past. The figure thus provides a complete picture of the double continuity of modes of running-off.

§ 11. Primal Impression, [*Urimpression*] and Retentional Modification

The "source-point" with which the "generation" of the enduring Object begins is a primal impression. This consciousness is engaged in continuous alteration. The actual [*leibhafte*] tonal now is constantly changed into something that has been; constantly, an ever fresh tonal now, which passes over into modification, peels off. However, when the tonal now, the primal impression, passes over into retention, this retention is itself again a now, an actual existent. While it itself is actual (but not an actual sound), it is the retention of a sound that has been. A ray of meaning [*Strahl der Meinung*] can be directed toward the now, toward the retention, but it can also be directed toward that of which we are conscious in retention, the past sound. Every actual now of consciousness, however, is subject to the law of modification. The now changes continuously from retention to retention. There results, therefore, a stable continuum which is such that every subsequent point is a retention for every earlier one. And every retention is already a continuum. The sound begins and steadily continues. The tonal now is changed into one that has been. Constantly flowing, the *impressional* consciousness passes over into an ever fresh *retentional* consciousness. Going along the flux or with it, we have a continuous series of retentions pertaining to the beginning point. Moreover, every earlier point of this series shades off [*sich abschattet*] again as a now in the sense of retention. Thus, in each of these retentions is included a continuity of retentional modifications, and this continuity is itself again a point of actuality which retentionally shades off. This does not lead to a simple infinite regress because each retention is in itself a continuous modification which, so to speak, bears in itself the heritage [*Erbe*] of the past in the form of a series of shadings. It is not true that lengthwise along the flux each earlier retention is merely replaced by a new one, even though it is a continuous process. Each subsequent retention, rather, is not merely a continuous modification arising from the primal impression but a continuous modification of the same beginning point. . . .

§ 14. Reproduction of Temporal Objects [*Objekten*]—Secondary Remembrance

We characterized primary remembrance or retention as a comet's tail which is joined to actual perception. Secondary remembrance or recollection is completely different from this. After primary remembrance is past [*dahin*], a new memory of this motion or that melody can emerge. The difference between the two forms of memory, which we have already touched on, must now be explained in detail. . . . Let us consider an example of secondary remembrance. We remember a melody, let us say, which in our youth we heard during a concert. Then it is obvious that the entire phenomenon of memory has, *mutatis mutandis*, exactly the same constitution as the perception of the melody. Like the perception, it has a favored point; to the now-point of the perception corresponds a now-point of the memory, and so on. We run through a melody in phantasy; we hear "as if" [*gleichsam*] first the first note, then the second, etc. At any given time, there is always a sound (or a tonal phase) in the now-point. The preceding sounds, however, are not erased from consciousness. With the apprehension of the sound appearing now, heard as if now, primary remembrance blends in the sounds heard as if just previously and the expectation (protention) of the sound to come. Again, the now-point has for consciousness a temporal halo [*Hof*] which is brought about through a continuity of memory. The complete memory of the melody consists of a continuum of such temporal continuities or of continuities of apprehension of the kind described. Finally, when the melody presentified has been run through, a retention is joined to this as-if hearing; the as-if heard still reverberates a while, a continuity of apprehension is still there but no longer as heard. Everything thus resembles perception and primary remembrance and yet is not itself perception and primary remembrance. We do not really hear and have not really heard when in memory or phantasy we let a melody run its course, note by note. In the former case, we really hear; the temporal Object itself is perceived; the melody itself is the object of perception. And, likewise, temporal periods, temporal determinations and relations are themselves given, perceived. And again, after the melody has sounded, we no longer perceive it as present although we still have it in consciousness. It is no longer a present melody but one just past. Its being just past is not mere opinion but a given fact, self-given and therefore perceived. In contrast to this, the temporal present [*Gegenwart*] in recollection is remembered, presentified. And the past is remembered in the same way, presentified but not perceived. It is not the primarily given and intuited past.

On the other hand, the recollection itself is present, originarily constituted recollection and subsequently that which has just been. It generates itself in a continuum of primal data and retentions and is constituted (better, re-constituted) jointly with an immanent or transcendent objectivity of duration (depending on whether it is immanently or transcendently oriented). On the other hand, retention generates no objectives of duration (whether originary or reproductive), but merely retains what is produced in consciousness and impresses on it the character of the "just past."

§ 19. The Difference between Retention and Reproduction (Primary and Secondary Remembrance or Phantasy)

... That a great phenomenological difference exists between representifying memory and primary remembrance which extends the now-consciousness is revealed by a careful comparison of the lived experiences involved in both. We hear, let us say, two or three sounds and have during the temporal extension of the now a consciousness of the sound just heard. ...

The modification of consciousness which changes an originary now into one that is *reproduced* is something wholly other than that modification which changes the now—whether originary or reproduced—into the *past*. This last modification has the character of a continuous shading-off; just as the now continuously grades off into the ever more distant past, so the intuitive consciousness of time also continuously grades off. On the other hand, we are not speaking here of a continuous transition of perception to phantasy, of impression to reproduction. The latter distinction is a separate one. We must say, therefore, that what we term originary consciousness, impression, or perception is an act which is continuously gradated. Every concrete perception implies a whole continuum of such gradations. Reproduction, phantasy-consciousness, also requires exactly the same gradations, although only reproductively modified. On both sides, it belongs to the essence of lived experiences that they must be extended in this fashion, that a punctual phase can never be for itself.

Naturally, the gradation of what is given originarily as well as of what is given reproductively indeed concerns the content of apprehension, as we have already seen. Perception is built upon sensations. Sensation which functions presentatively for the object forms a stable continuum, and in just the same way the phantasm forms a continuum for the representation [*Repräsentation*] of an Object of phantasy. Whoever assumes an essential difference between sensations and phantasms naturally may not claim the content of apprehension of the temporal phases just past to be phantasms, for these, of course, pass continually over into the content of apprehension of the moment of the now.

§ 21. Levels of Clarity of Reproduction

. . . The specific modes of vividness and lack of vividness, of clarity and lack of clarity of the presentification do not belong to what is presentified, or belong to it only by virtue of the modality of the presentification. They belong to the actual lived experience of the presentification.

§ 24. Protentions in Recollection

In order now to understand the disposition of this constituted unity of lived experience, "memory," in the undivided stream of lived experience, the following must be taken into account: every act of memory contains intentions of expectation whose fulfillment leads to the present. Every primordially constitutive process is animated by protentions which voidly [*leer*] constitute and intercept [*auffangen*] what is coming, as such, in order to bring it to fulfillment. However, the recollective process not only renews these protentions in a manner appropriate to memory. These protentions were not only present as intercepting, they have also intercepted. They have been fulfilled, and we are aware of them in recollection. Fulfillment in recollective consciousness is re-fulfillment [*Wieder-Erfüllung*] (precisely in the modification of the positing of memory), and if the primordial protention of the perception of the event was undetermined and the question of being-other or not-being was left open, then in the recollection we have a pre-directed expectation which does not leave all that open. It is then in the form of an "incomplete" recollection whose structure is other than that of the undetermined primordial protention. And yet this is also included in the recollection. There are difficulties here, therefore, with regard to the intentional analysis both for the event considered individually, and, in a different way, for the analysis of expectations which concern the succession of events up to the actual present. Recollection is not expectation; its horizon, which is a posited one, is, however, oriented on the future, that is, the future of the recollected. As the recollective process advances, this horizon is continually opened up anew and becomes richer and more vivid. In view of this, the horizon is filled with recollected events which are always new. Events which formerly were only foreshadowed are now quasi-present, seemingly in the mode of the embodied present.

§ 25. The Double Intentionality of Recollection

If, in the case of a temporal Object, we distinguish the content together with its duration (which in connection with "the" time can have a different place) from its temporal position, we have in the reproduction of an enduring being, and in addition to the reproduction of the filled duration, the intentions which affect the position, in fact, necessarily affect it. A duration is not imaginable, or better, is not positable unless it is posited in a temporal nexus, unless the intentions of the temporal nexus are there. Hence it is necessary that these intentions take the form of either past or future intentions. To the duality of the intentions which are oriented on the fulfilled duration and on its temporal position corresponds a dual fulfillment. The entire complex of intuitions which makes up the appearance of past enduring Objects has its possible fulfillment in the system of appearances which belong to the same enduring thing. The intentions of the temporal nexus are fulfilled through the establishment of the fulfilled nexuses up to the actual present. In every presentification, therefore, we must distinguish between the reproduction of the consciousness in which the past enduring Object was given, i.e., perceived or in general primordially constituted, and that consciousness which attaches to this reproduction as constitutive for the consciousness of "past," "present" (coincident with the actual now), and "future."

Now is this last also reproduction? This is a question which can easily lead one astray. Naturally, the whole is reproduced, not only the then present of consciousness with its flux but "implicitly" the whole stream of consciousness up to the living present. This means that as an essential *a priori* phenomenological formation [*Genese*] memory is in a continuous flux because conscious life is in constant flux and is not merely fitted member by member into the chain. Rather, everything new reacts on the old; its forward-moving intention is fulfilled and determined thereby, and this gives the reproduction a definite coloring. An *a priori,* necessary retroaction is thus revealed here. The new points again to the new, which, entering, is determined and modifies the reproductive possibilities for the old, etc. Thereby the retroactive power of the chain goes back, for the past as reproduced bears the character of the past and an indeterminate intention toward a certain state of affairs in regard to the now. It is not true, therefore, that we have a mere chain of "associated" intentions, one after the other, this one suggesting the next (in the stream). Rather, we have an intention which in itself is an intention toward the series of possible fulfillments.

But this intention is a non-intuitive, an "empty" intention, and its objectivity is the Objective temporal series of events, this series being the dim surroundings of what is actually recollected. Can we not characterize the non-general "surroundings" as a unitary intention which is based on a multiplicity of interconnected objectivities and in which a discrete and manifold givenness comes gradually to fulfillment? Such is also the case with the spatial background. And so also, everything in perception has its reverse side as background (for it is not a question of the background of attention but of apprehension). The component "unauthentic perception" which belongs to every transcendent perception as an essential element is a "complex" intention which can be fulfilled in nexuses of a definite kind, in nexuses of data.

The foreground is nothing without the background; the appearing side is nothing without the non-appearing. It is the same with regard to the unity of time-consciousness—the duration reporduced is the foreground; the classifying intentions make us aware of a background, a temporal background. And in certain ways, this is continued in the constitution of the temporality of the enduring thing itself with its now, before, and after. We have the following analogies: for the spatial thing, the ordering into the surrounding space and the spatial world on the one side, and on the other, the spatial thing itself with its foreground and background. For the temporal thing, we have the ordering into the temporal form and the temporal world on the one side, and on the other the temporal thing itself and its changing orientation with regard to the living now.

§ 26. The Difference between Memory and Expectation

We must further investigate whether memory and expectation equal each other. Intuitive remembrance offers me the vivid reproduction of the expiring duration of an event, and only the intentions which refer back to the before and forward to the living now remain unintuitive.

In the intuitive idea of a future event, I now have intuitively the productive "image" of a process which runs off reproductively. Joined thereto are indeterminate intentions of the future and of the past, i.e., intentions which from the beginning of the process affect the temporal surroundings which terminate in the living now. To that extent, expectational intuition is an inverted memorial intuition, for the now-intentions do not go "before" the process but follow after it. As empty environmental intentions, they lie "in the opposite direction." How do matters stand now with the mode of givenness of the process itself. Does it make any essential difference that in memory the content of the process is determinate? Moreover, the memory can be intuitive but still not very determinate, inasmuch as many intuitive components by no means have real memorial character. With "perfect" memory to be sure, everything would be clear to the last particular and properly characterized as memory. But, ideally, this is also possible with expectation. In general, expectation lets much remain open, and this remaining-open is again a characteristic of the components concerned. But, in principle, a prophetic consciousness (a consciousness which gives itself out as prophetic) is conceivable, one in which each character of the expectation, of the coming into being, stands before our eyes, as, for example, when we have a precisely determined plan and, intuitively imagining what is planned, accept it lock, stock, and barrel, so to speak, as future reality. Still there will also be many unimportant things in the intuitive anticipation of the future which as makeshifts fill out the concrete image. The latter, however, can in various ways be other than the likeness it offers. It is, from the first, characterized as being open.

The principal differences between memory and expectation, however, are to be found in the manner of fulfillment. Intentions of the past are necessarily fulfilled by the establishment of nexuses of intuitive reproductions. The reproduction of past events permits, with respect to their validity (in internal consciousness) only the confirmation of the uncertainties of memory and their improvement by being transformed in a reproduction in which each and everything in the components is characterized as reproductive. Here we are concerned with such questions as: Have I really seen or perceived this? Have I really had this appearance with exactly this content? All this must at the same time dovetail into a context of similar intuitions up to the now. Another question, to be sure, is the following: Was the appearing thing real? On the other hand, expectation finds its fulfillment in a perception. It pertains to the essence of the expected that it is an about-to-be-perceived. In view of this, it is evident that if what is expected makes its appearance, i.e., becomes something present, the expectational situation itself has gone by. If the future has become the present, then the present has changed to the relatively past. The situation is the same with regard to environmental intentions. They are also fulfilled through the actuality of an impressional living experience.

Notwithstanding these differences, expectational intuition is something primordial and unique exactly as is intuition of the past.

§ 27. Memory as Consciousness of Having-Been-Perceived

§ 28. Memory and Figurative Consciousness—Memory as Positing Reproduction

§ 29. Memory of the Present

With regard to the sphere of the intuition of external time and objectivity there is yet another type of immediate reproductive intuition of temporal objects to be considered. (Indeed, all our explanations are restricted to the immediate intuition of temporal objects, and the question of mediate or non-intuitive expectations and memories is left alone.)

Whether on the basis of earlier perceptions or on the basis of a description, etc., I can also represent to myself something present as now existing without having it now embodied before me. In the first case, I certainly have a memory, but to what is remembered I grant duration up to the actual now, and for this duration I have no internal, remembered "appearances." The "memory-image" serves me, but I do not posit what is remembered as such, what is objective in the internal memory, in the duration proper to it. What is posited is the enduring as the self-exhibiting in this appearance. We posit the appearing now and the ever-fresh now, etc., but we do not posit it as "past."

We know that the "past" in the case of memory also does not imply that in the present act of remembrance we form an image of the earlier one and others of like construction. Rather, we simply posit the appearing, the intuited, which in conformity with its temporality is naturally intuitable only in modes of temporality. And to what appears thereby we give, in the mode of remembrance and by means of environmental intentions, position with regard to the now of actuality. . . . The environmental intention always produces for the "possible" appearances themselves a halo of intentions. Such also is the case with the intuition of enduring being. I now perceive this being and posit it as having been before without having perceived it before and remembering it now. In addition, I posit it as continuing to be in the future.

§ 30. The Preservation of the Objective
[*gegenständlichen*] Intention in the
Retentional Modification

§ 31. Primal Impressions and Objective
[*objektiver*] Individual Temporal Points

Seemingly, we have been led here to an antinomy. The Object, in sinking back, constantly alters its temporal position, and yet in sinking back is said to preserve its temporal position. In truth, however, the Object of primary remembrance constantly being shoved back does not alter its temporal position but only its interval from the actual now, specifically, because the actual now is accepted as an ever new objective temporal point, whereas the past temporal thing remains what it is. But the question now is, how does it happen that, despite the phenomenon of the continuous alteration of the consciousness of time, there is the consciousness of Objective time, and above all the consciousness of identical temporal positions? Very closely bound to this is the question of the constitution of the Objectivity of individual temporal objects and processes. All Objectification takes place in time-consciousness, and without a clarification of the identity of temporal position no clarification of the identity of an Object in time can be given.

Stated more precisely, the problem is the following: The now-phases of perception constantly undergo a modification. They are not preserved simply as they are. They flow. Constituted therein is what we have referred to as sinking back in time. The tone sounds now and immediately sinks into the past, as the same tone. This affects the tone in each of its phases and, therefore, the whole tone also. Now, through our previous observations, this sinking away is, in some measure, comprehensible. But how is it that despite the sinking away of the tone, we still say, as our analysis of reproductive consciousness has shown, that it has a fixed position in time, that temporal points and temporal positions may be identified in repeated acts? The tone, as well as every temporal point in the unity of the enduring tone, indeed has its absolutely fixed place in "Objective" (or even in immanent) time. Time is motionless and yet it flows. In the flow of time, in the continuous sinking away into the past, there is constituted a non-flowing, absolutely fixed, identical Objective time. This is the problem.

Let us first consider somewhat more closely the state of affairs with regard to the sinking away of the same tone. Why do we say it is the same tone which sinks away? The tone is built up in a temporal flux through its phases. Of every phase, that of an actual now, for example, we know that although subject to the law of constant modification, it still must appear as objectively the same, as the same tonal point, so to speak. This is true because a continuity of apprehension is present here which is governed by the identity of sense and exists in continuous coincidence. This coincidence concerns the extra-temporal matter which even in flux preserves the identity of the objective sense. This holds true for every now-phase. But every new now is precisely that, a new one, and is phenomenologically characterized as such. Even if every tone continues completely unaltered, in such a way that not the least alteration is visible to us—even if every new now, therefore, possesses exactly the same content of apprehension as regards moments of quality, intensity, and the like, and carries exactly the same apprehension—nevertheless, a primordial difference still exists, one which pertains to a new dimension. And this difference is a constant one. From the point of view of phenomenology, only the now-point is characterized as an actual now, that is, as new. The previous temporal point has undergone its modification, the one before that a continuing modification, etc. The continuum of modifications in the content of apprehension and the apprehensions based thereon produce the consciousness of the extension of the tone with the continuous sinking down into the past of what is already extended.

But despite the phenomenon of the continuous alteration of time-consciousness, how does the consciousness of Objective time and, above all, of identical temporal place and temporal extension come about? The answer is that in contrast to the flux resulting from being shoved back in time, i.e., the flux of modifications of consciousness, the Object which appears to be shoved back remains preserved even in absolute identity, that is, the Object together with the positing as a "this" experienced in the now-point. The continuous modification of apprehension in the constant flux does not affect the "as what" of the apprehension, i.e., the sense. It intends no new Object or Object-phase; it yields no new temporal point, but always the same Object with the same temporal points. Every actual now creates a new temporal point because it creates a new Object, or rather a new Object-point which is held fast in the flux of modifications as one and the same individual Object-point. And the constancy in which again and again a new now is constituted shows us that in general it is not a question of "novelty" but of a constant moment of individuation, in which the temporal position has its origin. It is part of the essence of the modifying flux that this temporal position stands forth as identical and necessarily identical. The now as the actual now is the givenness of the actual present of the temporal position. As a phenomenon moves into the past, the now acquires the character of a past now. It remains the same now, however. Only in relation to the momentarily actual and temporally new now does it stand forth as past.

The Objectivation of temporal Objects rests, therefore, on the following moments. The content of sensation which belongs to the different actual now-points of the Objects can qualitatively remain absolutely unaltered, but even with so far-reaching an identity with regard to content it still does not have true identity. The same sensation now and in another now has a difference, in fact, a phenomenological difference which corresponds to the absolute temporal position. This difference is the primal source of the individuality of the "this" and therewith of the absolute temporal position. Every phase of the modification has "in essence" the same qualitative content and the same temporal moment, although modified. Furthermore, each phase has in itself the same temporal moment in such a way that precisely by means of it the subsequent apprehension of identity is made possible: this on the side of sensation, or of the foundation of apprehension. The different moments sustain different parts of the apprehension, of the true Objectivation. One aspect of the Objectivation finds its support purely in the qualitative content of the material of sensation. This yields the temporal matter, e.g., the sound. This matter is held identically in the flux of the modification of the past. A second aspect of the Objectivation arises from the apprehension of the representatives of the temporal positions [Zeitstellenrepräsentanten]. This apprehension is also continuously retained in the flux of modification.

To recapitulate: the tonal point in its absolute individuality is retained in its matter and temporal position, the latter first constituting individuality. To this must be added, finally, apprehension which belongs essentially to the modification and which, while retaining the extended objectivity with its immanent absolute time, allows the continuous shoving-back into the past to appear. . . .

Every perceived time is perceived as a past which terminates in the present, the present being a boundary-point. Every apprehension, no matter how transcendent it otherwise may be, is bound to this regularity. If we perceive a flight of birds, a squadron of cavalry at a gallop, and the like, we find the described distinctions in the underlying basis of sensation—ever new primal sensations, their temporal character, which provides their individuation, being carried with them; and on the other side, we find the same modes in the apprehension. Precisely by this means, the Objective iiself appears, the flight of birds as primal givenness in the now-point, as complete givenness, though in a continuum of the past which terminates in the now, while the continuously preceding in the continuum of the past is moved ever further back. The appearing event always has the identical, absolute temporal value. Since, following the segment of time that has expired, the event is shoved ever further back into the past, it is shoved back with its absolute temporal position and hence with its entire temporal interval into the past, i.e., the same event with the same absolute temporal extensity continually appears (as long as it appears at all) as identically the same. Only the form of its givenness is different. On the other hand, in the living source-point of the now there also wells up ever fresh primal being, in relation to which the distance from the actual now of the temporal points belonging to the event is constantly increased. Accordingly, the appearance of sinking back, of withdrawing, arises.

§ 32. The Part of Reproduction in the Constitution of the One Objective [objektiven] Time

With the preservation of the individuality of the temporal points in their sinking back into the past, we still do not have, however, consciousness of a unitary, homogeneous, Objective time. In the occurrence of this consciousness, reproductive memory (in its intuitive capacity, as in the form of empty intentions) plays an important role. Every temporal point which has been shoved back can, be means of reproductive memory, be made the null-point of an intuition of time and be repeated. The earlier temporal field, in which what is presently shoved back was a now, is reproduced, and the reproduced now is identified with the temporal point still vivid in recent memory. The individual intention is the same. The temporal field that is reproduced extends further than that actually present. If we take a point of the past in this temporal field, the reproduction, by being shoved along with the temporal field in which this point was the now, provides a further regress into the past, and so on. Theoretically, this process is to be thought of as capable of being continued without limit, although in practice actual memory soon breaks down. It is evident that every temporal point has its before and after, and that the points and intervals coming before cannot be compressed in the manner of an approximation to a mathematical limit, as, let us say, the limit of intensity. If there were such a boundary-point, there would correspond to it a now which nothing preceded, and this is obviously impossible. A now is always and essentially the edge-point [*Rand-punkt*] of an interval of time. And it is evident that this entire interval must sink back and therby its entire magnitude, its entire individuality, is preserved. To be sure, phantasy and reproduction do not make possible an extension of the intuition of time in the sense that the extent of the real, given temporal gradations in simultaneous consciousness is increased. With reference to this, one may perhaps ask: How, with this successive stringing together of temporal fields, does the one Objective time with the one fixed order come to be? The answer proffers the continuous shoving along of the temporal fields, which in truth is no mere temporal stringing together of temporal fields. The segments being shoved along are individually identified in connection with the intuitively continuous regress into the past. If, starting from any actual lived and experienced temporal point—i.e., any one which is originarily given in the temporal field of perception or one which reproduced a distant past—we go back into the past, along, so to speak, a well-established chain of Objectivities which are interconnected and always identified, then the question arises: How is the linear order there established? In such an order every temporal interval, no matter which—even the external continuity with the actual temporal field reproduced—must be a part of a unique chain, continuing to the point of the actual now. Even every arbitrarily phantasied time is subject to the requirement that if one is able to think of it as real time (i.e., as the time of any temporal Object) it must subsist as an interval within the one and unique Objective time.

§ 33. Some *A priori* Temporal Laws

Obviously, this *a priori* requirement is grounded in the recognition of the immediately comprehensible and fundamental temporal certainties which become evident on the basis of intuitions of the data of temporal position.

If, to begin with, we compare two primal sensations, or correlatively two primal data, both really appearing in one consciousness as now, then they are distinguished from one another through their matter. They are, however, simultaneous; they have identically the same temporal position; they are both now, and in the same now they necessarily have the same value with regard to temporal position. They have the same form of individuation and both are constituted in impressions which belong to the same impressional level. These data are modified in this identity and always retain it in the modification of the past. A primal datum and a modified datum of like or dissimilar content necessarily have different temporal positions—the same if they arise from the same now-point, different if from the different now-points. The actual now is *a* now and constituted *a* temporal position. No matter how many Objectivities are constituted separately in the now, they all have the same temporal present and retain their simultaneity in flowing off. That the temporal positions have differences, that these are magnitudes, and the like, can here be seen as evident. Also evident are additional truths such as the law of transitivity, namely, the law that if A is earlier than B then B is later than A. It is part of the *a priori* essence of time that the latter is sometimes identified with a continuity of temporal positions, sometimes with the changing Objectivities which fill it; and that the homogeneity of absolute time is necessarily constituted in the flow of the modifications of the past and in the continual welling-forth of a now, of the creative temporal point, of the source-point of temporal positions in general.

Furthermore, it belongs to the *a priori* essence of the state of affairs that sensation, apprehension, position-taking, all share in the *same* temporal flux and that Objectified absolute time is necessarily the same as the time which belongs to sensation and apprehension. Pre-Objectified time, which pertains to sensation, necessarily founds the unique possibility of an Objectivation of temporal positions which corresponds to the modification of the sensation and the degree of this modification. To the Objectified temporal point in which, let us say, a peal of bells begins, corresponds the temporal point of the matching sensation. In the beginning phase, the sensation has the same time, i.e., if subsequently it is made into an object, it necessarily maintains the temporal position which coincides with the corresponding temporal position of the bell-peal. In the same way, the time of the perception and the time of the perceived are necessarily the same. The act of perception sinks back in time in the same way as the perceived in the appearance, and in reflection each phase of the perception must be given identically the same temporal position as the perceived.

27

Minkowski:

THE PRESENCE OF THE PAST

Eugène Minkowski (1885–   ) was one of the first and foremost of French phenomenologists. His prime focus has been on the interrelationships of philosophy and psychopathological studies. His first book, applying temporal notions derived from Bergson to studies of schizophrenia, was published in 1927. The selection is taken from the chapter entitled "The Past," * which is the final chapter of the first half of his *Lived Time* (entitled "Essay on the Temporal Aspect of Life"); the second half of the book carries the themes developed forward into a study of the "Spatiotemporal Structure of Mental Disorders."

If, as we have done for the future,[1] we pose the question, "How do we live the past?" the answer will seem simple. The phenomenon of memory seems to give us this answer directly. But let us not be hasty.

We must set aside the biological notion of memory, which considers it a preservation of material traces capable of influencing the reactions of living matter. As we place ourselves on the phenomenological level, phenomena as they appear in evolved consciousness ought to serve as our point of departure. We will thus give memory itself priority over biological memory, taking care not to consider the former simply a result of the evolution of the latter. Further, we ought to inquire what

---

* Eugène Minkowski, *Lived Time: Phenomenological and Psychopathological Studies,* trans. Nancy Metzel (Evanston: Northwestern University Press, 1970). By permission.

1. The reader will perhaps be surprised to find that I do not deal with the past immediately after the future. But, after all, the past could not have been understood without the phenomenon of death, and that is why it seemed natural for me to speak first of death.

distortion of the original phenomenon was required in order for this transformation of memory into the biological domain to occur. It is clear that mnemonic traces cannot be deduced directly from the data of consciousness related to memory. I had the opportunity to insist on this point in one of my earlier works. When I recognize an object, I deduce from this experience, if indeed there is an actual deduction, that I have seen this object before; I do not deduce that it has left traces in me capable of causing this act of recognition; indeed, I do not see how I could deduce anything else. The reiteration of similar stimuli which is involved in the conception of the biological memory is, after all, only an external projection of our faculty of recognition; and it appears paradoxical, at the least, since it has the result of rendering that faculty itself problematic. According to biological reasoning, because I recognize an object that I am seeing for the second time, it must have left traces at the time of the first perception. We would like to oppose to this the following givens of consciousness: I see it for the second time only because I recognize it; or, perhaps better, to recognize is nothing else but to see for the second time, so that it is completely natural that I recognize what I see again, just as it is completely natural that I find new what I have before my eyes for the first time, that is, that I have the notion of "never or not yet seen" in its presence.

But the common notion of memory also needs to be examined more closely, above all if we consider it as the only means of putting us in touch with the past. From this point of view it very soon appears to be insufficient.

It comes to be associated much too easily, we think, with utility. We find it useful in life, in a more or less explicit way, to be able to remember at a desired time things previously experienced or learned. Memory creates our experiences and determines the degree of our learning. In anticipation of the future we are even forced to adopt a mnemonic attitude in the present and to retain certain things with which we find ourselves in relation. Here Janet's "delayed act" takes place. The failure of memory always lessens our powers in these conditions. It is always tiring to have forgotten a name, a date, or an errand. . . . Forgetting could not exist if memory were absolute—an ideal which seems thinkable in this conception—or if there weren't any memories at all. Nothing is more logical. Forgetting is only the negation of memory. It must necessarily give way to memory and remain subordinate to it, as negation does logically in relation to affirmation. Our research, however, has taught us to be mistrustful of relations that are too logical. Whenever the phenomena of life agree too well with the demands of discursive thought, we should inquire whether, through misapprehension, we have not substituted an aspect more or less distorted by reason for the true nature of these phenomena. Be that as it may, it is useful to remember here that the "photographic" memory seen in certain patients constitutes an impediment and represents more a monstrosity than anything else. . . .

It appears that all memory contains *knowledge*. To recognize someone is the equivalent of *knowing* that we have seen him before, just as to remember an event is equivalent to knowing that this event has taken place in the past. Nothing is altered by the fact that memory can reproduce a past event with its emotional tonality, since it will still be a question, above all, of a knowledge related to this tonality, which is evidently not controverted by the fact that memory itself can be a function of that more or less agreeable or painful tonality. On the whole, here, everything is reduced to a knowledge relative to the past. Our memory seems destined above all to enrich our knowledge in the broadest sense of the word. That is the essential thing. But we would find ourselves thus in the presence of a completely different situation from the one which our studies of the phenomena of the temporal order have made familiar to us. In their irrational character these phenomena were far from being related to knowledge, and it was the phenomenon of death that first brought a kind of knowledge to us. Memory, on the contrary, tells us what we know about the past. Certainly we would be wrong to exclude a priori the thought that the only way of living the past consists in knowing something about it. But in that case the past would be a dead thing, a veritable tomb, scarcely deserving to be known, not to mention the fact that its temporal nature would be annulled by the same stroke. Being the domain of our knowledge, the past would have nothing to do with lived time. As a matter of fact, however, the past does not really appear to us as devoid of life; expressions such as "living in the past" or "reliving the past" are witness to this.

Another difficulty soon arises. In memory two things are noticed. "I remember an event" means: I remember it *and* it has really taken place. But if memory is the only means of putting us in relation with the past, it seems equally as justified to affirm that memory produces the past as it is to say that it reproduces it. In other words, the two affirmations "We remember that which has been" and "That which has been has been only because we remember it" are equally legitimate, and we can see no reason for preferring one over the other. Besides, it is easy to realize that all verification of our memories can only depend on a new act of memory. To cite an example: if I find an object in my drawer that I remember having put there yesterday, what counts is not the finding of the object in the drawer but finding it again, that is, the recognition of it as something that I remember having put there. Any memory can be confirmed only by other memories, a circumstance which only emphasizes the equivalence of the two affirmations of which we have just spoken. As a matter of fact, however, we opt for the first, and without the least hesitation. That memories reproduce the past and do not create it with all that it has in it seems completely evident, and this evidence is irreducible. Whatever errors memory contains, they never make us doubt the reproductive character of our memories. As for reducing this evidence to an agreement of memories, we could not take this seriously; apart from the fact that this concordance can be constituted only with the aid of new memories, that is, that it always rests of the thesis that it is attempting to demonstrate, this opinion leads us directly to the conception that these memories are "mere hallucinations" of memory, a formula which appears as completely unacceptable to us as the one which tends to present our perceptions as mere hallucinations. Both are counter to the immediate givens of consciousness.

In a word, the manner in which we live the past doesn't seem to be reducible to either a memory or a sum of memories. The idea of a primitive intuition of the past, completely independent of concrete memories, which in turn come to furnish this past, begins to come to the surface. A breach appears in the purely rational conception of the phenomena of memory; a way leading toward a phenomenology of the past opens before us.

We will be able to contribute only a few fragments to this phenomenology. . . .

. . . [W]e see that the phenomena of memory are really complicated and become singularly rich. In regard to its positive manifestations, we see already that memory cannot be reduced merely to an image or a representation of a fact. The recall, as Pichon says, can have depth; not only can we represent to ourselves a slice of our past, but we can actually relive it; and sometimes the past penetrates as if by intrusion in all its anguishing power into the present and occupies it entirely; or, as we said earlier, the present, with its fluid and extendible limits, is capable of including within itself an indeterminate slice of the past. But, what is more, all these so-called positive manifestations of memory plunge into the "mass of the forgotten," as Mignard says. Whatever the slice of the recalled past may be, whatever the depth of the recall that characterizes it, there is always a vast but obscure zone around it, from which it emerges and which serves as its support. There is always behind it an inscrutable "before," which softly, without commotion, becomes lost in the infinite. It is this "mass of the forgotten," it is the forgotten, which seems to be the first intuition of the past, to constitute the essential basic material upon which memory comes to embroider the remembrances of isolated events. Forgetting is thus not simply memory failure. It appears to us now in its positive value. This value evidently has nothing to do with the fact that it is useful in everyday life to forget insignificant things, minute details, in order to retain only that which is important. Life itself does not comprehend these subtleties or, more precisely, these coarse distinctions that the "principle of simplicity" or any other principle of the same order imposes on it. There is no such thing as minute detail for it; moreover, life doesn't comprehend the difference between what is negligible and what is important. It has no need to comprehend this difference, for everything that it touches becomes by that very fact of capital importance. It is life that tells us that forgetting has nothing to do with having a good or bad memory but that it marks out this obscure mass behind us which gives us our first intuition of the past. From this point of view, the vision that everything is destined to be forgotten seems much more natural, much more appeasing, than the fact that it can be reproduced again as an isolated event. How, in effect, can we conceive of an isolated event that not only can exist but can even outlive itself in face of becoming?

We have had occasion more than once to insist on the fact that phenomena, according to their positive or negative value, insert themselves in life in an asymmetrical manner. Here we find this asymmetry again.

We are not considering ethical acts, which generate the future. They do not comprehend the past. They disappear, as they came, without leaving a trace except for the luminous ray which they constantly project before us. They have no history. They have nothing precise to tell us about the past. We will have to come down a rung on the ladder. There we find the work. The work is integrated with ambient becoming but always remains, because of its perfectibility, virtually unfinished. Works thus are spread out over the length of a life without ever having a natural end, some of them always attempting to surpass others. They owe to that their vibrant and dynamic character and force us thus to look ahead of ourselves. Sometimes we feel tempted to look behind in order to contemplate what we have done, but there we discover particular relations. To spread out these works before oneself or before others is to see them fixed, shrunken, tarnished, extinguished; it is at the same time a proof of presumption; it is to adopt an attitude which not only is displeasing but which seems contrary to life. In reality, as long as a breath of life is in us, we see past works synthesizing into a compact mass which seems to have only one end: that of making us go further. When we try to stop in order to search the past from this point of view, we feel an irresistible desire germinating in us to go on again, to push forward. The past does not unfold before our eyes in successive stages, each one having its value independent of the other; on the contrary, it turns back upon itself, condenses to the maximum, without for all that losing anything of its force. We have before us a past which is concentrated, gathered together, from which our *élan* surges anew to carry us toward the future. This is the role of this past, and it has no other. In this sense I would almost like to speak of *prospective memory,* in which the past, gathered into a single block, occurs above all in the form of the surpassed or of the "to be surpassed" and not in the current meaning of the word "past."

This characteristic of the surpassed seems to be something primitive and natural. As a result, we find it quite frequently when we are concerned with the past. We do not abandon it easily. Moreover, we have no reason to do so. When we come to line up the facts or events of the past, and when we attempt to establish a connection between them, that is, when we come to make a history, we always order these events almost instinctively from the point of view of the surpassed, especially when it is a question of our collective efforts, our institutions, or our beliefs. There is always a line which progresses and which, ending in the present, seems to have no other meaning than that of stimulating us to take up our *élan* in order to continue this journey forward. Indeed, we are inclined to say that this idea of progression is extracted from the facts themselves, accumulated in the past. In reality, however, nothing becomes a fact in the past, it seems, unless it can be integrated to this past from the point of view of the surpassed. This should not surprise us, since this state of affairs is in perfect accord with the essential dynamism of life and, more especially, with the factor of continuation, or the fabric of life, which emerges quite naturally from the *élan vital*.

This dynamism is much more in accord with the character of the surpassed than with that of the past. We marvel that a past can exist for us, that we can recall a memory, look back, relive past things distinctly, fix precise events in the past, or, to speak more concretely, that we find in life, which is only progression, the time to turn our eyes from that progression and look at what we have left behind us. Why do we do this? Well, someone will say, we detach ourselves from the present in order to plunge ourselves into the past only momentarily. Yet, even as a passing attitude, this retrospection must seem surprising in face of the propulsion of life. Memory being a synthetic whole, prospective memory would seem to be sufficient for life; from this point of view, the physiological aspect of memory, that is, memory as it serves to unite past experiences in the form of habit, which can then automatically determine our behavior in the future so that we are able to avoid dangers and to seek out agreeable situations—this aspect, I say, seems in certain respects much closer to the vital dynamism than the faculty of isolating and consciously fixing a precise event in the past. Shouldn't we be individuals capable of storing up the past without knowing it and of making our *élan* toward the future benefit from this instinctively?

Other phenomena in life give us the answer to this question. Evil, as we have said, is found on the same level as the work. More "material," more tangible than the good, it is united to becoming and leaves traces, bits of evidence, there. It aims at an isolated fact or action, much more so than does the good, in which there is always something impersonal. But we cannot apply the principle of prospective memory to these actions. That is evident. But evil cannot disappear without leaving traces in the past. That would deprive it of its nature. Here it is indispensable that something remain. The phenomenon of remorse occurs. Remorse concerns the past and can only concern the past. But further, it cuts out and isolates a fact in that past, a precise event; it fixes it and makes it survive. One might even say that this is the most natural way of isolating a precise fact in the past, which is primitively only the shadowy "mass of the forgotten." One could also say that this is the first step toward the unfolding of the synthetic form of the past, which the prospective memory of our *élan* includes in itself while giving our past the character of the surpassed.

Certainly we often utilize remorse for purposes other than those for which it was intended. We use it to "save face," to put on airs of virtue, and so forth. However, that is not true remorse. In spite of all counterfeit, we know what sincere remorse ought to be, and we take account of the role that it is called to play in life. From our point of view, we could say that it is the *primary memory*. In any case, remorse shows us, with all the desired evidence, not only the possibility but the necessity of conscious memory.

Also, remorse contains more than one factor of the past. We also find in it—and this justifies the attention that we have given it here—prospection. It ought to open the way again toward the future. It is not a question of marking time but of a rocky path which obstructs the horizon which is nevertheless destined to open up into the great road of life and to free the way again toward the blossoming of the personality and toward the search for ethical action. But there is retrospection in it—and this is the point that we have tried to show. It isolates a fact of the past, revives it, and forms the primary memory, as we said before.

Beneath remorse comes regret. Used more commonly, and applicable to events of a lesser seriousness, it is more commonplace than remorse. It is to remorse as fault or error is to sin. Moreover, regret, through a seemingly completely natural extension, concerns, above all, events that happen outside our direct intervention. We can thus think with regret of a past time. Nevertheless, regret shares in the essential characteristics of remorse. Like it, it has to do with a precise fact of the past and, at the same time, projects a ray toward the future. Certainly we can regret only what has happened; but all regret, if it is not to become completely sterile, contains an "It would have been better if," which is either to influence our future behavior or produces in us the hope that things will be different another time.

It is only below regret that we find the "remembrance," in the usual, or, if you prefer, in the "scientific" sense of the word, that is, when it is considered as a simple reminiscence or a simple reproduction of some fact of the past. Stripped of all meaningful content, completely detached from the future, this assumed base of memory can appear only as a paradoxical and incomprehensible phenomenon in relation to life's incessant journey forward. As such, it always sinks into the "mass of the forgotten."

The past exists for our personal *élan* only in a synthetic and global form, that is, in the form of the surpassed, as we said before. This is the only form which is compatible with the essentially dynamic nature of our *élan*. The precise and isolated fact, *cut out,* as it were, from the whole of becoming, does not free itself from this global form until evil enters life. Then this form of the past becomes even indispensable, since it is only through remorse that we conceive not only of the possibility but of the necessity of the conscious memory in life. The precise fact, confronted with the primitive dynamism of life, has a static character about it. In this sense we can say that the memory of evil is more static than the memory of the good; it is fixed longer on a fact of the past and obliges us to remain with it longer. This static nature, moreover, is completely relative, since phenomena such as remorse or regret remain at the same time oriented toward the future and pursue, because of their very nature, only one end, that of integrating the thing done, despite its negative meaning, into the forward journey of life toward the good.

The asymmetry between the positive and negative values is revealed equally in their relation to the past. We do not have a phenomenon equivalent to remorse for positive actions. We could not have one, as that would be contrary to the very meaning of life. To remain in admiration of a past action is to destroy it. It is proof of a narrow-mindedness, of pride, and of narcisism. As a result, if we consider the most elevated phenomenon of life, freedom, we easily see that this freedom is never concerned with the past. It is never born from a retrospective examination of the facts, but it emerges in us, in all its power, "once the thing is done," leaving it behind as an insignificant detail compared to the infinite horizon in the future that it opens before us. I never feel free or great in what I have done in the past. On the contrary, in the past, freedom seems always unreliable; for in its very nature it can concern only the future. I have not been free in what I have done, but what I have just done can liberate my freedom for the future. And it is remorse that then opens the door through which the past penetrates life.

In order to emphasize again the asymmetry between the good and the bad, it is useful perhaps to remember here—although this doesn't touch our subject directly—that rancor and resentment are much more persistent than the feeling of gratitude. We remember once and for all the evil that has been done to us. We always demand new proofs of the good. This is true because it is much more difficult to forget and to pardon than to remember the vexations that we suffered.

A certain structural analogy is established, however, between the past and future. As for the future, we had activity, desire, and ethical action or, again, anticipation, hope, and prayer, which were arranged one above the other; we have, in the same way, the impression of having before us three levels of the past: remorse, regret, and ordinary remembrance. The analogy is far from being complete. For the future, it is the superior levels which allow our gaze to embrace the future in all its power. For the past, on the contrary, it is only to the extent that we distance ourselves from the vital characteristic, that is, from the retrospective characteristic of the phenomena studied, finally to end with remembrance, that we come nearer to the synthetic form of the past. This is not the only difference; remembrance always emerges from the forgotten, a characteristic phenomenon of the past that has no equivalent in the context of the future. . . .

It is remorse that renders retrospection plausible to us in life. Second, retrospection, forgetful of its origins and also of the propulsion which it engenders primitively, seems to free itself and become an autonomous attitude.

It leads us toward the past of the forgotten, or if you like, toward the "past." When we look behind us, what we discover, first of all, is the general form of the past in its particular tonality; it is the realm of shadows, oblivion, and silence. We lost our bearings there, for it is dark. There is no clarity there, no horizon, but at most a perspective which becomes lost in the shadows of the infinite. It even seems that these shadows grow only in proportion as we search to pierce them; like a light fog at first, they change at last into an impenetrable night. Our gaze is buried as if in a solid mass but discovers nothing there. It is an obscure perspective without horizon, without limits, since our gaze, being quite unconcerned with our first memory and the fact of our birth (which we know only as a biological fact, admitted by analogy to other living beings), flees, without confronting any obstacle, toward the immensity of the infinite. In the absence of any perceptible transition, the individual past merges with the past in general.

Such is the general foundation that retrospection reveals to us. It is, as I have said, the *past of the forgotten.*

The forgotten here is not a memory failure. It has a positive meaning. The guiding principle of the past that retrospection reveals is, above all, the principle of the forgotten. Everything in the past submits to the wear of time; all is fatally destined to be forgotten. We have reconstructed the past, we have forged hypotheses spreading out over millions of years behind; but that does not unlock the primitive intuition of the past, an intuition which tells us that, when we look back, we see things, whatever their importance may be, climbing slowly toward the eternal silence of oblivion. Moreover, this accounts for the fact that the past is not nothingness or some kind of surrogate of a spatial order. There is movement, there is dynamism, there is time in it—to the extent that everything is destined to oblivion. Time thus safeguards its nature. It ends by taking back what it has given, it submerges fatally that which has been able to float on the surface for an instant—and this instant can be centuries. . . .

Isn't it completely natural that, from the moment that there is past in becoming, we are in a position to see it, just as we see the present or the future? There would have to be some premise such as, for example, that nothing outside the "now" can be given to us in an immediate manner, in order that the understanding of the "spection" of the past could become problematic. It contains no such premise in itself.

We see thus that the past has a completely different organization from the present and the future. We are trying to specify the essential characteristics of a particular manner of living time. Consequently, the past cannot be reduced to that part of time which precedes the present, as is customarily done. This is perhaps one of the most convenient ways of giving the past a more concrete and rational expression. But it is *only* that; just as the usual conception of forgetfulness (that is, as the forgetting of a name or a date) is only a particular case, and one of the least important aspects, of forgetting in general.

Thus, questions such as: "Is the present destined to become entirely past?" "Is there past in the present and in the future?," "Is there anything else in the past besides the past of history?," contradictory as they might seem, ought not to be immediately set aside.

The passage of the past into the present does not have a linear character. Try as I may to represent facts or events to myself and to establish a connection between them, either in the form of causality or in the more vibrant form of evolution or progress, and try as I may to breathe new life into these facts and to make them live again before me, it remains that only isolated facts will be related in this way, and the present, understood as the lived present, cannot thereby be deduced in any manner. For this present is not a part of time, ridiculously conceived to be between the past and the future; it is a completely different manner of living time from that which characterizes the past. It does not set apart or isolate, but integrates, unfolds, and radiates, in opening the horizon of the future before us. And these are perspectives that the past does not have at all. The present has one more dimension than the past. Present and past are incommensurable because of this. Thus, when I pass from the past to the present, I feel a radical change of attitude happening in me. Likewise, in the present, I am quite aware that this present is not destined to become entirely past—and not because there are facts of different importance in it, not because there are trifling details which don't deserve to be retained or because, when all is said and done, everything is destined to become forgotten, but because it is completely different from the past.

Compared to the lived present, the past always seems to us to be diminished, as amputated from something, because it is characterized by being cut up and carved out. Thus, however extended and detailed our historical knowledge may be, in looking back to gain a glimpse of the whole, we cannot repress a feeling of deception. How could life be only the continuation of this succession, at base contingent and devoid of meaning, sentenced to repeat to infinity the events recorded by history? . . . Not that we doubt the utility of history or that it will be able someday to give us a still more complete picture—one richer in detail, more gripping, more vibrant than we have today—but because, even if it should succeed in making us *re*live the *past,* it could never make the past *live* again, because the content of the past, being necessarily modeled on the essential characteristics of what is past, no longer lends itself to the constitution of a present. It is as if someone said that the present is reduced to the group of facts that I find recorded in the morning newspaper or to a range, even an infinite range, of facts of that order.

This last remark helps me to surmount a difficulty. If the present and the past are actually different from each other as far as their contexture is concerned, how do we come to establish an intelligible connection between them? I believe that that doesn't come from the fact that we see the present become past but because, on the contrary, the past eats into the present, just as it does with regard to the future. In other words, if in general we come to unite the three forms of time, it is because we introduce the past into the present and into the future. We do it precisely in distinguishing isolated facts or events in them. When I anticipate what I will do tomorrow, at base there is nothing of the lived future in this anticipation. There is only the past, or, more exactly, there is only what will be past after tomorrow. The lived future only begins further on. It is the same when we see the present broken down into "various facts," whatever their importance may be. This operation is only a reflection of the knowledge we have of facts in the past; or, if you prefer, the various facts in the present are only what they are because they are destined to be inscribed in the past. This does not mean that we necessarily preserve a remembrance of it, since the preservation of such a remembrance will depend on our faculty of memory and on the interest that the fact in question can have for us in daily life. It is still the case that an isolated fact is detached from the present only because the past, with its particular form, exercises its authority over it. The set of these facts is shown to be incapable of exhausting the present because they are precisely of the past and not of the present, or more exactly, of the past in the present and not the inverse. Even when we are conscious of participating in a great historical event, as was the case during the war, we feel clearly that this event, as it will be inscribed in history, will be only a part of what it has been for us in the present—and that, again, not because history will never be able to reproduce the war in all its details but because there is always something in the present which, without being forgotten, is nevertheless not inscribed at all in the past. When we participate in a historical event in the present, we know perfectly well that what is "historic" in it is only a part, only an aspect of what we do and of what we live.

Science with its forecasts is thus only an attempt on a large scale to build the future on the model of the past, and religious feeling some-times tries to project the ray of the future into the past; both set time—and especially the lived future—aside and remain beneath it.

With poignant acuity a new element penetrates the past: the past not only has been but *is no longer.*

Remembrance is born. Negation has penetrated time. From this moment, time rationalizes itself.

According to reason, certainly, the two judgments "That which has been is no longer" and "That which will be is not yet" are both equally negative by comparison to the present. But let us lean forward and listen to life itself. Haven't we yet learned to do this in the course of this long effort? Certainly. And it is thus that we discover nuances in negation, regardless of what logic says. The negation of the past seems much more categorical than the negation of the future. The latter is much less affirmative, as if, feeling out of place here, it remained here, against its will, only to the extent that the future borrows its charac-teristics from the past. If, in respect to the past, we can actually speak without inconvenience or constraint of an "It is no longer," it is not the same with the future. We have for the future—and we have said this more than once already—no phenomenon analogous to remem-brance; we are not prophets, and we don't want to be, and the future is not primitively "That which will be and that which is not yet" but "That which can be and above all that which ought to be." As such it springs entirely from the present, which it animates with its vivifying rays.

It is only from the moment when negation penetrates the past that the rational problem of memory arises. At bottom, it is the remem-brance which furnishes all the elements of memory. Remembrance puts us, in a direct way, in relation with a *mediate* past. There is no remembrance for the present which just disappeared, no remembrance for a past losing itself in the infinite. Remembrance always bears on an event which happened "some time ago"; and however short this "some time ago" may be, it nonetheless constitutes a lapse of time from the qualitative point of view, an open interval during which the past event has not been present to consciousness in any way. Remembrance thus juxtaposes, in an immediate way and by itself, two points of time, in the form of the present and of a point of the past separated from each other by an empty interval with respect to the event in question. In other words, in appearing to consciousness after an indeterminate interval, it tells us not only that an event of the past has been but, further, that it is no longer. And reason, which, like nature, has a horror of a void, is terrified in the presence of all of these givens full of contradiction and puts everything to work to suppress them. It appeals to mnemonic traces and other conceptions of the same order. We do not follow in this path. It is sufficient for now to have elucidated the principal characteristics of lived time.

28.

Heidegger:

THE PRIORITY OF THE FUTURE

In his thirty-seventh year, Martin Heidegger (1889–) completed the writing of his *magnum opus,* the published portions of *Being and Time.* It was intended, as he said, to show that whenever we seek to understand the meaning of Being, of what it means to be, we necessarily do so from within the perspective of time, that our conception of time and "the ordinary understanding of it, have sprung from temporality" and that it is in terms of temporality that human experience is structured. The selection,* drawn from Division Two ("Dasein and Temporality"), consists of extracts from sections 65, 66, 67, 68, 69, and all of 81. Unfortunately—but perhaps necessarily for good philosophic reasons—it is now almost a ritual to introduce Heidegger with an apology for his vocabulary. The reader who is not already familiar with it is invited to read, not an apology, but an apologia and explanation, beginning on p. 455 above. (N.B. Translator's footnotes are unidentified; Heidegger's own footnotes are signed "M.H." and those of the editor "C.M.S.")

¶ 65. Temporality as the Ontological Meaning of Care

In characterizing the 'connection' between care and Selfhood, our aim was not only to clarify the special problem of "I"-hood, but also to help in the final preparation for getting into our grasp phenomenally the totality of Dasein's structural whole. We need the *unwavering discipline* of the existential way of putting the question, if, for our ontological point of view, Dasein's kind of Being is not to be finally

---

* From *Being and Time* by Martin Heidegger, trans. John Macquarrie and Edward Robinson. Copyright (C) 1962 by SCM Press Ltd. Reprinted by permission of Harper & Row, Publishers, Inc., and Basil Blackwell & Mott, Ltd.

perverted into a mode of presence-at-hand, even one which is wholly undifferentiated. Dasein becomes 'essentially' Dasein in that authentic existence which constitutes itself as anticipatory resoluteness. Such resoluteness, as a mode of the authenticity of care, contains Dasein's primordial Self-constancy and totality. We must take an undistracted look at these and understand them existentially if we are to lay bare the ontological meaning of Dasein's Being. . . .

Dasein is either authentically or inauthentically disclosed to itself as regards its existence. In existing, Dasein understands itself, and in such a way, indeed, that this understanding does not merely get something in its grasp, but makes up the existentiell Being of its factical potentiality-for-Being. The Being which is disclosed is that of an entity for which this Being is an issue. The meaning of this Being—that is, of care—is what makes care possible in its Constitution; and it is what makes up primordially the Being of this potentiality-for-Being. The meaning of Dasein's Being is not something free-floating which is other than and 'outside of' itself, but is the self-understanding Dasein itself. What makes possible the Being of Dasein, and therewith its factical existence?

That which was projected in the primordial existential projection of existence has revealed itself as anticipatory resoluteness. What makes this authentic Being-a-whole of Dasein possible with regard to the unity of its articulated structural whole? Anticipatory resoluteness, when taken formally and existentially, without our constantly designating its full structural content, is *Being towards* one's ownmost, distinctive potentiality-for-Being. This sort of thing is possible only in that Dasein *can, indeed,* come towards itself in its ownmost possibility, and that it can put up with this possibility as a possibility in thus letting itself come towards itself—in other words, that it exists. This letting-itself-*come-towards*-itself in that distinctive possibility which it puts up with, is the primordial phenomenon of the *future as coming towards*. If either authentic or inauthentic *Being-towards-death* belongs to Dasein's Being, then such Being-towards-death is possible only as something *futural* [als *zukünftiges*], in the sense which we have now indicated, and which we have still to define more closely. By the term 'futural', we do not here have in view a "now" which has *not yet* become 'actual' and which sometime *will be* for the first time. We have in view the coming [Kunft] in which Dasein, in its ownmost potentiality-for-Being, comes towards itself. Anticipation makes Dasein *authentically* futural, and in such a way that the anticipation itself is possible only in so far as Dasein, *as being*, is always coming towards itself—that is to say, in so far as it is futural in its Being in general.

Anticipatory resoluteness understands Dasein in its own essential Being-guilty. This understanding means that in existing one takes over Being-guilty; it means *being* the thrown basis of nullity. But taking over throwness signifies *being* Dasein authentically *as it already was*. Taking over throwness, however, is possible only in such a way that the futural Dasein can *be* its ownmost 'as-it-already-was'—that is to say, its 'been' [sein "Gewesen"]. Only in so far as Dasein *is* as an "I-*am*-as-having-been", can Dasein come towards itself futurally in such a way that it comes *back*. As authentically futural, Dasein *is* authentically as *"having been"*. Anticipation of one's uttermost and ownmost possibility is coming back understandingly to one's ownmost "been". Only so far as it is futural can Dasein *be* authentically as having been. The character of "having been" arises, in a certain way, from the future.[1]

Anticipatory resoluteness discloses the current Situation of the "there" in such a way that existence, in taking action, is circumspectively concerned with what is factically ready-to-hand enviromentally. Resolute Being-alongside what is ready-to-hand in the Situation—that is to say, taking action in such a way as to let one encounter what *has presence* environmentally—is possible only by *making* such an entity *present*. Only as the *Present* [*Gegenwart*] in the sense of making present, can resoluteness be what it is: namely, letting itself be encountered undisguisedly by that which it seizes upon in taking action.

Coming back to itself futurally, resoluteness brings itself into the Situation by making present. The character of "having been" arises from the future, and in such a way that the future which "has been" (or better, which "is in the process of having been") releases from

1. 'Die Gewesenheit entspringt in gewisser Weise der Zukunft.' Here 'The character of having been' represents 'Die Gewesenheit' (literally, 'beenhood'). Heidegger distinguishes this sharply from 'die Vergangenheit' ('pastness'). We shall frequently translate 'Gewesenheit' simply as 'having been'.

itself the Present. This phenomenon has the unity of a future which makes present in the process of having been; we designate it as *"temporality"*. Only in so far as Dasein has the definite character of temporality, is the authentic potentiality-for-Being-a-whole of anticipatory resoluteness, as we have described it, made possible for Dasein itself. *Temporality reveals itself as the meaning of authentic care.*

The phenomenal content of this meaning, drawn from the state of Being of anticipatory resoluteness, fills in the signification of the term "temporality". In our terminological use of this expression, we must hold ourselves aloof from all those significations of 'future', 'past', and 'Present' which thrust themselves upon us from the ordinary conception of time. This holds also for conceptions of a 'time' which is 'subjective' or 'Objective', 'immanent' or 'transcendent'. Inasmuch as Dasein understands itself in a way which, proximally and for the most part, is inauthentic, we may suppose that 'time' as ordinarily understood does indeed represent a genuine phenomenon, but one which is derivative [ein abkünftiges]. It arises from inauthentic temporality, which has a source of its own. The conceptions of 'future', 'past' and 'Present' have first arisen in terms of the inauthentic way of understanding time. In terminologically delimiting the primordial and authentic phenomena which correspond to these, we have to struggle against the same difficulty which keeps all ontological terminology in its grip. When violences are done in this field of investigation, they are not arbitrary but have a necessity grounded in the facts. If, however, we are to point out without gaps in the argument, how inauthentic temporality has its source in temporality which is primordial and authentic, the primordial phenomenon, which we have described only in a rough and ready fashion, must first be worked out correctly.

If resoluteness makes up the mode of authentic care, and if this itself is possible only through temporality, then the phenomenon at which we have arrived by taking a look at resoluteness, must present us with only a modality of temporality, by which, after all, care as such is made possible. Dasein's totality of Being as care means: ahead-of-itself-already-being-in (a world) as Being-alongside (entities encountered within-the-world). When we first fixed upon this articulated structure, we suggested that with regard to this articulation the ontological question must be pursued still further back until the unity of the totality of this structural manifoldness has been laid bare. *The primordial unity of the structure of care lies in temporality.*

The "ahead-of-itself" is grounded in the future. In the "Being-already-in . . .", the character of "having been" is made known. "Being-alongside . . ." becomes possible in making present. While the "ahead" includes the notion of a "before", neither the 'before' in the 'ahead' nor the 'already' is to be taken in terms of the way time is ordinarily understood; this has been automatically ruled out by what has been said above. With this 'before' we do not have in mind 'in advance of something' [das "Vorher"] in the sense of 'not yet now—but later'; the 'already' is just as far from signifying 'no longer now—but earlier'. If the expressions 'before' and 'already' were to have a time-oriented [zeithafte] signification such as *this* (and they can have this signification too), then to say that care has temporality would be to say that it is something which is 'earlier' and 'later', 'not yet' and 'no longer'. Care would then be conceived as an entity which occurs and runs it course 'in time'. The *Being* of an entity having the character of Dasein would become something *present-at-hand*. If this sort of thing is impossible, then any time-oriented signification which the expressions we have mentioned may have, must be different from this. The 'before' and the 'ahead' indicate the future as of a sort which would make it possible for Dasein to be such that its potentiality-for-Being is an issue. Self-projection upon the 'for-the-sake-of-one-self' is grounded in the future and is an essential characteristic of *existentiality. The primary meaning of existentiality is the future.*

Likewise, with the 'already' we have in view the existential temporal meaning of the Being of that entity which, in so far as it *is*, is already something that has been thrown. Only because care is based on the character of "having been", can Dasein exist as the thrown entity which it is. 'As long as' Dasein factically exists, it is never past [vergangen], but it always is indeed as already having *been*, in the sense of the "I *am*-as-having-been". And only as long as Dasein is, *can* it *be* as having been. On the other hand, we call an entity "past", when it is no longer present-at-hand. Therefore Dasein, in existing, can never establish itself as a fact which is present-at-hand, arising and passing away 'in the course of time', with a bit of it past already.

Dasein never 'finds itself' except as a thrown Fact. In the *state-of-mind in which it finds itself,* Dasein is assailed by itself as the entity which it still is and already was—that is to say, which it constantly *is* as having been. The primary existential meaning of facticity lies in the character of "having been". In our formulation of the structure of care, the temporal meaning of existentiality and facticity is indicated by the expressions 'before' and 'already'.

On the other hand, we lack such an indication for the third item which is constitutive for care—the Being-alongside which falls. This should not signify that falling is not also grounded in temporality; it should instead give us a hint that *making-present,* as the *primary* basis for *falling* into the ready-to-hand and present-at-hand with which we concern ourselves, remains *included* in the future and in having been, and is included in these in the mode of primordial temporality. When resolute, Dasein has brought itself back from falling, and has done so precisely in order to be more authentically 'there' in the 'moment of *vision'* as regards the Situation which has been disclosed.

Temporality makes possible the unity of existence, facticity, and falling, and in this way constitutes primordially the totality of the structure of care. The items of care have not been pieced together cumulatively any more than temporality itself has been put together 'in the course of time' ["mit der Zeit"] out of the future, the having been, and the Present. Temporality 'is' not an *entity* at all. It is not, but it *temporalizes* itself. Nevertheless, we cannot avoid saying, 'Temporality "is" . . . the meaning of care', 'Temporality "is" . . . defined in such and such a way'; the reason for this can be made intelligible only when we have clarified the idea of Being and that of the 'is' in general. Temporality temporalizes, and indeed it temporalizes possible ways of itself. These make possible the multiplicity of Dasein's modes of Being, and especially the basic possibility of authentic or inauthentic existence.

The future, the character of having been, and the Present, show the phenomenal characteristics of the 'towards-oneself', the 'back-to', and the 'letting-oneself-be-encountered-*by*'. The phenomena of the "towards . . .", the "to . . .", and the "alongside . . .", make temporality manifest as the ἐκστατικόν pure and simple. *Temporality is the primordial 'out-side-of-itself' in and for itself.* We therefore call the phenomena of the future the character of having been, and the Present, the *"ecstases"* of temporality.[2] Temporality is not, prior to this, an entity which first emerges from *itself;* its essence is a process of temporalizing in the unity of the ecstases. What is characteristic of the 'time' which is accessible to the ordinary understanding, consists, among other things, precisely in the fact that it is a pure sequence of "nows", without beginning and without end, in which the ecstatical character of primordial temporality has been levelled off. But this very levelling off, in accordance with its existential meaning, is grounded in the possibility of a definite kind of temporalizing, in conformity with which temporality temporalizes as inauthentic the kind of 'time' we have just mentioned. If, therefore, we demonstrate that the 'time' which is accessible to Dasein's common sense is *not* primordial, but arises rather from authentic temporality, then, in accordance with the principle, *"a potiori fit denominatio"*, we are justified in designating as *"primordial time"* the *temporality* which we have now laid bare.

In enumerating the ecstases, we have always mentioned the future first. We have done this to indicate that the future has a priority in the ecstatical unity of primordial and authentic temporality. This is so, even though temporality does not first arise through a cumulative sequence of the ecstases, but in each case temporalizes itself in their equiprimordiality. But within this equiprimordiality, the modes of temporalizing are different. The difference lies in the fact that the nature of the temporalizing can be determined primarily in terms of the different ecstases. Primordial and authentic temporality temporalizes itself in terms of the authentic future and in such a way that in having been futurally, it first of all awakens the Present. *The primary phenomenon of primordial and authentic temporality is the future.* The priority of the future will vary according to the ways in which the

---

2. The root-meaning of the word 'ecstasis' (Greek, ἔκστασις; German, 'Ekstase') is 'standing outside'. Used generally in Greek for the 'removal' or 'displacement' of something, it came to be applied to states-of-mind which we would now call 'ecstatic'. Heidegger usually keeps the basic root-meaning in mind, but he also is keenly aware of its close connection with the root-meaning of the word 'existence'.

temporalizing of inauthentic temporality itself is modified, but it will still come to the fore even in the derivative kind of 'time'. . . .

In our thesis that temporality is primordially finite, we are not disputing that 'time goes on'; we are simply holding fast to the phenomenal character of primordial temporality—a character which shows itself in what is projected in Dasein's primordial existential projecting.

The temptation to overlook the finitude of the primordial and authentic future and therfore the finitude of temporality, or alternatively, to hold *'a priori'* that such finitude is impossible, arises from the way in which the ordinary understanding of time is constantly thrusting itself to the fore. If the ordinary understanding is right in knowing a time which is endless, and in knowing only this, it has not yet been demonstrated that it also understands this time and its 'infinity'. What does it mean to say, 'Time goes on' or 'Time keep passing away?' What is the signification of 'in time' in general, and of the expressions 'in the future' and 'out of the future' in particular? In what sense is 'time' endless? Such points need to be cleared up, if the ordinary objections to the finitude of primordial time are not to remain groundless. But we can clear them up effectively only if we have obtained an appropriate way of formulating the question as regards finitude and in-finitude. Such a formulation, however, arises only if we view the primordial phenomenon of time understandingly. The problem is not one of *how* the *'derived'* [*"abgeleitete"*] infinite time, 'in which the ready-to-hand arises and passes away, becomes *primordial* finite temporality; the problem is rather that of how *in*authentic temporality arises out of finite authentic temporality, and how inauthentic temporality, *as in*authentic, temporalizes an *in*-finite time out of the finite. Only because primordial time is *finite* can the 'derived' time temporalize itself as *infinite*. In the order in which we get things into our grasp through the understanding, the finitude of time does not become fully visible until we have exhibited 'endless time' so that these may be contrasted.

Our analysis of primordial temporality up to this point may be summarized in the following theses. Time is primordial as the temporalizing of temporality, and as such it makes possible the Constitution of the structure of care. Temporality is essentially ecstatical. Temporality temporalizes itself primordially out of the future. Primordial time is finite.

However, the Interpretation of care as temporality cannot remain restricted to the narrow basis obtained so far, even if it has taken us the first steps along our way in viewing Dasein's primordial and authentic Being-a-whole. The thesis that the meaning of Dasein is temporality must be confirmed in the contrete content of this entity's basic state, as it has been set forth.

¶ 66. Dasein's Temporality and the Tasks Arising Therefrom of Repeating the Existential Analysis in a more Primordial Manner

... The ontological structure of that entity which, in each case, I *myself* am, centres in the Self-subsistence [Selbständigkeit] of existence. Because the Self cannot be conceived either as substance or as subject but is grounded in existence, our analysis of the inauthentic Self, the "they", has been left wholly in tow of the preparatory Interpretation of Dasein. Now that Selfhood has been *explicitly* taken back into the structure of care, and therefore of temporality, the temporal Interpretation of Self-constancy and non-Self-constancy acquires an importance of its own. This Interpretation needs to be carried through separately and thematically. However, it not only gives us the right kind of insurance against the paralogisms and against ontologically inappropriate questions about the Being of the "I" in general, but it provides at the same time, in accordance with its central function, a more primordial insight into the *temporalization-structure* of temporality, which reveals itself as the historicality of Dasein. The proposition, "Dasein is historical", is confirmed as a fundamental existential ontological assertion. This assertion is far removed from the mere ontical establishment of the fact that Dasein occurs in a 'world-history'. But the historicality of Dasein is the basis for a possible kind of historiological understanding which in turn carries with it the possibility of getting a special grasp of the development of historiology as a science.

By Interpreting everydayness and historicality temporally we shall get a steady enough view of primordial time to expose it as the condition which makes the everyday experience of time both possible and necessary. As an entity for which its Being is an issue, Dasein *utilizes itself* primarily *for itself* [*verwendet sich ... für sich selbst*], whether it does so explicitly or not. Proximally and for the most part, care is circumspective concern. In utilizing itself for the sake of itself, Dasein 'uses itself up'. In using itself up, Dasein uses itself—that is to say, its time. In using time, Dasein reckons with it. Time is first discovered in the concern which reckons circumspectively, and this concern leads to the development of a time-reckoning. Reckoning with time is constitutive for Being-in-the-world. Concernful circumspective discovering, in reckoning with its time, permits those things which we have discovered, and which are ready-to-hand or present-at-hand, to be encountered in time. Thus entities within-the-world become accessible as 'being in time'. We call the temporal attribute of entities within-the-world *"within-time-ness"* [die *Innerzeitkeit*]. The kind of 'time' which is first found ontically in within-time-ness, becomes the basis on which the ordinary traditional conception of time takes form. But time, as within-time-ness, arises from an essential kind of temporalizing of primordial temporality. The fact that this is its source, tells us that the time 'in which' what is present-at-hand arises and passes away, is a genuine phenomenon of time; it is not an externalization of a 'qualitative time' into space, as Bergson's Interpretation of time—which is ontologically quite indefinite and inadequate—would have us believe.

In working out the temporality of Dasein as everydayness, historicality, and within-time-ness, we shall be getting for the first time a relentless insight into the *complications* of a primordial ontology of Dasein. . . .

¶ 67. The Basic Content of Dasein's Existential Constitution, and a Preliminary Sketch of the Temporal Interpretation of it

Our preparatory analysis has made accessible a multiplicity of phenomena; and no matter how much we may concentrate on the foundational structural totality of care, these must not be allowed to vanish from our phenomenological purview. Far from excluding such a multiplicity, the *primordial* totality of Dasein's constitution *as articulated* demands it. The primordiality of a state of Being does not coincide with the simplicity and uniqueness of an ultimate structural element. The ontological source of Dasein's Being is not 'inferior' to what springs from it, but towers above it in power from the outset; in the field of ontology, any 'springing-from' is degeneration. If we penetrate to the 'source' ontologically, we do not come to things which are ontically obvious for the 'common understanding'; but the questionable character of everything obvious opens up for us. . . . Being-in-the-world was first characterized with regard to the phenomenon of the world. And in our explication this was done by characterizing ontico-ontologically what is ready-to-hand and present-at-hand *'in'* the environment, and then bringing within-the-world-ness into relief, so that by this the phenomenon of worldhood in general could be made visible. But understanding belongs essentially to disclosedness; and the structure of worldhood, significance, turned out to be bound up with that upon which understanding projects itself—namely that potentiality-for-Being *for the sake of which* Dasein exists.

The temporal Interpretation of everyday Dasein must start with those structures in which disclosedness constitutes itself: understanding, state-of-mind, falling, and discourse. The modes in which temporality temporalizes are to be laid bare with regard to these phenomena, and will give us a basis for defining the temporality of Being-in-the-world. This leads us back to the phenomenon of the world, and permits us to delimit the specifically temporal problematic of worldhood. This must be confirmed by characterizing that kind of Being-in-the-world which in an everyday manner is closest to us—circumspective, falling concern. The temporality of this concern makes it possible for circumspection to be modified into a perceiving which looks at things, and the theoretical cognition which is grounded in such perceiving. The temporality of Being-in-the-world thus emerges, and it turns out, at the same time, to be the foundation for that spatiality which is specific for Dasein. We must also slow the temporal Constitution of deseverance and directionality. Taken as a whole, these analyses will reveal a possibility for the temporalizing of temporality in which Dasein's inauthenticity is ontologically grounded; and they will lead us face to face with the question of how the temporal character of everydayness—the temporal meaning of the phrase 'proximally and for the most part', which we have been using constantly hitherto—is to be understood. By fixing upon this problem we shall have made it plain that the clarification of this phenomenon which we have so far attained is insufficient, and we shall have shown the extent of this insufficiency.

The present chapter is thus divided up as follows: the temporality of disclosedness in general (Section 68); the temporality of being-in-the-world and the problem of transcendence (Section 69); the temporality of the spatiality characteristic of Dasein (Section 70); the temporal meaning of Dasein's everydayness (Section 71).

¶ 68. The Temporality of Disclosedness in General

Resoluteness, which we have characterized with regard to its temporal meaning, represents an authentic disclosedness of Dasein—a disclosedness which constitutes an entity of such a kind that in existing, it can be its very 'there'. Care has been characterized with regard to its temporal meaning, but only in its basic features. To exhibit its concrete temporal Constitution, means to give a temporal Interpretation of the items of its structure, taking them each singly: understanding, state-of-mind, falling, and discourse. Every understanding has its mood. Every state-of-mind is one in which one understands. The understanding which one has in such a state-of-mind has the character of falling. The understanding which has its mood attuned in falling, Articulates itself with relation to its intelligibility in discourse. The current temporal Constitution of these phenomena leads back in each case to that *one* kind of temporality which serves as such to guarantee the possibility that understanding, state-of-mind, falling, and discourse, are united in their structure.

(a) The Temporality of Understanding

With the term "understanding" we have in mind a fundamental *existentiale,* which is neither a definite *species of cognition* distinguished, let us say, from explaining and conceiving, nor any cognition at all in the sense of grasping something thematically. Understanding constitutes rather the Being of the "there" in such a way that, on the basis of such understanding, a Dasein can, in existing, develop the different possibilities of sight, of looking around [Sichumsehens], and of just looking. In all explanation one uncovers understandingly that which one cannot understand; and all explanation is thus rooted in Dasein's primary understanding.

If the term "understanding" is taken in a way which is primordially existential, it means *to be projecting towards a potentiality-for-Being for the sake of which any Dasein exists.* In understanding, one's own potentiality-for-Being is disclosed in such a way that one's Dasein always knows understandingly what it is capable of. It 'knows' this, however, not by having discovered some fact, but by maintaining itself in an existentiell possibility. The kind of ignorance which corresponds to this, does not consist in an absence or cessation of understanding, but must be regarded as a deficient mode of the projectedness of one's potentiality-for-Being. Existence can be questionable. If it is to be possible for something 'to be in question' [das "In-Frage-stehen"], a disclosedness is needed. When one understands oneself projectively in an existentiell possibility, the future underlies this understanding, and it does so as a coming-towards-oneself out of that current possibility as which one's Dasein exists. The future makes ontologically possible an entity which is in such a way that it exists understandingly in its potentiality-for-Being. Projection is basically futural; it does not primarily grasp the projected possibility thematically just by having it in view, but it throws itself into it as a possibility. In each case Dasein *is* understandingly in the way that it can be. Resoluteness has turned out to be a kind of existing which is primordial and authentic. Proximally and for the most part, to be sure, Dasein remains irresolute; that is to say, it remains closed off in its ownmost potentiality-for-Being, to which it brings itself only when it has been individualized. This implies that temporality does not temporalize itself constantly out of the authentic future. This inconstancy, however, does not mean that temporality sometimes lacks a future, but rather than the temporalizing of the future takes various forms.

To designate the authentic future terminologically we have reserved the expression *"anticipation"*. This indicates that Dasein, existing authentically, lets itself come towards itself as its ownmost potentiality-for-Being—that the future itself must first win itself, not from a Present, but from the inauthentic future. If we are to provide a formally undifferentiated term for the future, we may use the one with which we have designated the first structual item of care—the *"ahead-of-itself"*. Factically, Dasein is constantly ahead of itself, but inconstantly anticipatory with regard to its existentiell possibility.

How is the inauthentic future to be contrasted with this? Just as the authentic future is revealed in resoluteness, the inauthentic future, as an ecstatical mode, can reveal itself only if we go back ontologically from the inauthentic understanding of everyday concern to its existential-temporal meaning. As care, Dasein is essentially ahead of itself. Proximally and for the most part, concernful Being-in-the-world understands itself in terms of that with *which* it is concerned. Inauthentic *understanding* projects itself upon that with which one can concern oneself, or upon what is feasible, urgent, or indispensable in our everyday business. But that with which we concern ourselves is as it is for the sake of that potentiality-for-Being which cares. This potentiality lets Dasein come towards itself in its concernful Being-alongside that with which it is concerned. Dasein does not come towards itself primarily in its ownmost non-relational potentiality-for Being, but it *awaits this* concernfully *in terms of that which yields or denies the object of its concern.* Dasein comes towards itself from that with which it concerns itself. The inauthentic future has the character of *awaiting.* One's concernful understanding of oneself as they-self in terms of what one does, has its possibility 'based' upon this ecstatical mode of the future. And *only because* factical Dasein *is* thus *awaiting* its potentiality-for-Being, and *is awaiting* this potentiality in terms of that with which it concerns itself, can it *expect* anything and wait for it [*erwarten* und warten auf . . .]. In each case some sort of awaiting must have disclosed the horizon and the range from which something can be expected. *Expecting is founded upon awaiting, and is a mode of that future which temporalizes itself authentically as anticipation.* Hence there lies in anticipation a more primordial Being-towards-death than in the concernful expecting of it.

Understanding, as existing in the potentiality-for-Being, however it may have been projected, is *primarily* futural. But it would not temporalize itself if it were not temporal—that is, determined with equal primordiality by having been and by the Present. The way in which the latter ecstasis helps constitute inauthentic understanding, has already been made plain in a rough and ready fashion. Everyday concern understands itself in terms of that potentiality-for-Being which confronts it as coming from its possible success or failure with regard to whatever its object of concern may be. Corresponding to the inauthentic future (awaiting), there is a special way of Being-*alongside* the things with which one concerns oneself. This way of Being-alongside is the Present—the "waiting-towards"; this ecstatical mode reveals itself if we adduce for comparison this very same ecstasis, but in the mode of authentic temporality. To the anticipation which goes with resoluteness, there belongs a Present in accordance with which a resolution discloses the Situation. In resoluteness, the Present is not only brought back from distraction with the objects of one's closest concern, but it gets held in the future and in having been. That *Present* which is held in authentic temporality and which thus is *authentic* itself, we call the *"moment of vision"*. This term must be understood in the active sense as an ecstasis. It means the resolute rapture with which Dasein is carried away to whatever possibilities and circumstances are encountered in the Situation as possible objects of concern, but a rapture which is *held* in resoluteness. The moment of vision is a phenomenon which *in principle* can *not* be clarified in terms of the *"now"* [dem *Jetzt*]. The "now" is a temporal phenomenon which belongs to time as within-time-ness: the "now" 'in which' something arises, passes away, or is present-at-hand. 'In the moment of vision' nothing can occur; but as an authentic Present or waiting-towards, the moment of vision permits us *to encounter for the first time* what can be 'in a time' as ready-to-hand or present-at-hand.

In contradistinction to the moment of vision as the authentic Present, we call the inauthentic Present *"making present"*. Formally understood, every Present is one which makes present, but not every Present has the character of a 'moment of vision'. When we use the expression "making present" without adding anything further, we always have in mind the inauthentic kind, which is irresolute and does not have the character of a moment of vision. Making-present will become clear only in the light of the temporal Interpretation of falling into the 'world' of one's concern; such falling has its existential meaning in making present. But in so far as the potentiality-for-Being which is projected by inauthentic understanding is projected in terms of things with which one can be concerned, this means that such understanding temporalizes itself in terms of making present. The moment of vision, however, temporalizes itself in quite the opposite manner—in terms of the authentic future.

Inauthentic understanding temporalizes itself as an awaiting which makes present [gegenwärtigendes Geswärtigen]—an awaiting to whose ecstatical unity there must belong a corresponding *"having been"*. The authentic coming-towards-oneself of anticipatory resoluteness is at the sametime a coming-back to one's ownmost Self, which has been thrown into its individualization. This ecstasis makes it possible for Dasein to be able to take over resolutely that entity which it already is. In anticipating, Dasein *brings* itself *again forth* into its ownmost potentiality-for-Being. If *Being*-as-having-been is authentic, we call it *"repetition"*.[3] But when one projects oneself inauthentically towards those possibilities which have been drawn from the object of concern in making it present, this is possible only because Dasein has *forgotten* itself in its ownmost *thrown* potentiality-for-Being. This forgetting is not nothing, nor is it just a failure to remember; it is rather a 'positive' ecstatical mode of one's having been—a mode with a character of its own. The ecstasis (rapture) of forgetting has the character of backing away *in the face of* one's ownmost "been", and of doing so in a manner which is closed off from itself—in such a manner, indeed, that this backing-away closes off ecstatically that in the face of which one is backing away, and thereby closes itself off too. *Having forgotten* [*Vergessenheit*] as an inauthentic way of having been, is thus related to that thrown *Being* which is one's own; it is the temporal meaning of that Being in accordance with which I *am* proximally and for the most part as-having-been. Only on the basis of such forgetting can anything be *retained* [*behalten*] by the concernful making-present which awaits; and what are thus retained are entities encountered within-the-world with a character other than that of Dasein. To such retaining there corresponds a non-retaining which presents us with a kind of 'forgetting' in a derivative sense.

Just as expecting is possible only on the basis of awaiting, *remembering* is possible only on that of forgetting, *and not vice versa;* for in the mode of having-forgotten, one's having been 'discloses' primarily the horizon into which a Dasein lost in the 'superficiality' of its object of concern, can bring itself by remembering. The *awaiting which forgets and makes present* is an ecstatical unity in its own right, in accordance with which inauthentic understanding temporalizes itself with regard to its temporality. The unity of these ecstases closes

3. [N.B. The translators have translated *wiederholen* and its variations as 'to repeat'; for reasons explained in my *Heidegger, Kant, and Time* (see footnote 25 to VII), p. 12 n., I suggest 'to retrieve' as more appropriate.—C. M. S.]

off one's authentic potentiality-for-Being, and is thus the existential condition for the possibility of irresoluteness. Though inauthentic concernful understanding determines itself in the light of making present the object of concern, the temporalizing of the understanding is performed primarily in the future.

(b) The Temporality of State-of-mind

Understanding is never free-floating, but always goes with some state-of-mind. The "there" gets equiprimordially disclosed by one's mood in every case—or gets closed off by it. Having a mood brings Dasein *face to face* with its thrownness in such a manner that this thrownness is not known as such but disclosed far more primordially in 'how one is'. Existentially, *"Being*-thrown" means finding oneself in some state-of-mind or other. One's state-of-mind is therefore based upon thrownness. . . .

(d) The Temporality of Discourse

When the "there" has been completely disclosed, its disclosedness is constituted by understanding, state-of-mind, and falling; and this closedness becomes Articulated by discourse. Thus discourse does not temporalize itself primarily in any definite ecstasis. Factically, however, discourse expresses itself for the most part in language, and speaks proximally in the way of addressing itself to the 'environment' by talking about things concernfully; because of this, *making-present* has, of course, a *privileged* constitutive function.

Tenses, like the other temporal phenomena of language—'aspects' and 'temporal stages' ["Zeitstufen"]—do not spring from the fact that discourse expresses itself 'also' about 'temporal' processes, processes encountered 'in time'. Nor does their basis lie in the fact that speaking runs its course 'in a psychical time'. Discourse *in itself* is temporal, since all talking about . . . , of . . . , or to . . . , is grounded in the ecstatical unity of temporality. *Aspects* have their roots in the primordial temporality of concern, whether or not this concern relates itself to that which is within time. The problem of their existential-temporal structure *cannot even be formulated* with the help of the ordinary traditional conception of time, to which the science of language needs must have recourse. But because in any discourse one is talking about entities, even if not primarily and predominantly in the sense of theoretical assertion, the analysis of the temporal Constitution of discourse and the explication of the temporal characteristics of language-patterns can be tackled only if the problem of how Being and truth are connected in principle, is broached in the light of the problematic of temporality. We can then define even the ontological meaning of the 'is', which a superficial theory of propositions and judgments has deformed to a mere 'copula'. Only in terms of the temporality of discourse—that is, of Dasein in general—can we clarify how 'signification' 'arises' and make the possibility of concept-formation ontologically intelligible.

Understanding is grounded primarily in the future (whether in anticipation or in awaiting). States-of-mind temporalize themselves primarily in having been (whether in repetition or in having forgotten). Falling has its temporal roots primarily in the Present (whether in making-present or in the moment of vision). All the same, understanding is in every case a Present which "is in the process of having been'. All the same, one's state-of-mind temporalizes itself as future which is 'making present'. And all the same, the Present 'leaps away' from a future that is in the process of having been, or else it is held on to by such a future. Thus we can see that *in every ecstasis, temporality temporalizes itself as a whole; and this means that in the ecstatical unity with which temporality has fully temporalized itself currently, is grounded the totality of the structural whole of existence, facticity, and falling—that is, the unity of the care-structure.*

Temporalizing does not signify that ecstases come in a 'succession.' The future is *not later* than having been, and having been is *not earlier* than the Present. Temporality temporalizes itself as a future which makes present in the process of having been.

Both the disclosedness of the "there" and Dasein's basic existentiell possibilities, authenticity and inauthenticity, are founded upon temporality. But disclosedness always pertains with equal primordiality to the entirety of *Being-in-the-world*—to Being-in as well as to the world. So if we orient ourselves by the temporal Constitution of disclosedness, the ontological condition for the possibility that there can be entities which exist as Being-in-the-world, must be something that may also be exhibited.

¶ 69. The Temporality of Being-in-the-world and the Problem of the Transcendence of the World

The ecstatical unity of temporality—that is, the unity of the 'out-side-of-itself' in the raptures of the future, of what has been, and of the Present—is the condition for the possibility that there can be an entity which exists as its "there". . . .

(a) The Temporality of Circumspective Concern

. . . When, in one's concern, one lets something be involved, one's doing so is founded on temporality, and amounts to an altogether pre-ontological and non-thematic way of understanding involvement and readiness-to-hand. In what follows, it will be shown to what extent the understanding of these types of Being as such is, in the end, also founded on temporality. We must first give a more concrete demonstration of the temporality of Being-in-the-world. With this is our aim, we shall trace how the theoretical attitude towards the 'world' 'arises' out of circumspective concern with the ready-to-hand. Not only the circumspective discovering of entities within-the-world but also the theoretical discovering of them is founded upon Being-in-the-world. The existential-temporal Interpretation of these ways of discovering is preparatory to the temporal characterization of this basic state of Dasein.

(b) The Temporal Meaning of the Way in which Circumspective Concern becomes Modified into the Theoretical Discovery of the Present-at-hand Within-the-world

When in the course of *existential ontological* analysis we ask how *theoretical* discovery 'arises' out of *circumspective* concern, this implies already that we are not making a problem of the *ontical* history and development of science, or of the factical occasions for it, or of its proximate goals. In seeking the *ontological genesis* of the theoretical attitude, we are asking which of these conditions implied in Dasein's state of Being are existentially necessary for the possibility of Dasein's existing in the way of scientific research. This formulation of the question is aimed at an *existential conception of science*. This must be distinguished from the 'logical' conception which understands science with regard to its results and defines it as 'something established on an interconnection of true propositions—that is, propositions counted as valid'. The existential conception understands science as a way of existence and thus as a mode of Being-in-the-world, which discovers or discloses either entities or Being. Yet a fully adequate existential Interpretation of science cannot be carried out until the *meaning of Being and the 'connection' between Being and truth* have been *clarified* in terms of the temporality of existence. The following deliberations are preparatory to the understanding of *this central problematic,* within which, moreover, the idea of phenomenology, as distinguished from the preliminary conception of it which we indicated by way of introduction will be developed for the first time. . . .

¶ 81. Within-time-ness and the Genesis of the
Ordinary Conception of Time

How does something like 'time' first show itself for everyday circumspective concern? In what kind of concernful equipment-using dealings does it become *explicitly* accessible? If it has been made public with the disclosedness of the world, if it has always been already a matter of concern with the discoveredness of entities within-the-world—a discoveredness which belongs to the world's disclosedness—and if it has been a matter of such concern in so far as Dasein calculates time in reckoning with *itself,* then the kind of behaviour in which 'one' explicitly regulates oneself *according to time,* lies in the use of clocks. The existential-temporal meaning of this turns out to be a making-present of the travelling pointer. By *following* the positions of the pointer in a way which makes present, one *counts* them. This making-present temporalizes itself in the ecstatical unity of a retention which awaits. To *retain* the 'on that former occasion' and to retain it by *making it present,* signifies that in saying "now" one is open for the horizon of the earlier—that is, of the "now-no-longer". To *await* the 'then' by *making it present,* means that in saying "now" one is open for the horizon of the later—that is, of the "now-not-yet". *Time is what shows itself in such a making-present.* How then, are we to define the *time* which is manifest within the horizon of the circumspective concernful clock-using in which one takes one's time? *This time is that which is* counted *and which shows itself when one follows the travelling pointer, counting and making present in such a way that this making-present temporalizes itself in an ecstatical unity with the retaining and awaiting which are horizonally open according to the "earlier" and "later".* This, however, is nothing else than an existential-ontological interpretation of Aristotle's definition of "time": τοῦτο γάρ ἐστιν ὅ χρόνος, ἀριθμὸς κινήσεως κατὰ τὸ πρότερον καὶ ὕστερον. "For this time: that which is counted in the movement which we encounter within the horizon of the earlier and later." [4] This definition may seem strange at first glance; but if one defines the existential-ontological horizon from which Aristotle has taken it, one sees that it is as 'obvious' as it at first seems strange, and has been genuinely derived. The source of the time which is thus manifest does not become a problem for Aristotle. His Interpretation of time moves rather in the direction of the 'natural' way of understanding Being. Yet because this very understanding and the Being which is thus understood have in principle been made a problem for the investigation which lies before us, it is only *after* we have found a solution for the question of Being that the Aristotelian analysis of time can be Interpreted thematically in such a way that it may indeed gain some signification in principle, if the formulation of this question in ancient ontology, with all its critical limitations, is to be appropriated in a positive manner.

Ever since Aristotle all discussions of the concept of time have clung *in principle* to the Aristotelian definitions; that is, in taking time as their theme, they have taken it as it shows itself in circumspective concern. Time is what is 'counted'; that is to say, it is what is expressed and what we have in view, even if unthematically, when the *travelling* pointer (or the shadow) is made present. When one makes present that which is moved in its movement, one says 'now here, now here, and so on'. The "nows" are what get counted. And these show themselves 'in every "now" ' as "nows" which will 'forwith be no-longer-now' and "nows" which have 'just been not-yet-now'. The world-time which is 'sighted' in this manner in the use of clocks, we call the *"now-time"* [*Jetzt-Zeit*].

When the concern which gives itself time reckons with time, the

---

4. [Cf. Aristotle, *Physica*, Δ 11, 219b 1 ff. M.H. (See p. 52 above.—C. M. S.)]

more 'naturally' it does so, the less it dwells at the expressed time as such; on the contrary, it is lost in the equipment with which it concerns itself, which in each case has a time of its own. When concern determines the time and assigns it, the more 'naturally' it does so—that is, the less it is directed towards treating time as such thematically—all the more does the Being which is alongside the object of concern (the Being which falls as it makes present) say unhesitatingly (whether or not anything is uttered) "now" or "then" or "on that former occasion". Thus for the ordinary understanding of time, time shows itself as a sequence of "nows" which are constantly 'present-at-hand', simultaneously passing away and coming along. Time is understood as a succession, as a 'flowing stream' of "nows", as the 'course of time'. *What is implied by such an interpretation of the world-time with which we concern ourselves?*

We get the answer if we go back to the *full* essential structure of world-time and compare this with that with which the ordinary understanding of time is acquainted. We have exhibited *datability* as the first essential item in the time with which we concern ourselves. This is grounded in the ecstatical constitution of temporality. The 'now' is essentially a "now that . . .". The datable "now", which is understood in concern even if we cannot grasp it as such, is in each case one which is either appropriate or inappropriate. *Significance* belongs to the structure of the "now". We have accordingly called the time with which we concern ourselves "*world*-time". In the ordinary interpretations of time as a sequence of "nows", both datability and significance are *missing*. These two structures are *not* permitted to 'come to the fore' when time is characterized as a pure succession. The ordinary interpretation of time *covers them up*. When these are covered up, the ecstatico-horizonal constitution of temporality, in which the datability and the significance of the "now" are grounded, gets *levelled off*. The "nows" get shorn of these relations, as it were; and as thus shorn, they simply range themselves along after one another so as to make up the succession.

It is no accident that world-time thus gets levelled off and covered up by the way time is ordinarily understood. But just *because* the every-day interpretation of time maintains itself by looking solely in the direction of concernful common sense, and understands only what 'shows' itself within the common-sense horizon, these structures must escape it. That which gets counted when one measures time concernfully, the "now", gets co-understood in one's concern with the present-at-hand and the ready-to-hand. Now so far as *this* concern with time comes back to the time itself which has been co-understood, and in so far as it 'considers' that time, it sees the "nows" (which indeed are also somehow 'there') within the horizon of that understanding-of-Being by which this concern is itself constantly guided. Thus the "nows" are in a certain manner *co-present-at-hand:* that is, entities are encountered, *and so too* is the "now". Although it is not said explicitly that the "nows" are present-at-hand in the same way as Things, they still get 'seen' ontologically within the horizon of the idea of presence-at-hand. The "nows" *pass away,* and those which have passed away make up the past. The "nows" *come along,* and those which are coming along define the 'future'. The ordinary interpretation of world-time as now-time never avails itself of the horizon by which such things as world, significance, and datability can be made accessible. These structures necessarily remain covered up, all the more so because this covering-up is reinforced by the way in which the ordinary interpretation develops its characterization of time conceptually.

The sequence of "nows" is taken as something that is somehow present-at-hand, for it even moves 'in to time'. We say: '*In every* "now" is now; *in* every "now" it is already vanishing.' In *every* "now" the "now" is now and therefore it constantly has presence *as something selfsame,* even though in every "now" another may be vanishing as it comes along. Yet at *this* thing which changes, it simultaneously shows its own constant presence. Thus even Plato, who directed his glance in this manner at time as a sequence of "nows" arising and passing away, had to call time "the image of eternity": ["But he decided to make a kind of moving image of the eternal; and while setting the heaven in order, he made an eternal image, moving according to number—an image of that eternity which abides in oneness. It is to this image that we have given the name of 'time'."] [5]

5. Heidegger cites the original Greek from *Timaeus,* 37d and offered *his* translation of Greek into German as a footnote. Presumably, the translators have translated from Heidegger's German translation and not the 'standard' German translation. Cf. pp. 43-44 above—C. M. S.

The sequence of "nows" is uninterrupted and has no gaps. No matter how 'far' we proceed in 'dividing up' the "now", it is always now. The continuity of time is seen within the horizon of something which is indissolubly present-at-hand. When one takes one's ontological orientation from something that is constantly present-at-hand, one either looks for the problem of Continuity of time or one leaves this impasse alone. In either case the specific structure of world-time must remain *covered up*. Together with datability (which has an ecstatical foundation) it has been *spanned*. The spannedness of time is not to be understood in terms of the horizontal *stretching-along* of the ecstatical unity of that temporality which has made itself public in one's concern with time. The fact that in every "now", no matter how momentary, it is *in each case already* now, must be conceived in terms of something which is 'earlier' *still* and from which every "now" stems: that is to say, it must be conceived in terms of the ecstatical stretching-along of that temporality which is alien to any Continuity of something present-at-hand but which, for its part, presents the condition for the possibility of access to anything continuous that is present-at-hand.

The principle thesis of the ordinary way of interpreting time—namely, that time is 'infinite'—makes manifest most impressively the way in which world-time and accordingly temporality in general have been levelled off and covered up by such an interpretation. It is held that time presents itself proximally as an uninterrupted sequence of "nows". Every "now", moreover, is already either a "just-now" or a "forthwith". If in characterizing time we stick primarily and exclusively *to such a sequence,* then in principle neither beginning nor end can be found in it. Every last "now", as *"now",* is always *already* a "forthwith" that is no longer [ein Sofort-nicht-mehr]; thus it is time in the sense of the "no-longer-now"—in the sense of the past. Every first "now" is a "just-now" that is not yet [ein Socben-noch-nicht]; thus it is time in the sense of the "not-yet-now"—in the sense of the 'future'. Hence time is endless 'on both sides'. This thesis becomes possible only on the basis of an orientation *towards a free-floating "in-itself" of a course of "nows" which is present-at-hand*—an orientation in which the full phenomenon of the "now" has been covered up with regard to its datability, its worldhood, its spannedness, and its character of having a location of the same kind as Dasein's, so that it has dwindled to an unrecognizable fragment. If one directs one's glance towards Being-present-at-hand and not-Being-present-at-hand, and thus 'thinks' the sequence of "nows" through 'to the end', then an end can never be found. In *this* way of *thinking* time through to the end, one *must* always *think* more time; from this one infers that time *is* infinite.

But wherein are grounded this levelling-off of world-time and this covering-up of temporality? In the Being of Dasein itself, which we have, in a prepartory manner, Interpreted as *care*. Thrown and falling, Dasein is proximally and for the most part lost in that with which it concerns itself. In this lostness, however, Dasein's fleeing in the face of the authentic existence which has been characterized as "anticipatory resoluteness", has made itself known; and this is a fleeing which covers up. In this concernful fleeing lies a fleeing *in the face of* death—that is, a looking-away *from* the end of Being-in-the-world. This looking-away from it, is in itself a mode of that Being-*towards*-the-end which is ecstatically *futural*. The inauthentic temporality of everyday Dasein as if falls, must, as such a looking-away from fini-tude, fail to recognize authentic futurity and therewith temporality in general. And if indeed the way in which Dasein is ordinarily under-stood is guided by the "they", only so can the self-forgetful 'repre-sentation' of the 'infinity' of public time be strengthened. The 'they' never dies because it *cannot* die; for death is in each case mine, and only in anticipatory resoluteness does it get authentically understood in an existentiell manner. Nevertheless, the "they", which never dies and which misunderstands Being-towards-the-end, gives a charac-teristic interpretation to fleeing in the face of death. To the very end 'it always has more time'. Here a way of "having time" in the sense that one can lose it makes itself known. 'Right now, this! then that! And that is barely over, when . . .' Here it is not as if the finitude of time were getting understood; quite the contrary, for concern sets out to snatch as much as possible from the time which still keeps coming and 'goes on'. Publicly, time is something which everyone takes and can take. In the everyday way in which we are with one another, the levelled-off sequence of "nows" remains completely unrecognizable as regards its origin in the temporality of the individual Dasein. How is 'time' in its course to be touched even the least bit when a man who has been present-at-hand 'in time' no longer exists? Time goes on, just as indeed it already 'was' when a man 'came into life'. The only time one knows is the public time which has been levelled off and which belongs to everyone—and that means, to nobody.

But just as he who flees in the face of death is pursued by it even as he evades it, and just as in turning away from it he must see it none the less, even the innocuous infinite sequence of "nows" which simply runs its course, imposes itself 'on' Dasein in a remarkably enigmatical way. Why do we say that time *passes away,* when we do not say with *just as much* emphasis that it arises? Yet with regard to the pure sequence of "nows" we have as much right to say one as the other. When Dasein talks of time's *passing away,* it understands, in the end, more of time than it wants to admit; that is to say, the *temporality* in which world-time temporalizes itself has *not been completely closed off,* no matter how much it may get covered up. Our talk about time's passing-away gives expression to this 'experience': time does not let itself be halted. This 'experiene' in turn is possible only because the halting time is something that we want. Herein lies an inauthentic *awaiting* of 'moments'—an awaiting in which these are already *forgotten* as they glide by. The *awaiting* of inauthentic existence—the awaiting which forgets as it makes present—is the condition for the possibility of the ordinary experience of time's passing away. Because Dasein is futural in the "ahead-of-itself", it must, in awaiting, understand the sequence of "nows" as one which *glides by* as it passes away. *Dasein knows fugitive time in terms of its 'fugitive' knowledge about its death.* In the kind of talk which emphasizes time's passing away, the *finite futurity* of Dasein's temporality is publicly reflected. And because even in talk about time's passing away, death can remain covered up, time shows itself as a passing-away 'in itself'.

But even in this pure sequence of "nows" which passes away in itself, primordial time still manifests itself throughout all this levelling off and covering up. In the ordinary interpretation, the stream of time is defined as an *irreversible* succession. Why cannot time be reversed? Especially if one looks exclusively at the stream of "nows", it is incomprehensible in itself why this sequence should not present itself in the reverse direction. The impossibility of this reversal has its basis in the way public time originates in temporality, the temporalizing of which is primarily futural and 'goes' to its end ecstatically in such a way that it 'is' already towards its end.

The ordinary way of characterizing time as an endless, irreversible sequence of "nows" which passes away, rises from the temporality of falling Dasein. *The ordinary representation of time has its natural justification.* It belongs to Dasein's average kind of Being, and to that understanding of Being which proximally prevails. Thus proximally and for the most part, even *history* gets understood *publicly* as happening *within-time.* This interpretation of time loses its exclusive and pre-eminent justification only if it claims to convey the 'true' conception of time and to be able to prescribe the sole possible horizon within which time is to be interpreted. On the contrary, it has emerged that *why and how world-time belongs to Dasein's temporality* is intelligible only in terms of that temporality and its temporalizing. From temporality the full structure of world-time has been drawn; and only the Interpretation of this structure gives us the clue for 'seeing' at all that in the ordinary conception of time something has been covered up, and for estimating how much the ecstatico-horizonal constitution of temporality has been levelled off. This orientation by Dasein's temporality indeed makes it possible to exhibit the origin and the factical necessity of this levelling off and covering up, and at the same time to test the arguments for the ordinary theses about time.

On the other hand, within the horizon of the way time is ordinarily understood, *temporality is inaccessible in the reverse direction.* Not only must the now-time be oriented primarily by temporality in the order of possible interpretation, but it temporalizes itself only in the inauthentic temporality of Dasein; so if one has regard for the way the now-time is derived from temporality, one is justified in considering temporality as the *time which is primordial.*

Ecstatico-horizonal temporality temporalizes itself *primarily* in terms of the *future.* In the way time is ordinarily understood, however, the basic phenomenon of time is seen in the *"now",* and indeed in that pure *"now"* which has been shorn in its full structure—that which they call the 'Present'. One can gather from this that there is in principle no prospect that *in terms of this kind of "now"* one can clarify the ecstatico-horizonal phenomenon of the *moment of vision* which belongs to temporality, or even that one can derive it thus. Correspondingly, the future as ecstatically understood—the datable and significant 'then'—does not coincide with the ordinary conception of the 'future' in the sense of a pure "now" which has not yet come along but is only coming along. And the concept of the past in the sense of the pure "now" which has passed away, is just as far from coinciding with the ecstatical "having-been"—the datable and significant 'on a former occasion'. The "now" is not pregnant with the "not-yet-now", but the Present arises from the future in the primordial ecstatical unity of the temporalizing of temporality.[6]

Although, proximally and for the most part, the ordinary experience of time is one that knows only 'world-time', it always gives it a *distinctive* relationship to 'soul' and 'spirit', even if this is still a far cry from a philosophical inquiry oriented explicitly and primarily towards the 'subject'. As evidence for this, two characteristic passages will suffice. Aristotle says: ["But if nothing other than the soul or the soul's mind were naturally equipped for numbering, then if there were no soul, time would be impossible."][7] And Saint Augustine writes: *"inde mihi visum est, nihil esse aliud tempus quam distentionem; sed cuius rei nescio; et mirum si non ipsius animl."*[8] Thus in principle even the Interpretation of Dasein as temporality does not lie beyond

6. "The fact that the traditional conception of 'eternity' as signifying the 'standing "now"' (*nunc stans*), has been drawn from the ordinary way of understanding time and has been defined with an orientation towards the idea of 'constant' presence-at-hand, does not need to be discussed in detail. If God's eternity can be 'construed' philosophically, then it may be understood only as a more primordial temporality which is 'infinite'. Whether the way afforded by the *via negativa et eminentiae* is a possible one, remains to be seen." M.H. [Cf. Kant as cited on p. 118 above. C. M. S.]

7. Heidegger cites the original Greek and offers his translation of Greek into German as a footnote: Aristotle, *Physica* Δ 14, 223 a 25; cf. *ibid.* 11, 218b 29-219a 4-6. M.H. [See pp. 60-61 & 51 above.—C. M. S.]

8. "Hence it seemed to me that time is nothing else than an extendedness; but of what sort it is an extendedness, I do not know; and it would be surprising if it were not an extendedness of the soul itself." Augustine, *Confessions* XI, 26. M.H. [See p. 92 above.—C. M. S.]

the horizon of the ordinary conception of time. And Hegel has made an explicit attempt to set forth the way in which time as ordinarily understood is connected with spirit. In Kant, on the other hand, while time is indeed 'subjective', it stands 'beside' the 'I think' and is not bound up with it.[9] The grounds which Hegel has explicitly provided for the connection between time and spirit are well suited to elucidate indirectly the foregoing Interpretation of Dasein as temporality and our exhibition of temporality as the source of world-time.

9. On the other hand, the extent to which an even more radical understanding of time than Hegel's makes itself evident in Kant, will be shown in the first division of the second part of this treatise. M.H. [Although never published as such, it apparently was incorporated into a separate book published, in English, under the title, *Kant and the Problem of Metaphysics.*—C. M. S.]

VIII.

The Open Agenda

We have been seeking out the nature of that time which forms or constitutes the structure of human experiencing. Insofar as any human investigation, regardless of what it is about, is necessarily structured as a temporal project, so this consideration of our developing conceptions of time has, in fidelity to the nature of its subject, been pursued in a temporal form.

From the beginning, we have seen the identification of time with meaningful or intelligent activity and with functioning rationality. In the attempt to comprehend the patterns of change and development—whether attributed to a transcendent divine or some kind of immanent source—time was seen as the frame for the reality of the encountered world and thereby, by implication, as a key to the comprehension of its meaning and significance.

Plato's vision of time as rational ordering, Aristotle's identification of it with self-awareness while limiting its function to that of measuring, Plotinus' grounding of processes in the permissibility of the temporal, and Augustine's quest for the truth of time in the outlook of the human mind—all have conjoined to testify to the developmental potential of the Heraclitean proposal that reason or *Logos*, and time are somehow conjoined in the reality of the world we experience. Setting out the frame of discussion for three thousand years, their continuing influence was effectively reaffirmed by those who sought new beginnings at the beginning of the modern era.

Following out Descartes' lead, Locke, Leibniz, and Kant, each pursuing the Cartesian program in his own way, each found an intimate connection between the time of the thinking subject and the thinking of the subject himself. Kant, more than the others, saw the immediate implications that ensued for the necessary temporalization of the rules of cognitive thinking and the new problems posed regarding the explanation of functioning moral reason.

Seeking to move without further ado from the nature of human thinking to the world in which and about which one claims to think, Hegel, Lotze, Bergson, and Alexander, each, in his own way, sought to make the jump from 'mind' to 'world' without really considering the legitimacy of doing so. Having served up too many conundrums on each of these attempts, something of a restaging appeared to be in order. The inquiry into the nature of experiential time proceeded separately on three more or less contemporary fronts that are not as isolated from each other as some of their protagonists like to claim.

We have been urged to clarify our logical and linguistic usages. Two types of linguistic temporal expressions have been analyzed and discerned. The resultant controversy concerning their priority seems to have been settled for some by treating human thinking as a lapse from the functioning of physical nature which is the object of their own cognitive studies. For others, it has been settled by the observation that the time of non-conscious natural becoming and that of conscious human experiencing are of two different orders but both are somehow real within the one world within which they both transpire; for these, however, it would seem that the topic is then thrown back onto its subjective ground, onto that individuality in which experience has customarily been analyzed and which has been taken by most modern thinkers as somehow irreducible.

Meanwhile, a new perspective on time and the individualism in which it had been seen as rooted was being shaped on a new shore. Finding the self in the commitments and creative activity in which the individual engages himself, it was seen that his experiential time could not be described or explained by the use of a geometer's pen drawing points on a line. The experiential present, in which all experiencing transpires, was found to be an essentially indeterminate spread of duration with no clear boundaries marking it off from future or from past. Finding the reality of the self in its community activities—whether in the raising of a frontier barn or the common effort of a scientific investigation—the individuality, which the received tradition had made the locus of thinking, time and truth, was seen to be intrinsically involved in networks of community relationships. But the import for temporal understanding of this central insight was not really pursued; yet the tradition emerging continued to note the preeminence of temporal continuity, its essential unification in futurity, and the temporal nature of the notion of meaning it made central to its outlook.

A new investigation, arising out of the heart of the old tradition, came to ask similar questions as it confronted the ground questions it had inherited. Seeking to systematize that introspection in which the assurances of individual existence are first encountered, it found every aspect of the revelation of the self to be marked by a temporality that would succumb neither to linear designation nor quantifiable description. Locating the originating source of experience in the experiential present, it rapidly discovered the necessity of transcending the ambiguously marked borders of that present in order to comprehend what it was about. Working its way into the time of lived experience, it found that experience to be fundamentally temporal in nature, oriented to what yet might be, intrinsically involving the not-self in the self-awareness of the self, and that the human understanding of its own temporality inherently necessitates the transcendence of any strict chronology. This investigation found the temporal undergirding any experience we might possibly have and permeating any actual experience we do have; it could only conclude that the ontology in terms of which we see the world and our own selves, and the languages we use to express such perception, is already structured in temporalizing terms.

However we may find ourselves in agreement or dissent regarding any aspects of this general development, we must first remember that the discussion cannot be closed and that we cannot discuss everything at once. The discussion cannot be closed, first, because time itself cannot, in the most vivid imagination, come to an absolute end, and, thereby, what is temporal is ongoing. But, second, in this, as in any focused inquiry, the spotlight while moving along to focus attention on certain elements of the terrain, necessarily thrusts other elements of that same terrain into a shadowy background. This investigation, as any other, is thereby inherently incomplete. Those other and obscured objects of inquiry must themselves be searched out, looked at, placed under deliberate scrutiny, and thus be explicated. This need ensues from the inherently temporal limitations of any investigation—we cannot look at everything at once. While we are striving for comprehensiveness, we do so with the knowledge that our perspectives are inherently finite, that *any* focus excludes while it includes selected aspects of what is being discussed, and that what is left out is also, somehow, part of what is to be investigated. Just because of the conviction that we must start with the instrument of inquiry, the focus here has been on how the human experience of time has itself been formed, for the instrument of inquiry is, first of all, ourselves.

As we look to a continuing exploration of the nature and meaning of experiential time, we keep ourselves open to a review of the ground covered in terms of other perspectives. However, we also seek out some of the leading themes which present themselves to us as leading issues propounded by the heritage which has brought us here, and which we might well face directly as we seek to move on.

From the beginning, time has been tied to nature, to our observations of natural processes and thus, with the increasing modern focus on epistemology, to time as an element in the knowing of what is to be deemed as knowledge. Only recently, in some pragmatic and existential work, has the suggestion been made that this has, perhaps, been a reversal of the true order. Perhaps Kant's intuition [1] about the priority of practical reason is worth pursuing—for the revelation of human time appears to come foundationally from the nexus of value conflicts which animate and mark our interactions with the world of our fellows and of the things which are revealed to us in the unfolding of a temporal field. It seems more than ironic that in considerations of ethics, of esthetics, of the human value-laden activities in which temporality is necessarily presupposed and so pervasively present, the question of time so rarely appears.[2] For most value-laden questions are concerned with the use of unfolding time; although Heidegger, in his notion of 'resoluteness', has, in fact, seen the temporally foundational character of such value conflicts and decisions, one serious criticism of his work is the fact that he never seems to have really pursued it.

From the beginning, time as tied to nature has also been tied to the natural sciences; and we have generally looked to the work of our scientists for clarification of the nature, if not the meaning, of time. We have tended to forget that our scientists are also men and subject to the confines of whatever constitutional outlook may be uniquely human. Even Kant, in raising the landmark question, 'how is science possible?', forgot to ask about the humans who create our sciences as human interpretations of the phenomena of natural appearances.

Also we have forgotten that physical scientists are concerned to study the world of inanimate matter along the lines of a more or less mechanistic model. If there be any truth in Leibniz's contention that *all* existents 'reflect' the nature of the universe in radically peculiar ways, we cannot afford to forget that the time of the non-conscious provides a wholly different kind of 'reflection' than the temporality of conscious, purposeful, anticipatory beings. The reduction of the conscious to the unconscious is a category mistake of first magnitude. If Whitehead's notion, which seems to convey something of this Leibnizian insight in more contemporary dress, has any merit, we might then wonder about the efficacy of subordinating human to inanimate 'experience' and of seeking out of the study of the non-conscious physical, a key to the outlook of conscious human experience which is also part of the existent reality of the world.

From the beginning, time has been thought essentially in terms of chronology; yet such a completely neutral way of seeing its pervasive cast has never really rung true. For experiential time, as Bergson insisted, is not even-handed; it has depths and shallows; it runs sometimes wide and sometimes deep. One might well invoke the song of Ecclesiastes about temporal propitiousness and explore the crucial distinction which Paul Tillich, for one, marked out quite well: time as *chronos,* and time as *kairos:* for both are certainly present in any situation in which humans find themselves; but these two different perspectives on the temporal speak quite differently about the face of time, and it might well be argued that it is the latter sense which guides the human approach to the former.

From the beginning, the notion of introspection or self-examination has been seen as essentially individual; even Aristotle, not ordinarily considered an introspective thinker, urged an identification of self-awareness and individual time-perception. By and large, we have followed the developing modern journey, which started from the Cartesian postulation of the atomic individual apart from the historically social. His *je pense* may indeed depend, as Descartes urged, on the *je suis,* but it depends even more crucially on the *nous sommes.* We might wonder whether his atomic view of momentary time was really unrelated to his view of absolutely atomically separate individuals. Yet, in the background, the thought has hovered that this could not finally be good enough: perhaps, we have too long ignored the metaphysical truth embedded in Aristotle's *Politics* and Rousseau's *Du contrat social*—that the individual, in his activity, his individuality, his time, and his being, ultimately depends upon and necessarily presupposes his free membership in his community. Stimulated, perhaps, by Fichte's insight that the 'I' always needs the 'not-I' for its own awareness and fulfillment, it would seem that the pragmatic tradition had first seen that true individuality necessarily involves an essentially social involvement. Essential to an existential notion of Being, the essential sociality of the individual seems to have been suggested by Heidegger in his notion of 'being-in-the-world' and sketched out by Sartre in his focus on the 'other'. Significantly they are often criticized for not having pursued any further this insight into the essentially social nature of the individual—in whom time and truth have been located. But an authentic phenomenology of experiential time would then go beyond them in contextual fashion, demonstrating the dialectic of the individual and of the social in the temporal developing of each.

When we pursue this last point, we turn to a consideration of human time that takes this essential sociality of the individual as a fundamental datum; we can find no better person of this century to initiate a continuing discussion along these lines than that neglected British thinker, Robin S. Collingwood. A good part of his work can be seen as a fairly sustained attack on what he pointed up as a persistently perverse prejudice of much contemporary thought—the deliberate ignoring of the temporal, and thereby the historical nexus in which human problems arise and human thought transpires. The tradition has generally seen human time in the individual sense of memory. But seen in societal terms, time is not merely memory but history, as history is the collective memory and sense of being of a time-binding community. Indeed, the development of thought which this book has portrayed has been not out of the individual memory of any of us, but out of the collective memory, the history, of a cultural tradition which brings its past into its present as it presses on. Collingwood has, indeed, argued that all the human sciences are historical sciences: each portrays a chain of development and reflects the concerns of the age from which it emerged. The doctrines of a particular science at a particular time are not isolated propositions of a new revelation but the expression of the presuppositions consciously or unconsciously invoked by the investigators of the time, as they asked specific questions embodying their presuppositions and formulated the answers then developed in terms of them.

In examining some of the 'perplexities' which he finds in most discussions of time, he has summed them up in a conclusion about our usual use of temporal language which only rarely can be taken as implying anything but metaphoric truth: *"All statements ordinarily made about time seem to imply that time is something which we know it is not, and make assumptions about it which we know to be untrue."* As we persist in trying to find ways out of the puzzlements in which we find ourselves enmeshed, we find ourselves raising the crucial metaphysical queries—the relation between the actual, potential, and possible, between the contingent and the necessary, the existential and the real. But, we find ourselves raising these anew within a distinctively historical or temporal framework of inquiry. Hopefully, pursuing these old topics under the lead of this ground question of time and history, our researches and inquiries may open up new frontiers instead of merely recircling well-worn ground.

One other way of carrying the investigation forward has been suggested by that preeminent American teacher of philosophy, Richard McKeon. He urged on our attention, as a mode of facing the various facets of our question of time, not so much questions or issues as topics or "commonplaces," which have in the past and might for us again provide guidance as we seek to develop greater insight into the ways in which the events that provide the content of our experience and the time of our experiencing join to light up the significance of what is transpiring. Although he would favor taking the traditional priority of the stance of the world in this new time-oriented investigation, there is no reason why the "commonplaces" he suggests cannot be approached from the viewpoint of the experienc*er* as the tradition hailing from Kant and Heidegger would insist.

It is, indeed, of some symbolic significance that Collingwood wrote his prime book while on a round-the-world ocean cruise; and, that McKeon's call for a new but continuing discussion was delivered as an opening statement at a philosophy conference called to discriminate and transcend the differences between East and West. Our heritage has here been traced in terms of the Western tradition; but it is not the only tradition of human thought or of human thinking about its time. We need, at last, to heed Kant's call to write our ongoing historical development in cosmopolitan terms.

Too long we have permitted ourselves to be bound by a parochial convention—that all that needs to be said is being said in that culture which has derived from the 'glory that was Greece'. Without suggesting any lesser respect for it, without suggesting that we can do without it, it surely is becoming clear that we can no longer do only with it. In large measure, perhaps, the globalization of human thinking has been engendered by the outreach of our developed Hellenic civilization. We must bring it, which has brought us here, to where we are. But we can only go further if we bring it with us while seeking to assimilate other cultural outlooks and philosophic traditions. We can only approach the new situation facing us in this new age with the perspectives which we have brought with us and which have brought us to where we are: We need to take up the questions which have brought us to this uniquely new era when we must, with the strength of our philosophic past, begin to look beyond its own borders. We can do this with perspective and with wisdom if we bring to a broadened outlook the questions which our own historical past has impelled us to pose. We must do so in the subjective existential conviction that the West is not the only human tradition that has sought to face the question of human time. As we continue to wrestle with this ground question, which accompanies and forms all human concerns, we must begin by being insightful about the outlook which our own tradition has imposed upon us as we face the wisdoms and the foolishnesses of other cultures and seek to assimilate them with our own.

Man may not be the only being consciously bound by time. But, among the conscious creatures we know, man is the only being capable of determining *how to use* the time that has been given to him. Man—not just Western man—is essentially temporal in his being; man—not just Western man—is, insofar as we can tell, "the one being who can determine the order and content of his time. Freedom means to him essentially freedom to dispose of his time." [3] In this statement is enshrouded every problematic of value and possible course of commitment, and, thereby, of every question of significance or meaning within the human perspective. The question of time, then, is not merely a problematic for an esoteric and inherently pedantic conceptual understanding; the question of time is that of the meaning of the present and of the possibility of the human future. It is the framework within which, and perhaps also the 'matter' out of which, the forms of the continuity of the human community and of the life that shares man's terrestrial home with him shall be salvaged. It is within the perspective of the temporal that we can find its inherently meaning-giving freedom ensured, and, thereby, the possibilities which we may still make our own.

29

Collingwood:

"SOME PERPLEXITIES ABOUT TIME"

Robin George Collingwood (1889–1943) insisted that we can only understand what a thinker is doing if we see his work in terms of what he was trying to accomplish. Just so, he urged, no propositions are true except in terms of the questions they pretend to answer; and the questions themselves depend upon the presuppositions which give them meaning. So with the problem of time, which, he saw as the problem of collective memory or historical consciousness. His address, which was originally delivered on 15 February 1926, has, so far as can be ascertained, never before been reprinted. Originally entitled, "Some Perplexities about Time: With an Attempted Solution," * he points us to the historical nexus of any individual outlook and to the metaphysical issues where are inherently involved.

I.

By way of preface, I will enumerate certain points with which I do not propose to deal.

(*a*) There are difficulties attaching to the idea of time in itself, or abstract time, considered apart from all events which are said to occupy or differentiate it. With these I am not here concerned. I am not sure that I think them very important, except from a dialectical point of view; for everyone who finds himself entangled in them may fairly be accused of complaining that he cannot see through the dust he has himself raised. For it is he that has made the false abstraction of time from the temporal events of which alone he has experience; and if he finds the abstraction unintelligible, the remedy is in his own hands.

---

* Reprinted from the *Proceedings of the Aristotelian Society,* by courtesy of the Editor of The Aristotelian Society "(C)" 1926. The Aristotelian Society.

Another set of difficulties attaches to the question how time in itself is related to temporal events: the question what is meant by *occupying* time. These are no less formidable than the others; but these, too, I intend to ignore. My reason is that these, too, depend on a false abstraction. First we abstract time in itself from temporal events, and then we perplex ourselves about the relation between time and events. But the answer is simple: the problem is one which we have created by making this false abstraction and setting it alongside the facts from which we have abstracted it as if it were another fact.

It may not be a waste of space to point out a parallel to these two false problems: (i) What is the State, in itself, quite apart from its members? The answer is, Nothing: and that answer is the right answer to all the questions which people ask about the State in abstraction from the persons whose political activities and passivities make them a State. (ii) What is (or ought to be) the relation between the State and the individual? The answer is, there can be no possible relation between a nonentity and a person; and that, I think, disposes of another large group of pseudo-problems that encumber political theory.

My perplexities, then, attach wholly to "full" time, as opposed to "empty" time. They are all, so far as I can see, of such a kind that they do *not* disappear when we bring time into relation with space or stuff it full of events or call it an illusion or do any of the other things which are generally held to draw the teeth of its paradoxes.

(*b*) I shall allow myself to speak of an event, and events in the plural; but I shall not raise the problem of the relation between the continuity of events and their plurality. We are told by mathematicians that this problem has been solved by the notion of a compact series, which is at once discrete, because a series, and continuous, because compact. I do not see that this is really so. If, between any two terms of a series, there is still a third term, it appears to me that we are still as far from continuity as ever; for *ex hypothesi* the series *as actually counted* always has gaps between all its terms, and however many terms are interpolated the gaps always remain. We are told that these gaps are filled by uncounted terms; but no evidence is brought that these terms, however numerous, *fill* the gaps. At most, they will only *occur in* the gaps, and serve to delimit their extent; and even this is not guaranteed by the definition of a compact series, for that definition does not stipulate that all the gaps between the terms are to be of the same size, or that an interpolated term is to be interpolated at one point in the gap rather than another. The theory of compact series, in a word, does not meet the facts; and it is only advanced in the interests of what I take to be a logical error, namely logical atomism, which in its application does not differ very widely from the sensational atomism of Hume, and is amenable to all the same criticisms. I shall therefore assume that an event takes time and is always (i) part of an event which takes more time, (ii) divisible into events that take less; and that events are in no sense composed of instants or point-instants but always of events. Time is therefore, I shall assume, to be sought *within* events, not in the relations *between* events, except so far as these relations fall within larger events. Thus, I should call the murder of Caesar and the battle of Actium two events forming parts of one event, namely, the fall of the Roman Republic; and in this use of words I should confidently claim support from the language of everybody who is not distorting English to defend a thesis.

(*c*) It is possible to torpedo all inquiries about such a subject as time by saying, "Time is just time; it is an ultimate fact, a thing *sui generis*, and in seeking to clear up these difficulties about it you are merely trying to explain it in terms of something else, which cannot be done: it is simple, and cannot be rendered intelligible by analysing it into simpler elements: hence there *cannot* be a theory of time, for a theory would be just such an analysis; nor *need* there be, for we all understand time perfectly well, and all perplexities about it are, in fact, cases of raising a dust and complaining that we cannot see."

The force of this objection is derived from its ambiguity. (i) It may mean that time is something perfectly intelligible, in which case we can reasonably retort that to understand a thing implies either seeing no difficulties about it or being able intelligently to dispose of them when they arise. (ii) It may mean that time is not intelligible at all, but simply an object of immediate feeling or intuition like a toothache or a blue colour. Such objects, I suppose, are wholly and adequately apprehended by feeling or intuiting them, and the question of understanding them, in any further sense, does not and cannot arise. There is nothing to understand; and whereas, if you feel or intuite them, you thereby know all that there is to know in them, if you do not, the way to acquire the knowledge is not to reason but to feel or to look. But neither of these accounts holds in the case of time. (i) The confessed inability of the objector to dispose of the difficulties shows that he does not really find time intelligible; (ii) the admitted fact that difficulties do arise shows that it is not sufficiently grasped by feeling or intuition.

No doubt time, like knowledge and goodness and number, is *sui generis;* but it does not follow that there cannot be a "theory" of it, if that means a reasoned discussion of the difficulties which are encountered when we try to think about it. And that is the only sense in which I ask for a theory of time or of anything else.

II.

My central difficulty is this:—*All statements ordinarily made about time seem to imply that time is something which we know it is not, and make assumptions about it which we know to be untrue.*

(*a*) Thus, we say that time flies. But what is the air in which, or the ground over which, it flies? Nothing, surely, but a system of reference, a temporal system of reference; in fact, time itself. The movement of time can only be a movement relative to something that is itself time—time regarded as stationary and existing *totum simul.* That, relatively to which time moves, cannot be space; for what moves relatively to space can only be a spatial object or body. We have, therefore, two times, a moving and a stationary; and since to be stationary implies permanence in time, we have a third time in which the stationary time remains stationary, and so *ad infinitum.* If I am told to accept this result in a spirit of natural piety, I reply that I cannot, because it had contradicted the thesis on which it depends: for it now appears that time as such does *not* fly, but that some times fly while others remain at rest. If I am told that my difficulty comes from taking a popular metaphor literally, I gratefully accept the confession that to speak of the flight, lapse, movement, &c., of time is a mere metaphor, and that in using it we are saying what we know to be untrue.

The difficulty is not removed if we say that events move "in" time. Here either (i) the time is regarded as moving with the events, in which case the difficulty recurs in the same form; or (ii) events are regarded as moving past a stationary frame of temporal references: in which case it recurs in a new form. For events must carry their own time-determinations with them—*e.g.*, an hour's journey, however far it recedes into the past, remains an hour long; and thus we have once more two times, one moving and one stationary, with results as before. It is not merely events but times that move in time; which is absurd. (The same difficulty arises in the conception of a body moving in space; the fact that it recurs there does not make it less serious here.)

(*b*) Again, we say that time can be measured. But how can it be measured? We measure (not abstract space, but) bodies by laying measuring-rods against them: that is, by juxtaposing two bodies and thus measuring one by reference, through the other, to a third. This could not be done unless we could move a rod from one place to another; and the hypothesis of the Lorenz-Fitzgerald contraction brings home the fact that the constancy of the length of the rod is an assumption and no more. But in the sense, which is the natural sense of measuring, we cannot measure temporal events at all. We can observe a rough simultaneity between the beginnings and ends of two events (*e.g.*, the rotation of a minute-hand and a journey), but only if the two events are going on at the same time; or, I ought to add, appear to the measurer to be going on at the same time. But we cannot then move the rotation of the minute-hand to another part of time and thus compare the length of the journey with the length of a symphony. For that, we have to use a different rotation of the hand. It is as if we were unable to move our measuring-rods at all, but were compelled to use a fresh rod for every fresh measurement. Then how could the rods be standardized? Obviously, they could not. But our clocks *are* standardized. Does the difficulty, therefore, vanish? No; for we can only standardize two clocks by observing that *at a given time* their hands are travelling at the same pace, and this does not prove that they will travel at the same pace at any other time. And in the nature of the case there can be no possible means of showing either that they will or that they will not; in other words, there can be no possible method of measuring the time taken by one event relatively to the time taken by another event not simultaneous with it.

If I am told in reply that, after all, it is reasonable to assume that a clock, wound up and kept in order, continues to move uniformly, and that pragmatic results justify the assumption that one "hour"—meaning one rotation of the minute-hand—is the same length as the next, I agree; but a pragmatic assumption is not quite the same thing as a measurement, and once more it appears that we have said one thing while meaning another.

(c) Suppose, again, we say that time is continuous. What, in saying this, are we denying? Presumably, that time is discrete. If it had been discrete, it would have had gaps in it. But what would the gaps have been made of? Nothing but time: any gap in a series of events must be a gap consisting of time, for if there is no time in the gap there is no gap. Clearly, then, time is continuous.—But this does not seem to follow. That which must be either discrete or continuous must be a quantity. But if time cannot be measured,[1] it is doubtful whether we ought to call it a quantity; it is, in any case, an unmeasurable quantity, and in this phrase it is reasonable to suspect a *contradictio in adjecto*. We must inquire further.

When one event in time is said to be continuous with the next, the statement is either meaningless or false. Meaningless, if "the next" is simply "that with which it is continuous"; false, if it is assumed that events in time are really packed side by side with no intervals between them. In actual history, events overlap; you cannot, except by a confessed fiction, state the point at which the event called the Middle Ages ends and the event called the Modern Period begins. This is not because our notions of the distinction between the mediaeval and the modern worlds are vague and confused. There is no sequence of events, however clearly conceived, that does not show the same overlap; and it is only when our knowledge of events is superficial and our account of them arbitrary that we feel able to point out the exact junction between them, or rather, feel that there *is* an exact junction if only we knew it. In the actual history of events there is, as the theory of compact series insists, no nextness; not so much because there is always something between (though in a sense the facts may be put that way—you may distinguish an intermediate period between the Middle Ages and the Modern Period, and so *ad infinitum*) as because there is no clear beginning or ending.

But if it is said that, whatever is true of concrete historical events, the parts of *abstract* time are continuous, a new difficulty arises. To know that two bodies are continuous, we must know that their sepa-

---

1. There may, evidently, be quantities which *we* cannot measure. But the difficulty in measuring time is not of this kind.

rate lengths are together equal to their length overall. This we can only know by measuring and adding the measurements; and this, as we have seen, cannot be done in the case of time. To put this objection differently: the continuity of any two things presupposes a system of reference other than themselves, by appeal to which we can assure ourselves of the absence of a gap. In the case of spatial bodies, this system consists of the marks on a measuring-rod. In the case of temporal events, it consists of continuous time with its clock-marked divisions. But in the case of time itself, there can be no system of reference except another time, concurrent with but distinct from the first. This second time, assumed to be continuous, can guarantee the continuity of the first by a one-one correspondence of its parts. But how do we know that *it* is continuous? There is no way of *even assuming* that it is, unless we suppose the existence of yet another time, and so *ad infinitum*. Hence, far less prove, we cannot even assume the continuity of time without surreptitiously assuming another time as a background to it, and assuming the continuity of this other.

(*d*) A difficulty of the same kind attaches to the statement that time is infinite. This presumably means that time is temporally infinite, *i.e.*, everlasting. But to say that something is everlasting means that it lasts all the time; it implies two terms—that which endures, and the time in which it endures. Hence to say that time is everlasting is to say that, in addition to the time which goes on, there is another time during which it goes on; and to say that it goes on always means that it goes on as long as this other time goes on. Once more, we are involved in the fallacy of a reduplicated time-series.

This cannot be avoided by pleading that when time is called infinite we only mean that after any given part of time another always follows. For to say this is already to say that one part of time follows after another, and this implies a system of reference by appeal to which we can say that a change or lapse has taken place. We are, in fact, back in the perplexities that arose out of the notion of time as flying or moving.

If, instead of saying that abstract time is infinite, we say that temporal events are infinite, it must again be asked whether we mean infinite in number as succeeding each other in a series of mutually external terms. If so, it must be pointed out that events are not related to one another in this way, as has already been shown; and the infinity of time seems from this point of view to be only a metaphorical phrase to describe the infinite complexity of that one event which is the history of the world.

III.

It is hardly necessary to pursue further the quest for a statement about time that shall be anything but a conscious and more or less deliberate falsehood; we should be better employed asking what it is in all these statements that makes them false. Perhaps part of the answer lies in the habit denounced by Bergson (and others before him) as the "spatializing" of time. We imagine time as a straight line along which something travels. Without inquiring too closely what it is that does the travelling, we may ask whether time is at all like a line; and, obviously, it is not. "Thought of as a line, it would only possess one *real* point—namely, the present. From it would issue two endless but imaginary arms, Past and Future" (Lotze, *Metaphysic,* Section 138 [2]). It is difficult to uproot from one's mind the illusion that somehow the past and the future exist, or that the past somehow exists, even if the future does not. Have we not, in memory and in historical inference, *knowledge* of the past? Have we not, in scientific prediction, knowledge of the future? And is it not self-evident that what we know must be real? This seems to be the argument on which we rely when we try to bolster up our belief in the reality of past and future against the attacks of the obvious common-sense reflection that what has been, and what is to be, do not in any sense exist at all. No doubt, the present would not be what it is if the past had not been and if the future were not to be; but it is a childish confusion of thought to argue that therefore the past and the future are now real. On the contrary, they are just therefore *not* real. It is just because I have left Euston and hope to get to Carlisle that I am at Crewe and not now in any sense whatever either at Euston or at Carlisle. Euston and Carlisle still exist, but they are not past or future events; the past event of my leaving the one and the future event of arrival at the other are not happening, and an event when it is not happening is just nothing. It is

2. [See p. 194 above.—C. M. S.]

true that the whole of which they are parts is happening, and that the parts, as we said, overlap one another in the structure of the whole; but this does not mean that the past *as past* continues to exist. What does continue to exist is the contribution it has made to the present.

The point may be illustrated by the way in which many theories of memory have broken down through confusing my present memory of a past event with the present effect of that past event on my bodily or psychic organism. To go about short of a leg is not the same thing as remembering the loss of it, and to suffer a neurotic disability as the result of fear is not the same thing as remembering the fear.[3] Memory is a kind of knowledge, if it is knowledge, having this peculiarity, if it is a peculiarity, that its objects have no existence of any kind whatever, and that they are known to have no existence. This may seem strange to people who believe that all thought is of a real object existing independently of the thought of it; but the alternative, to take literally the fairy-tale of the place where all the old moons are kept, is surely a good deal stranger.

To spatialize time is to fall into the illusion of thinking that past and future exist but are not "present to us" at the moment. And this fallacy seems to underlie all the ordinary statments about time—that it has one dimension, that it lapses uniformly, that it is continuous and divisble and measurable and infinite and so on—all of which rest on the assumption that a great deal of time, if not the whole of it, exists at any given moment and that we can somehow "go over" it in the same kind of way in which we go over a spatial object with a foot-rule: which we obviously could not do were it not present to us, as a whole, *now*. If this were not so, time could not be a quantity, for a quantity must exist somewhere, somehow, at some time; and when we say that from 1800 to 1900 is a hundred years, we are assuming (what we know to be untrue) that these dates exist now and that we can measure the interval between them. Nor is it better to say "from 1800 to 1900 *was* a hundred years." Was *when?* Obviously, at no identifiable time: in fact, never.

The first condition of clearing up our conception of time, then, is to

3. To give the name "mnemic phenomena" to occurrences arising out of the present effect of past experiences is to encourage, if not to betray, the confusion to which I refer.

stop thinking of it as a special kind of one-dimensional space and to think it as what it is—a perpetually changing present, having somehow bound up with it a future which does not exist and past which does not exist. Poetic imagination may think of the future as lying unrevealed in the womb of time and of the past as hidden behind some screen of oblivion; but these are metaphors, and the plain fact, obvious to anyone who will open his eyes and look at it, is that both future and past, consisting as they do of events that are not happening, are wholly unreal.

No doubt we remember the past and expect the future. But (i) these are not the same. "It is a mere accident," says a distinguished philosopher, "that we have no memory of the future"; but, accident or no accident, it is a fact. In expecting the future we may, and often do, regard it as necessarily implied in the present, but we invariably regard it as now non-existent. In remembering the past we may regard it as the necessary precondition of the present, but we invariably regard it as non-existent. Yet, though so far similar, memory and expectation are obviously and recognizably different. And merely to say that one is of the past and the other of the future does not serve to state the difference, because it leaves unanswered the crucial question, "What constitutes the difference between the past (having happened) and the future (being about to happen), granted that both are non-existent?" Attempts have been made to account for the obvious difference by a theory of the universe as an infinite midden or rubbish-heap, in which the outworn states of the present are conserved: the past is thus real and the future unreal, but unfortunately the distinction between past and present has now vanished. (ii) We do not remember *the* past, but only *our* past; and we do not expect *the* future but only *our* future. Hence the appeal to memory and expectation as guarantees of the reality of past and future proves the very opposite of what it is meant to prove: for the past and future which alone they guarantee are a purely subjective past and future.

Our attitude to *the* past is not memory but historical judgment; and this is also our attitude towards what we remember, in so far as we believe our memory to be trustworthy. Our attitude to *the* future is not expectation but something else for which, I think, we have no name, but which we distinguish from expectation somewhat as we distinguish history from recollection. I do not *remember* the battle of Waterloo, and in the same sense I do not *expect* the Aristotelian Society to survive my death; I hope and believe it will survive, and distinguish this, as a different kind of attitude, from my expectation of attending future meetings in my lifetime, even if I call it by the same name.

What we know must, I suppose, really exist. And if that is so we cannot really know either the past or the future. I am inclined to accept this consequence, and indeed to embrace it with some satisfaction as explaining why, in common with other students of history, I have found in my historical inquiries that I can never determine the exact truth about any historical fact, but have to be content with an account containing a large and unverifiable account of what I know to be conjecture. The only possible object of knowledge, I submit, is something that is real now. Of the past as past and the future as future we can have only conjecture, better or worse grounded. But, certainly, our conjectures about the past and the future are not on the same level, and I cannot dismiss the difference as accidental. Nobody's forecast of European history in the next ten years can possibly be so complete, so detailed, so well grounded, as any half-educated man's narrative of the last hundred years must necessarily be; and the difference is not, as this illustration might on hasty inspection suggest, a difference of degree. It is difficult to resist the conclusion that the future is as such not only unreal but indeterminate, belongs to the region of possibilities; and this in spite of astronomical predictions, which are, after all, hypothetical in their very essence and assume the absence of catastrophic or other disturbances. I may be reminded that such disturbances, too, are theoretically capable of prediction; but this only means that they would be capable of prediction under conditions that can never be fulfilled. *Our* knowledge of the future, at any rate, is a knowledge of the possible. The past, on the other hand, is equally unreal but is wholly determinate; it has its being in the region of the necessary but not actual. The past and the future, therefore, both baffle our endeavours to know them, but in different ways and for different reasons.

IV.

We have tried to clear up our ordinary conception of time, and have found it somewhat unstable. We certainly begin by thinking of it "spatially," as something existing *totum simul;* and we also, no less certainly, begin by thinking of it as a ceaseless flow in which the present, perpetually changing, sloughs off its states into the abyss of the past and acquires new states from the abyss of the future. And realizing, as we all do realize when we think of it, that the past and the future are non-existent, we find these two conceptions of time in mutual conflict. Yet we cannot very well abandon either. We cannot say that all time eternally exists and that the lapse from one to another is illusory, because that is nonsense. It implies that a given subject simultaneously possesses all the conflicting predicates that it ever has possessed and ever will possess, and fails to explain either how this is possible or why it certainly seems to possess now one, now the opposite. Nor can we say that time is an absolute flux from nothing into nothing through the mathematical point of the present, because that cuts the heart out of the very problem we have raised, namely, the perplexing fact that in some sense we do and must regard this flux not as a pure flux but as a series whose terms, however much they overlap or interpenetrate, really do succeed one another. For if time is a flux, it is at least a flux having determinate character and changing from something definite into something else equally definite. But if the whole truth were the reality of the present and the unreality of the past and future, the present, reduced to a mathematical point, would vanish entirely, and the wholly unreal past and future could contain no determinations of any kind. There would be no past, no present, no future, and no time. The terms of our problem, therefore, demand that *in some sense* we should restore to the past and future their actuality, in order that the present may not be exhausted of all its content. It is essential to the very being of the present that it should be in constant change. But in a purely momentary existence there can be no change; the whole universe is destroyed utterly at every instant and at every instant a wholly new universe is created. There is no *durée*, no continuity, no permanence; and where there is no permanence there is no change, for there is nothing that changes.

My concern is to justify our ordinary view of time; and my perplexity arises out of the fact that this ordinary view contains the two contradictory elements I have described. To think of time "spatially" may be a vice if pursued *à outrance;* but I hope I have shown that to eradicate this tendency by a one-sided adhesion to the opposite is no less vicious. The problem, therefore, is to find a conception of time which will justify both these tendencies of unreflective thought. And it cannot be solved psychologically, for that would lead straight into the sloughs of subjective idealism and scepticism.

The conception which I would suggest is that we should begin by distinguishing *being* from *existing*, and recognize that there are other modes of being beside existence, as well as what is generally called "subsistence," or the mode of being ascribed to the essences or attributes of the existent. Within being I would distinguish the actual from the ideal. The ideal is the non-existent, but not every non-existent is ideal; *e.g.* a square circle is non-existent, but it is not ideal. The ideal is that which is thought, but not thought as real or existing; and in this class fall the future, which is possible but not necessary, and the past, which is necessarily but not possible. The real is the present, conceived not as a mathematical point between the present and the past, but as the union of present and past in a duration or permanence that is at the same time change: the possible parting with its unnecessariness and the necessary parting with its impossibility in an actuality which is at once possible *and* necessary, not (like the abstract mathematical present) neither. Within this present there are, as really as you like, two elements (necessity and possibility), each of which taken singly or in isolation characterizes a being which is not real but ideal—the past and future respectively. Thus the past *as past* and the future *as future* do not exist at all, but are purely ideal; the past as living in the present and the future as germinating in the present are wholly real and indeed are just the present itself. It is because of the presence of these two elements *in* the present (not merely psychologically or illusorily, as in the doctrine of the specious present) that the present is a concrete and changing reality and not an empty mathematical point.

That which is ideal is for a mind, and has no other being except to be an object of mind. But the ideal and the real are not mutually exclusive. A thing may be ideal and also real. An example of this would be a duty, which is absolutely real in spite of the fact that it only exists for mind. But some things are merely ideal, and under this head fall the past and the future; unlike a duty, which exists only for thought but, for thought, really does exist, they have being for thought, but, even for thought, have no existence. Hence, if there were no mind, there would at any given moment be no past and no future; there would only be a present in which the past survived transformed and in which the future was present in germ. The past *as past* and the future *as future,* in contradistinction from their fusion in the present, have being for mind and only so. We do call the past, *as such,* into being by recollecting and by thinking historically; but we do this by disentangling it out of the present in which it actually exists, transformed, and re-transforming it in thought into what it was. Hence time, as succession of past, present and future, really has its being *totum simul* for the thought of a spectator, and this justifies its "spatialized" presentation as a line of which we can see the whole at once; it also justifies, so far as they go, subjectivist views of time like that of Kant. But time, as the ceaseless change of the present, is "transcendentally real," and the logical presupposition of any thought whatever; and this justifies the "pure flux" view of time and its treatment in philosophies like that of Bergson and Mr. Alexander.

But this conception, though the only one I can discover which gives any hope of escape from my perplexities about time, is only open to a logic which conceives the real as a synthesis of opposites and a metaphysic which has abandoned the hopeless attempt to think of all objects of thought as existent. If we must regard the real as a collocation of elements each of which is real by itself and in its own right, we must give up the solution which I have attempted to sketch and find another, if we can.

30

McKeon:

"TIME AND TEMPORALITY" *

Richard McKeon (1900–   ), Professor of Philosophy at the
University of Chicago, is one of the most distinguished of
American teachers. Long identified with a continuing fidelity to
the Aristotelian tradition, it is of significance that he was asked to
keynote an international conference on "Time and Temporality." Pointing out some of the continuing issues, topics, or "commonplaces" of discussion, his address * effectively calls upon us
to transcend the parochialism of an exclusively Western background while pointing out the 'topics' that must be brought into
the broader perspectives of an historically emerging world
culture.

The likeness and differences reported in comparative philosophy,
like all comparisons, are suspect because their perception is in the eye
of the beholder and their expression is in the vocabulary of the
reporter. The resulting distortions are somewhat mitigated when the
comparison is made, not by characterizing and naming philosophies
and doctrines, but by defining and distinguishing concepts used or
theses advanced. The concept of time is well suited for such purposes
of comparison, since it conveys a sense of concreteness and definiteness and suggests a possibility of escaping the inclusive abstractness
and ambiguous indefiniteness of terms, which are usually chosen as
more philosophical, like "being," or "reality," or "God," or "truth."
Yet "time" shares the same difficulties and dangers, for we approach
the task of comparing and contrasting concepts of time or doctrines of
time with convictions concerning what time really is and what it is
truly said to be and with supporting snippets of the history of false

* Reprinted from *Philosophy East and West*, vol. 24, No. 2, [1974]. Reprinted by
permission of the University Press of Hawaii.

conceptions and theories refuted and discarded. It is difficult to avoid the established clichés, which are the product of the accepted and repeated history of the periods and turns in the development of thought and the conventional geography of its transitions from place to place. Everybody knows, since he has learned and relearned, in school and in books, that the Middle Ages in the West was otherworldly, and, therefore, had no sense of time and no knowledge of or acquaintance with history. It is generally believed that one reason why East and West will never meet is that the Indians had no history until Greek historians taught them how to mark off historical periods by dates and how to trace consequences to causes and so transform poetical and mythical accounts of the Indian past into histories, and that the Chinese who, although they had histories which recorded the past and clocks which measured the lapse of time, had no knowledge of the nature of time and developed no science of mechanics.

Nonetheless time is not an entity which is encountered, or a concept which is perceived, in isolation. Indeed the history of thought about time includes theories which deny the existence of time and the perception of time. Even if it exists and can be experienced, time comes into being and knowledge only in connection with something else. "Time and temporality" is a formula to designate time in its circumstances, substantive and cognitive, and it may be used as a device by which to develop and examine the variety of circumstances in which "time" acquires its variety of meanings in the context of a variety of problems, philosophical in nature but with consequences detectable and traceable in history, science, art, and social and cultural structures. It is a device which the ancients called a "commonplace" or "topic" and used to discover arguments and relations among ideas and arguments. It was the device used and developed both in the arts of invention and in the arts of memory.

Time and temporality, time and the circumstances in which time is perceived as a problem or as a structure, is a formula which takes many forms and particularizations. They are easily recognized in the pairs of terms used in the treatments of time by philosophers, past and present, Eastern and Western—"time and eternity," "time and motion," "time and duration," "time and space." The list could be extended indefinitely to bring in other problems of time and other definitions of time, but they are all interrelated, and the pattern that a few of them form provides a structure for comparative analysis and consequential inquiry.

"Time and eternity" opens up the topic of time and change in the context of unchanging being. If time is, as Plato defined it, the moving image of eternity, time reflects in the sequences of change a structure of intelligibility, of intelligence, and of being. Cosmic order and cosmic motion are prior to, and order, the motions of things and the sequences of phenomena both as occurrences and as appearances. Moreover, the motion of the world-soul, which combines the motion of the same and the motion of the other, is prior in nature to the motions of bodies. Time enters into the analysis of the making of the world by the way of reason; space enters into the construction of bodies by the way of necessity. Substance and eternity, according to Spinoza, are conceived by intellect alone; quantity and duration may be conceived concretely and adequately by the intellect or abstractly and superficially by the imagination. And when we abstract quantity from substance, and duration from the mode by which it flows from eternal things, time and measure result—time to determine duration, and measure to determine quantity.

"Time and motion" places time in the context of change. In this commonplace some philosophers have identified time with the motion of the universe, or with the universe itself, or with motion, while others have held that it is distinct from but associated with motion as the measure of motion. As measure, time is involved in problems and paradoxes which place it in further commonplaces. One is the paradox of the measure and the measured: time may be said to measure motion, or motion may be said to measure time. Another is the paradox of the continuous and the discrete, of time and the moment or the now. The now provides the boundaries, the beginnings and the ends, of time which mark off discrete periods of time; the now also provides the continuity of time, joining the past to the future. Time is a continuity divided paradoxically into three parts: the past which is a continuity but no longer exists, the future which is a continuity but does not yet exist, and the present, the now, the moment, which exists but is discrete. There are such consequent logical paradoxes in the modalities of time as the sea fight tomorrow, which Aristotle propounded and which is still the subject of debate among logicians today. It is necessary that it either take place or not tomorrow, but it is not necessary that it take place tomorrow and it is not necessary that it not take place tomorrow. If there is a real alternative in future events, it is necessary by the principle of contradiction that one or the other take place, but the necessity is not determinative of which shall in fact occur. If the future is contingent, the transition from the future to the present is a transition from contingency to actuality, and the transition from the present to the past is a transition from actuality to necessity. Laws of motion which project from present observations of positions and motions will be laws of probability determined by a principle of indeterminacy or contingency. If future events are necessary, the same modality applies to statements about past, present, and future; but time is then inseparably attached, not to dimensions of space, but to specific motions, things, and causes. From these places have been developed distinct antagonistic theories of time based on the examination, respectively, of the necessities of what has been, the actualities of existence, and the possibilities of what will be.

Further paradoxes develop from the topic of time and what is in time. When what is in time is kinds, or collections, or sequences of individual things, persons, and events, the order of things and the account of their characteristics, relations, or sequences are histories. Histories of discrete things, and their species, and genera may be without time, as in the classifications of natural history, or with time, as in theories of the evolution of species. Histories of continuous sequences of things, events, persons, or institutions make use of time in a variety of ways constituting different kinds of histories embodying different theories of time. When what is in time is particulars identified by the universal characteristics they embody as particulars or by universal characteristics of the sequences in which they occur and are known, it is bodies with powers and functions acting and reacting in positions, or places, or space, and time in an ordering principle in the resulting physics, or physiology, or psychology. Finally, time may be conceived as an order separate from what is ordered in time, as Leibniz defined it, "an order of succession." What succeeds presents a further paradox: it may be a series of things or events or ideas, and paradoxically whichever one is chosen accounts for the others—a succession of things accounts for events and ideas, a succession of events produces things and ideas, and a succession of ideas identifies and explains things and events.

"Time and duration" is a topic which has operated in the first two commonplaces. Time and eternity were distinguished by use of the faculties of the mind by which they are perceived or conceived. Eternity is conceived by the intellect or understanding; time is perceived and differentiated by sensation and imagination and opinion. Eternity is of being; time is of becoming or existence. Duration may be perceived by the intellect as it flows from being, or it may be perceived by sense and imagination as time which measures duration. Time and motion, the second commonplace, was shown to involve the further topic of time and what is in time and thus to lead to the differentiation of successions of ideas from, or as a kind of, succession of things. Even philosophers who relate time to the motions of things acknowledge that there would be no time if there were no mind to observe and measure time. They sometimes distinguish a numbering number, supplied by the numbering mind, from a numbered number, found in things and motions; and they make time a numbered number to associate it more closely with the motion of things and to separate it from the intrusions of motions of minds. Other philosophers insist no less strongly and no less repeatedly that there would be no measuring by time unless there was something to measure, and they distinguish between duration which is perceived and lived before it is cut into moments and parts by time and measuring devices. Even if the succession of ideas is chosen in preference to the succession of things, however, a further topic remains to determine what mind or soul endures, the world mind or the individual mind. According to Plotinus, time began when the universal soul entered into movement, and it arose as a measurement of the activity of the soul. It is the life of the soul, and it consists in the movement by which the soul passes from one stage of its life to another. It is an image of eternity. According to St. Augustine, the mind in which time is measured is the individual mind. Time is a "distention of the mind," and the measurement of time is in the mind. There are three actions of the mind: memory, attention, and anticipation. They exist only as present actions. Past, present, and future exist therefore in the present—the past is the present memory of the past, the present is present perception of the present, and the future is present anticipation of the future. According to Bergson, the time of science is a spatialized time, whereas the duration of consciousness is a flowing current that has two characteristics: absolute novelty at each instant, and infallible and total conservation of all the past.

Many of the characteristics attributed to time, which seem to be substantive and not merely semantic, follow from the choice of commonplaces which determine inquiry and are not the products of warranted knowledge. If time is the measure of motion, time is continuous and without beginning or end; if motion is the measure of time, time begins with the beginning or creation of the world. Aristotle held that matter, motion, and time are eternal; Plato, Plotinus, and St. Augustine held that time began with the beginning of the world and of the life of the world soul. The choice moreover determines different characteristics for time: for the former time is associated with particular motions and neither space nor time are empty or characterless; for the latter time flows evenly and smoothly.

"Time and space" is a topic involved in the measurement of time, both in the theoretic and the chronometric measurement. Motion was defined and analyzed in terms of time and space even in antiquity. Galileo's construction of equations for motion and acceleration involved, to use Bergson's terms, the spatialization of time, the representation of both time and distance as straight lines. He conducted experiments to determine the acceleration of falling bodies using a crude clock, water dripping from the hole of a container. He made imaginary experiments with the pendulum. It was not until Huyghens perfected the pendulum and constructed a precision clock that the measurement of the accelerations of rolling balls or swinging pendulums was possible, but the experiments were then unnecessary since the pendulum clock was at once the measuring device and the experiment. When Newton distinguished absolute motion from relative motion, absolute time was distinguished from relative time. Absolute time, like Galileo's time, flows evenly. In relativity physics the dimensions of time and space are inseparable, and motion is conditioned by the characteristics of space-time to follow the paths of geodesics. The theoretic measurement of time is inseparable from the technological measurement of time. Artificial measuring devices have been constructed from antiquity to the present, in the East and in the

West. The use of natural measuring devices has had a similarly long history, observation of the positions of the planets, to physical and biological rhythms of vibration, to the rate of disintegration of radioactive elements. Similar natural measuring devices have been sought for historical time—geological formations, fossil remains, the rings of trees—and they have been supplemented by the artificial devices of archaeology and the extensions of human history from local histories and travel accounts of foreign places, to the universal histories of empires and religions, to the universal histories of arts and sciences and of culture and mankind.

If these variations in the meanings and instances of time were presented as an account of doctrines or of statements alleged to be true, they would each be in contradictory and incompatible opposition to the others. Since they have been presented as a pattern of commonplace possibilities for analysis, inquiry, and application, they stand instead in the relation of alternatives which focus on different aspects of time brought to the attention by different temporalities from which time takes its meanings. As alternatives they open up the way to progress in the investigation of time and the way to intelligibility in the comparison of doctrines of time developed in different philosophies in different cultures and at different times in each tradition. They afford a means of transition from the times of histories to the times of sciences, to the times of arts, to the times of experiences; and they suggest that there is a variety of kinds of histories, of sciences, of arts, and of experiences which deserve to be distinguished and examined as functions of time. They suggest that the commonplace used as the title of an influential modern book, *Being and Time,* might make the analysis to time developed in that topic clearer, if the topic itself were considered in contrast to alternative topics which it displaces—phenomena and time, motion and time, existence and time, nature and time. They suggest, in turn, topics in which time is related not only to objects which are determined and constituted in time, but also to faculties of the mind by which time is perceived and comprehended—sensation and time, memory and time, imagination and time, discursive reason and time, understanding and time, intuition and time. These commonplaces suggest the need to examine the commonplaces of living processes by which the processes of the observer and thinker are related to the processes observed and thought. They are devices by which comparative philosophy may uncover, in its inquiries and analyses, not rigid structures of dead, past philosophies or of opaque alien philosophies but living relations that animate past and unfamiliar inquiries and controversies and that are still relevant and operative in contemporary problems oriented to the emergence of new conceptions of time in the formulation and testing of new solutions to new problems.

NOTES

I. FOREWORD: IN THE BEGINNING

1. Henri Yaker, "Time in the Biblical and Greek Worlds," in Yaker et al., *The Future of Time* (New York: Doubleday & Company, Inc., Anchor Books, 1972), p. 32.
2. Kathleen Freeman, trans., *Ancilla to the Pre-Socratic Philosophers: A complete translation of the Fragments in Diels,* Fragmente der Vorsokratiker (Cambridge: Harvard University Press, 1957), p. 19.
3. Cf. Plato, *Phaedo,* 70–73.
4. Cf. Yaker, op. cit., p. 17.
5. George Boas, "The Acceptance of Time," *University of California Publications in Philosophy,* vol. XVI, no. 12, 1950, p. 250.

II. TIME AND MOTION

1. See Plato, *Phaedo,* 73 and *Theatetus,* 178–79.
2. J. F. Callahan, *Four Views of Time in Ancient Philosophy* (Cambridge: Harvard University Press, 1948), p. 18.
3. Plato, *Timaeus,* 48.
4. Cf. F. M. Cornford, *Plato's Cosmology* (London: Routledge & Kegan Paul, Ltd., 1937, 1948), p. 103.
5. Plato, *Timaeus,* 47.
6. Cf. Plato, *Timaeus,* 53b.
7. Cf. Aristotle, 251b10.
8. Aristotle, 449a 20–25.
9. Aristotle, 448b 25–30.
10. Aristotle, 436a 15–20.
11. See Aristotle, e.g., 251b 10–15, 252b 3, 267b20–26, 1071b 6–10.
12. W. Windelband, *A History of Philosophy,* trans. James H. Tufts (New York: The Macmillan Company, [1893], 1901), p. 215.
13. The essential structure of Plotinus' universe can be described as four-layered. At the pinnacle is the completely transcendent, primordial reality. Usually referred to as 'The One', it can, in other contexts, be described as 'The Good', 'The Absolute', 'The Simple', 'The Infinite', 'The Unconditioned'. The ultimate ground of all that is, it is so completely transcendent that no human predicates can really be ascribed to it; it is, therefore, inherently unknowable. Dependent on this ultimate source of all being are three knowable levels or "emanations" of reality. First is what can be variously described as existent 'Thought', 'Mind',

'*Nous*'; somewhat analogous to Plato's Demiurge, it is the first sign of multiplicity; as it produces plurality out of its own higher unity it incorporates the Ideas and may be regarded as the intellectual principle, as the Intelligible; it is then the first knowable. The second emanation is the Soul which may be patterned after the World-Soul of Plato's *Timaeus*. More directly the cause of movement, it is the eternal cause of the fourth level, the cosmos or material world which is its own emanation. The Soul is the connecting link between the supersensual and the sensual, the togetherness of individual souls which are individually real yet bound together. The three aspects of the Divine—the One, the Intelligible, the Soul—are not three separate entities but three hypostases of one unitary Supreme Being.

III. TIME AND UNDERSTANDING

1. René Descartes, "Meditations on First Philosophy," in *The Philosophical Works of Descartes*, trans. E. S. Haldane and G.R.T. Ross (Cambridge: Cambridge University Press, 1968), vol. I, p. 168.
2. "Arguments Demonstrating the Existence of God," (Addendum to Reply to Objection II), *Axiom II*, op. cit., vol. II, p. 56.
3. "Reply to Objections, V," op. cit., vol. II, p. 219, par. 9.
4. Because of the central import of this fiftyseventh Principle, it is important to compare it with a similar rendering from the authorized French version:

   That there are attributes which pertain to the things to which they are attributed, and that there are other attributes which depend upon our thinking.
   Of these *qualities* or attributes, there are some which are in things themselves, and others which are only in our thinking. Thus time, for example, which we distinguish from duration taken in general, and which we say is the number of motion, is nothing other than a certain *manner* in which we think about that duration, because we do not conceive that the duration of things which are moved is other than the duration of things which are not moved: as is evident from the fact that, if two bodies are moved for an hour, the one quickly and the other slowly, we do not count more time in the one than in the other, although we suppose more motion in the one of these two bodies. But, in order to comprehend the duration of all things under

the same measure, we ordinarily make use of the duration of *certain* regular motions which form the days and the years, and having thus compared it, we call it time; although in effect what we name in this way is nothing, *over and above* the true duration of things, except *a manner* of thinking.

Especially translated by John J. Blom from the original Latin and French texts as given in:Descartes, *Ouevres,* Adam & Tannery, (Paris: Vrin, 1964), VIII-1, pp.26-27, and IX-2, pp. 49-50.

5. For the sake of comparison one should also note the companion definition of absolute space:

Absolute space, in its own nature, without relation to anything external, remains always similar and immovable. Relative space is some movable dimension or measure of the absolute spaces; which our senses determine by its position to bodies; and which is vulgarly taken for immovable space; such is the dimension of a subterraneous, an aerial, or celestial space, determined by its position in respect of the earth. Absolute and relative space are the same in figure and magnitude; but they do not remain always numerically the same. For if the earth, for instance, moves, a space of our air, which relatively and in respect of the earth remains always the same, will at one time be part of the absolute space into which the air passes; at another time it will be another part of the same, and so, absolutely understood, it will be continually changed. [Motte, trans., "Scholium to Definition VIII," *Philosophiae Mathematica,* parts I–II; quoted in *The Leibniz–Clarke Correspondence,* ed. H. G. Alexander (Manchester: Manchester University Press, 1956), pp. 152–153.]

6. See *Essay on the Human Understanding,* ed. A. C. Fraser (New York: Dover Publications, n.d.), Bk. II, chap. XV, sec. 11.

7. Op. cit., Bk. II, chap. XV, sec. 12.

8. Op. cit., Bk. II, chap XV, sec. 12. cf. re Alexander, p. 178 below.

9. See G. W. Leibniz, *The New Essays on the Human Understanding,* trans. A. G. Langley (LaSalle: Open Court Publishing Co., 1949), p. 156.

10. We should take note in passing of the import of a change of metaphor, as it reverses the whole priority of the 'flow' and of the 'moment'; for in the stream analogy, the 'drop' is an abstraction from the flow of water as the 'moment' is from the flow of time.

11. Leibniz, "Fifth Paper," *The Leibniz–Clarke Correspondence,* p. 85.

12. G. W. Leibniz, *Discourse on Metaphysics,* trans. P. G. Lucas and L. Grint (Manchester: Manchester University Press, 1953), p. 23.

13. G. W. Leibniz, "Principles of Nature and of Grace," trans. and ed. L. E. Loemker, *Leibniz: Philosophical Papers and Letters* (Chicago: University of Chicago Press, 1956), vol. II, p. 1040.

14. For example, when I see an animal which I call a 'horse', I do not only see a generally brown shape; I become aware of it together with my conceptual notion of 'horse' and with whatever behaviorial connotations that that notion holds for me. This empirical concept of 'horse' is structured in a certain way that, abstractly can be set out as four groups of categories or "pure concepts"; these are a priori, Kant argues, and necessarily present in *all* empirical concepts.

15. Cf., Immanuel Kant, *Critique of Pure Reason,* trans. N. K. Smith, (London: Macmillan & Co., Ltd., 1933, 1968), A98–A103, pp. 131–133.

16. See Immanuel Kant, *Inaugural Dissertation and Early Writings on Space,* trans. J. Handyside (Chicago: The Open Court Publishing Company, 1929), sec. 28, pp. 78–81.

17. Kant, *Critique of Pure Reason,* A80 = B106, p. 113.

18. Immanuel Kant, *Prolegomena to Any Future Metaphysics,* trans. and ed. P. Carus (LaSalle: The Open Court Publishing Company, 1902, 1947), sec. 5, p. 26.

19. Ibid., sec. 10, p. 36.

20. Kant, "An Inquiry into the Distinctness of the Principles of Natural Theology and Morals," trans. and ed. L. W. Beck; Kant, *Critique of Practical Reason: And Other Writings in Moral Philosophy* (Chicago: The University of Chicago Press, 1949), cf., especially pp. 263, 266, 269.

21. Kant, *Critique of Practical Reason,* p. 235.

22. Ibid., p. 239.

IV. TIME AND REALITY

1. G. W. F. Hegel, *The Phenomenology of Mind,* trans. J. B. Baillie, Second Edition (London: George Allen & Unwin, Ltd., 1931, 1955), p. 104.

2. Alexandre Kojève, *Introduction to the Reading of Hegel,* trans. J. H. Nichols, Jr., ed. Allan Bloom (New York and London: Basic Books, Inc., 1969), p. 133.

3. See Hegel, op. cit., p. 800; J. N. Findlay, *Hegel: A Re-Examination* (London: George Allen & Unwin, Ltd., 1958), p. 146.

4. See Kojève, ibid.

5. G. W. F. Hegel, *The Philosophy of History,* trans. J. Sibree, Revised Edition (New York: The Colonial Press, 1899), p. 77.

6. See Hegel, *Phenomenology*, pp. 149 ff.

7. Aristotle, 1071b, quoted in Hegel, *Phenomenology*, p. 807 n.

8. Hegel, op. cit., p. 800. Cf. Kant, *Critique of Pure Reason*, A292 = B348, p. 295.

9. Kojève, ibid. (N.B. It is interesting to note that much of Kojève's inter-pretation of the significance of some aspects of Hegel's conception of time in the *Phenomenology* seem to incorporate a reading drawn from, or inspired by, Heidegger's *Being and Time;* cf. chap. VII and sel. 28 below, and "Editor's Introduction" to Kojève, p. x.)

10. Nicolas Berdyaev, *The Beginning and the End*, trans. R. M. French, (New York: Harper and Brothers Publishers, 1957) p. 000.

11. Etienne Gilson, *Being and Some Philosophers* (Toronto: Pontifical In-stitute of Mediaeval Studies, 1949), p. 141.

12. John Passmore, *A Hundred Years of Philosophy* (London: Gerald Duckworth & Co., Ltd., 1957), p. 49. (N.B. For a clear and concise overview of Lotze's metaphysics and influence see the excellent "In-troduction" by P. G. Kuntz to George Santayana, *Lotze's System of Philosophy* (Bloomington and London: Indiana University Press, 1971).

13. Hermann Lotze, *Metaphysic*, trans. T. H. Green and ed. B. Bosanquet (Oxford: The Clarendon Press, 1887), vol. I, p. 119, par. 47.

14. Harald Höffding, *A History of Modern Philosophy*, trans. B. E. Meyer (New York: The Humanities Press, 1950), vol. II, p. 509.

15. Hermann Lotze, *Microcosmus: An Essay Concerning Man and His Relations to the World*, trans. E. Hamilton and E. E. C. Jones (Edin-burgh: T. & T. Clark, 1885), vol. I, pp. 231, 230.

16. Augustine, *The Confessions of St. Augustine;* see p. 83 above.

17. N.B. This phrase is the English translation of the original French title, *Essai sur les données immédiates de la conscience*, which was translated into the English under the title, *Time and Free Will.*

18. Henri Bergson, *Creative Evolution*, trans. A. Mitchell (New York: Henry Holt and Company, 1911, 1937), pp. 344, 369–70.

19. Samuel Alexander, *Space, Time, and Deity* (New York: The Humanities Press, 1920, 1950), vol. II, pp. 38, 44; cf. p. 368.

20. Alexander, vol. I, p. 143; cf. p. 103 above.

V. THE ANALYSIS OF TEMPORAL CONCEPTS

1. Passmore, op. cit., p. 75n.
   N.B. By far the most extensive commentary has been, of course, that by his editor and Cambridge colleague, C. D. Broad, which is suitably entitled *An Examination of McTaggart's Philosophy.*

2. Cf. McTaggart, *The Nature of Existence,* ed. C. D. Broad (Cambridge: Cambridge University Press, 1927), vol. II, pp. 4, 273.
3. Cf. ibid., pp. 243–244.
4. Cf. ibid., Book VII, pp. 345–480, especially chap. LXI, "The Futurity of the Whole."
5. David Pears, in a carefully reasoned essay entitled "Time, Truth and Inference" (in Flew, *Essays in Conceptual Analysis,* London, 1963, pp. 228–252) has argued that McTaggart, like most 'logicists', has been "preoccupied with adjectives and neglected verbs." Finding a prime source of McTaggart's logicism in his failure to recognize the essential temporality which the verb imposes on any meaningful assertion (cf. p. 462, Pears has provided an essay which would seem to be an essential prelude to any serious discussion of temporal language.

 We might also note that McTaggart's logical rigor has enabled him to provide us with an important lesson though not one he intended. For his clearly developed logicism is a paradigm case of what may be expected when we forget that logical rules are conceptual tools developed by time-bound men. Once we subsume temporal experience under the strictures of a presumably non-temporal logic, then time *must* appear as unreal, as must every mark of change and of life—just because it does not measure up to the presumably non-temporal transcendent standard being applied and is obviously discordant with it. Indeed, Zeno seems to have offered a similar demonstration some two millenia ago. It would seem that the paradoxes revealed by such imposition of the allegedly timeless upon the essentially temporal would raise the question about the propriety of doing so. At least the question ought to be faced in candor by the thinker who attributes to his logic all the marks of a transcendent truth and then persists in this quixotic attempt to subsume time, life, and change under the lifeless frozen forms of an allegedly atemporal mode of thought, without extending us the courtesy of explaining by what right he does so. If it were remembered that human thinking is never experienced except in a temporal frame, we might come to see these systems of non-changing logics as the conceptual tools they are, created for human purposes out of human intellectual experiences by timebound men. Tools are not to govern but to be used when and as appropriate; most carpenters and mechanics are taught to respect their tools by not misusing them; perhaps philosophers someday will be as well.
6. For an excellent presentation of some original material on both sides of this debate, see Richard M. Gale, ed., *The Philosophy of Time: A Collection of Essays* (Garden City: Doubleday & Company, Inc., Anchor Books, 1967), especially secs. II and III.
7. Russell, quoted, Passmore, op. cit., p. 273.

8. It is somewhat startling to note the frequency with which those who are known for insisting on logical rigor concede arbitrary personal bias as the point of departure. One might well pursue the particular point here by examining a pamphlet which, to the best of my knowledge, has not been reprinted, viz., B. Russell, *The Philosophy of Bergson with a Reply by H. Wildon Carr and a Rejoinder by Mr. Russell* (London: Macmillan & Co., Ltd., 1914).

9. W. V. Quine, *Word and Object* (Cambridge: The M.I.T. Press, 1960), p. 170.

10. Quine, ibid., p. 171.
    N.B. One needs to investigate this whole issue of the representation of time in spatial terms which is a theme, as we have seen, with a long history of its own. One particularly helpful discussion is to be found in a paper by Nathaniel Lawrence entitled "Time Represented as Space" in *The Monist*, vol. 53. no. 3 (July, 1969), pp. 447-456.

11. N.B. One presumption also running through philosophic history, which is implicit in Reichenbach's point here, is the association of time and change. One discussion which may, indeed, open some now ground on this issue is the article entitled "Time Without Change" by Sydney Shoemaker, *The Journal of Philosophy*, vol. LXVI, no. 12 (June 19, 1969), pp. 363-381.

12. Hans Reichenbach, *The Direction of Time*, ed. Maria Reichenbach (Berkeley: University of California Press, 1971), p. 269.

13. Werner Heisenberg, *Physics and Beyond: Encounters and Conversations*, Trans. A.J. Pomerans, (New York: Harper & Row, Publishers, 1971), p. vii.

14. At the age of seventy-three, Whitehead still acknowledged the continuing import of Bergson, among others, to a philosophy of time; see J. G. Brennan, "Whitehead on Time and Endurance," *The Southern Journal of Philosophy*, vol. XII, no. 1 (Spring 1974), pp. 117-126.

VI. THE SIGNIFICANCE OF EXPERIENTIAL TIME

1. Too little notice has been taken of the decisive impact of German Idealism on the formation of American thought, first through Ralph Waldo Emerson and New England Transcendentalism, and then through 'technical' philosophy itself. Hardly alluded to in most current books on the history of American philosophic thought, various aspects of this development are spelled out in the unusually outstanding book by H. S. Thayer, *Meaning and Action: A Critical History of Pragmatism*, (Indianapolis, New York: The Bobbs-Merrill Co., 1968); cf. esp. pp. 47, 52-58, 74, 278, 456.

2. For an excellent critical discussion of this topic, see Bruce Wilshire, *William James and Phenomenology* (Bloomington and London: Indiana University Press, 1968), esp. chaps. 5–8. Beyond the immediate topic of this book, it is a most important contribution to the largely unexplored question of the relationship between American pragmatism and existential phenomenology.

3. William James, *A Pluralistic Universe* (New York and London: Longmans, Green & Co., Ltd., 1909, 1943), p. 254.

4. For a discussion of this last point, see Gerald E. Myers, "William James on Time Perception," *Philosophy of Science*, vol. 38, no. 3 (September 1971).

5. The continuity of this identification of knowledge, meaning and action, under the aegis of futurity, as a fundamental theme binding otherwise diverse spokesmen of the pragmatic tradition together, may be seen in the work of C.I. Lewis, a more contemporary pragmaticist thinker. In his late book, *An Analysis of Knowledge and Evaluation* (1946), for example, his opening discussion, "Knowledge, Action, and Evaluation," is offered under the heading "To know is to apprehend the future as qualified by values which action may realize."

6. The ignored affinity of Peirce for Royce is indicated quite clearly by a 1903 note by Peirce in which, protesting some Jamesian interpretation of his work, he wrote that he "continues to acknowledge, not indeed the Existence, but yet the Reality, of the Absolute, nearly as it has been set forth, for example, by Royce in his *The World and the Individual,* a work not free from faults of logic, yet valid in the main." Quoted in Max Fisch, ed., *Classic American Philosophers* (New York: Appleton–Century–Crofts, Inc., 1951), p. 54 n. That this was not a passing fancy is indicated by Peirce's statement in his 1891 essay, "The Architecture of Theories," that "The one intelligible theory of the universe is that of objective idealism, that matter is effete mind, inveterate habits becoming physical laws."

7. Bruce Kuklick, *Josiah Royce: An Intellectual Biography* (Indianapolis: The Bobbs-Merrill Company, Inc., 1972), p. 4.

8. Gabriel Marcel, *Royce's Metaphysics,* trans. V. and G. Ringer (Chicago: Henry Regnery Company, 1956), p. xii.

9. Kant, *Critique of Pure Reason*, A51 = B75, p. 93.

10. The relation between Bergson and Royce is one that merits more consideration than it has generally received. One might well refer to Royce's paper, "The Reality of the Temporal," *The International Journal of Ethics*, (April 1910), pp. 257–271. Some Roycean themes may be seen as having been anticipated by Bergson already in his first book—although it is not known just when Royce became familiar with Bergson's work; see, e.g., Bergson's *Time and Free Will,* trans. F. L. Pogson (London: George

Allen & Unwin, Ltd., 1910), esp. pp. 175, 183, 200, 233. Also see the excellent article by Milic Capek, "Time and Eternity in Royce and Bergson," *Revue internationale de philosophie*, vol. 79, pp. 22–45.

11. Quoted by J. Loewenberg, "Introduction," in Josiah Royce. *Fugitive Essays* (Cambridge: Harvard University Press, 1925), pp. 31, 33–34. (I am indebted to Prof. Rickard Donovan for bringing these passages, as well as the Capek paper, referred to in note 10, to my attention.)

12. Henry D. Aiken, "Introduction, Pragmatism and America's Philosophical Coming of Age," Barrett and Aiken, *Philosophy in the Twentieth Century* (New York: Harper & Row, 1971), vol. I. p. 52.

13. See pp. 113 ff above.

14. John Dewey, *Experience and Nature* (New York: W. W. Norton & Company, 1929), pp. 110–111.

VII. THE STRUCTURE OF EXPERIENTIAL TIME

1. See, e.g., Jean Piaget, *Insights and Illusions of Philosophy*, trans. Wolfe Mays (New York and Cleveland: The World Publishing Co., 1971). p. 131.

2. Quoted by H. Spiegelberg, *The Phenomenological Movement* (The Hague: Martinus Nijhoff, 1960), vol. I, pp. 87–88.

3. See Ibid., p. 88.

4. Jean Piaget, *Genetic Epistemology*, trans. E. Duckworth (New York and London: Columbia University Press, 1970), p. 9.

5. Piaget's stricture against reducing epistemology to logical formalism applies, with special force, to the concept of time:

The first reason is that there are many different logics, and not just a single logic. This means that no single logic is strong enough to support the total construction of human knowledge. But it also means that, when all the different logics are taken together, they are not sufficiently coherent with one another to serve as the foundation for human knowledge. . . .

The second reason is found in Gödel's theorem. It is the fact that there are limits to formalization. . . . logic is a formalization, an axiomatization of something, but of what exactly? What does logic formalize? This is a considerable problem. There are even two problems here. Any axiomatic system contains the undemonstrable propositions or the axioms, at the outset, from which the other propositions can be demonstrated, and also the undefinable, fundamental notions on the basis of which the other notions can be defined. Now in the case of logic what lies underneath the undemonstrable axioms and the undefinable notions? This is the problem of structuralism in logic, and it is a problem that shows the inadequacy of formalization as the fundamental basis. It shows the necessity for considering thought itself as well as considering ax- iomatized logical systems, since it is from human thought that the logical systems develop and remain still intuitive. [Ibid., pp. 10–11.]

6. Ibid., p. 15.
7. Jean Piaget, *The Child's Conception of Time*, trans., A. J. Pomerans (New York: Basic Books, Inc., 1969), p. x.
8. Ibid., p. 3.
9. Ibid., p. 60.
10. Ibid., p. 2, 197; cf. p. 103 and p. 178 above (Alexander and Locke).
11. *Genetic Epistemology*, pp. 64–65.
12. This suggests that one of McTaggart's problems is that even his A series, which came closest to experiential time, was still conceived as a *series*—of points on a line which (if the somewhat extraneous observer were removed) would really more accurately be described (in themselves and not from the viewpoint of an interpretive experiential structure) in terms of an unalterable B series of before-and-after. For his A series-subject was merely a *passive* and extraneous observer of a B-series set of events, certainly not an active participant in the perceptual relationship. It is, then, little wonder that, using a logically static B-series model, his fictional subject appeared logically discordant with what he was alleged to experience and disappeared into logical paradoxes while a timeless logicism was invoked to soothe the philosopher's brow.
13. Edmund Husserl, *The Phenomenology of Internal Time-Consciousness*, ed. Martin Heidegger, trans. James S. Churchill, intro. Calvin O. Schrag (Bloomington: Indiana University Press, 1964), p. 28.
14. Kant, *Critique of Pure Reason*, A65 = B90, p. 103.
15. Husserl, op. cit., pp. 43-44.
16. M. Merleau-Ponty, *Phenomenology of Perception*, trans. Colin Smith (New York: The Humanities Press, 1962, 1966), pp. 416–417. (Italiciza- tion of last sentence added—C.M.S.)
17. Husserl, op. cit., p. 23.
18. See H. Bergson, *Time and Free Will*, trans. F. L. Pogson, (London: George Allen & Unwin, Ltd., 1910, 1950), p. 210.
19. Spiegelberg, op. cit., vol. II, p. 399; cf. p. 403.
20. One who is impressed by *Zeitgeist* theories would note that Husserl and Bergson were both born in 1859 and that Bergson's first book was published in 1889, the year in which both Heidegger and Collingwood were born.

21. Nancy Metzel, "Translator's Introduction," in E. Minkowski, *Lived Time: Phenomenological and Psychopathological Studies,* (Evanston, Ill.: Northwestern University Press, 1970), p. xxxv.
22. See note 18 above.
23. Metzel, op. cit., p. xxvi.
24. See Minkowski, pp. 31–32.
25. Minkowski, p. 83.
26. Cf. Kant, *Critique of Pure Reason,* Bxiii, p. 20. For a defense of the thesis that Heidegger is really carrying Kant's Critical philosophy forward from a transcendental into a phenomenologically grounded existential ontology, see my *Heidegger, Kant and Time* (Bloomington and London: Indiana University Press, 1971): with regard to this particular reference, see especially pp. 67–96, 222–246.

N.B. In this connection one might well note the way in which Kant, himself, seems to have intimated something fairly close to Heidegger's concept of the temporal field:

Men are more interested in having foresight than any other power, because it is the necessary condition of all practical activity and of the ends to which we direct the use of our powers. Any desire includes a (doubtful or certain) foresight of what we can do by our powers. We look back on the past (remember) only so that we can foresee the future by it; and as a rule we look around us, in the standpoint of the present, in order to decide on something or prepare ourselves for it.

Empirical foresight is *anticipation of similar cases (exspectatio casuum similium)* and requires no rational knowledge of causes and effects . . .

Anthropology From a Pragmatic Point of View, trans. M.J. Gregor, (The Hague: Martinus Nijhoff, 1974), pp. 59–60.

27. The following table is taken from Michael Gelven's *A Commentary on Heidegger's "Being and Time"* (New York, Evanston and London: Harper & Row, 1970), p. 19:

	Being (Sein)	Entity (Das Seiende)
Object of Inquiry:	Being (Sein)	Entity (Das Seiende)
Type of Inquiry:	ontological	ontical
Terms of Inquiry:	existentials	categories
Status of Occurrence in Inquiry:	factical	factual
Type of Self-awareness in Inquiry	existential	existentiell

28. M. Heidegger, *Being and Time*, trans. John Macquarrie and Edward Robinson (New York: Harper & Row, 1962), pp. 63 (39), 374 (327). Cf. Kant, "Letter to Christian Garve," in *Kant: Philosophical Correspondence 1759–1799*, ed. A. Zweig (Chicago: University of Chicago Press, 1967), p. 101.

29. See Aristotle, *De Interpretatione*, 16a19–20, 16b6–7. N.B. In the German language, the word for 'verb' is *Zeitwort*, or 'time-word'—the answer to the question, 'when?'; the English 'verb' is derived from the Latin *verbum* or 'word' while the English 'noun' is derived from the Latin word for 'name'—thus perhaps indicating that it is merely one kind of word which arises from the act of naming and is not as fundamental as its normal use in English would seem to suggest.

30. Heidegger, op. cit., p. 456 (404).

31. Ibid., p. 464 (411).

32. Cf. ibid., p. 497, n. iii.

33. G. De Ruggiero, *Existentialism: Disintegration of Man's Soul* (New York: Social Science Publishers, 1948), p. 86.

34. See Heidegger, *Kant and the Problem of Metaphysics*, trans. and introd. James S. Churchill (Bloomington: Indiana University Press, 1962), p. 249; cf. Sherover, op. cit., pp. 241 ff.

35. Heidegger, "The Way Back Into the Ground of Metaphysics," in Kaufman (ed.), *Existentialism from Dostoevsky to Sartre* (New York: Meridian Books, 1958), p. 215.

VIII. THE OPEN AGENDA

1. For example, see Richard Kröner, *Kant's Weltanschauung*, trans. John E. Smith (Chicago: The University of Chicago Press, 1956), esp. chap. V., "Primacy of the Practical."

2. As but one example of the way in which this theme may be carried forward, see my essay, "Time and Ethics: How is Morality Possible?," J. T. Fraser & N. Lawrence (eds.), *The Study of Time*, vol. II., (Heidelberg and New York, 1975).

   A prime and stimulating example of what might be done in the application of temporal insights to esthetics is to be seen in the significant paper by Joan Stambaugh, "Music as a Temporal Form," *The Journal of Philosophy*, vol. LXI, no. 9 (April 23, 1964). Also, William Barrett's *Time of Need* (New York: Harper & Row, 1972) is a fine example of a time approach to literary themes.

3. Friedrich Kümmel, "Time as Succession and the Problem of Duration," in J. T. Fraser (ed.), *The Voices of Time* (New York: George Braziller, Inc., 1966), p. 32.

Index

Date Due

FE 13 '76			
JE 15 '78			
NO 6 '81			
JE 26 '87			
NO 30 '87			